DEEP FREEZING

DEEP FREEZING

*A Comprehensive Guide
to its Theory and Practice*

PAT M. COX, B.Sc.

*Recipe section compiled
in collaboration with*
PATRICIA MITCHELL, B.Sc.

FABER AND FABER
London

First published in 1968
by Faber and Faber Limited
24 Russell Square London WC1
Reprinted 1969, 1970
Printed in Great Britain by
W & J Mackay & Co Ltd, Chatham
All rights reserved

ISBN 0 571 08523 7

Preface

THIS book on deep freezing is written for the modern housewife and student of home economics alike.

It is believed that the first part of this book dealing with the theoretical and technical aspects of deep freezing will be of particular interest to the home economics student. At the same time it is hoped this section will explain the reasons underlying recommended practices to the interested freezer owner.

The principal function of a freezer for many users is the freezing of garden produce, carcass meat, fish and dairy produce. Therefore no book on deep freezing would be complete without detailed instructions for the preparation and treatments to be applied to these foods. Such instructions are given in Part Two of this book.

In Part Three, a comprehensive selection of recipes is given, together with detailed instructions for packaging, freezing and ultimate use. It is hoped that these recipes will illustrate the tremendous value of a freezer for storing pre-cooked and prepared foods which may not be fully appreciated by existing and potential freezer owners.

PAT M. COX

Acknowledgements

HELP and advice in the preparation of this book has been received from research institutes, libraries, freezer manufacturers—in particular Lec and English Electric—frozen food producers, suppliers of packaging materials, as well as a great many individuals. However, I would like to thank specifically Miss Margaret Leach of the Long Ashton Research Station and Miss Wendy Matthews of the Flour Advisory Bureau for their kind advice.

I am especially grateful to Dr Pat Lewis for her helpful guidance on the Microbiological section.

Above all, I am greatly indebted to Mrs Patricia Mitchell, without whose co-operation this book would not have been written. Not only has she selected and tested the majority of the recipes in Part Three of this book, but has also read the manuscript, making many helpful suggestions.

Contents

11

CONTENTS

12

CONTENTS

CONTENTS

14

Plates

15

PLATES

All photographs were taken by Bill Warner, except where otherwise stated

Figures and Tables

Figure 2 by courtesy of Birds Eye Ltd.

17

Part One

Part One

1

Introduction

A Look at the Past

FROM the time when our early ancestors ceased to move about in search of their food and attempted to settle and begin a primitive form of cultivation, so began the need to preserve food. The skill with which they were able to do this was an important factor in the development and progress of early man. Many methods in current use today are, in fact, developments of methods used by our ancestors. Drying, salting, smoking, and not least freezing, are by no means new techniques, but have been applied for thousands of years where natural resources and climatic conditions permitted.

It is interesting to speculate whether it was the result of mere chance or intelligent observation which led to the adoption of these very effective practices; for without knowing anything of the agencies which caused his food to decay, primitive man was taking very effective measures to control them. In this way he countered the problems of securing an adequate diet during winter, during ill health, and during disasters to his food sources. However, he must have had a strong stomach and not a little courage to eat some of the less appetizing foods during the long winter months in those ancient times.

Some of the first foods to be preserved commercially by freezing also left much to be desired. Some of the fish frozen in mixtures of salt and ice during the nineteenth century could not have been entirely agreeable, and even when mechanical refrigeration was introduced towards the end of that century the quality was very inferior to that which is achieved today.

The credit for the vastly improved quality of frozen foods is due in no small measure to an American, Clarence Birdseye. In Labrador in the 1920s he had tasted fish which had been frozen by the fur

21

trappers with whom he was working. The quality of the fish was superior in every way to that offered for sale in the markets of his home country. He contrasted the speedy preparation, and the rapid low-temperature freezing of the fish frozen in the Arctic conditions of Labrador with the often delayed preparation and slow rate of freezing of the fish prepared for the home market. It was common practice then to pack fish, frequently of indifferent quality, into wooden barrels and put them into the cold store to freeze, when freezing might take two to three days. The fish which emerged from those barrels would compare most unfavourably with the quick-frozen individually packed fillets available in the supermarkets today. Clarence Birdseye appreciated these points and pioneered the way to improved freezing techniques.

PRESENT TRENDS

With the passage of time the freezing of food ceased to be merely a method of providing food for survival; instead it developed into a means of providing food of better quality and, more important. as a means of distributing foods to an ever-increasing population, These functions are vital and valid in the problems of our present civilization. However, in recent years a new application has arisen, in that frozen foods are making a significant contribution to the changing social pattern in the more civilized countries of the world.

It is an accepted fact today that a woman's place should be in the home and not always in the kitchen. Her freedom has been secured to a large measure by the development of labour-saving materials and equipment; the washing machine, the refrigerator, the vacuum cleaner, detergents, plastic goods, drip-dry materials, central heating, etc. etc., have all contributed to her enfranchisement. Now a further contribution has arisen from the vast new range of convenience foods available today—many of which are frozen products. In some cases the housewife has bought her own freedom by taking work outside the home to help purchase this labour-saving equipment. This in turn has resulted in a greater appreciation and reliance on the ready-to-eat frozen foods.

Eating has always been a source of enjoyment, but in recent years there has been an increasing interest in food. Perhaps this is because more people take holidays abroad, and more people find dining out a pleasant way of spending an evening—but whatever the cause, there is much less conservatism about food than formerly. Frozen-food producers have not been slow to realize this, and are continually experimenting with new dishes to promote their sales. All this has created a greater interest in frozen-food products.

There are, however, many housewives who genuinely enjoy cooking and see it as the only creative part of their domestic duties. Of particular value to such people is the possession of a freezer, so that they are able to indulge their hobby when it is convenient, and at the same time plan for greater leisure time. There is a wide range of domestic freezers on the market to meet the varying requirements of large and small families, of country and town dwellers, and of kitchen garden and shooting enthusiasts.

Advances in the world of plastics have resulted in greatly improved packaging materials for frozen foods. The frozen-food producer is able to present his product cheaper, more attractively, and more securely, while the user of a domestic freezer finds the newer materials more manageable and convenient.

Thus while the technical role of frozen foods for feeding the world's population continues, the current advance is the application of frozen foods to the domestic scene for greater freedom and leisure.

A Look at the Future

In future years frozen foods will play an increasing part in both large- and small-scale catering. Where it is necessary to feed a large number of people from a factory, school, or hospital, in a limited time, the resources of most kitchens and kitchen staff are seriously strained. Increasing use of frozen foods will greatly reduce the time and space which must be allowed for preparation. In particular the future trend will be the increased use of microwave ovens in canteen kitchens, in cafés and motels. Prepared and precooked foods placed into a microwave oven would be ready to eat in less than a minute. A freezer provides the best means of storing foods at the present time, although most foods would need some thawing prior to being placed in a microwave oven if uneven heating is to be prevented. Thawing in a domestic refrigerator provides one answer to this problem, although some form of two stage oven, one section for thawing and another for heating, might prove the more successful solution. This development, together with vending machines and disposable plates

23

and utensils, will thus lead to virtually staffless canteens.

Perhaps the future trend in the home will also be towards this form of instant catering. In the future modern kitchen a freezer and microwave oven will stand side by side. To serve a meal it may ultimately be possible to transfer an appropriately packaged pre-cooked dish from the freezer into a suitably developed microwave oven at the appropriate setting and in a minute or so the meal may be served.

It is thus obvious that frozen foods have secured a firm place in the feeding techniques of the more civilized countries, and that future developments will see an even greater application of frozen foods.

2

The Physical Principles of Food Freezing and the Nutritional Value of Frozen Foods

THE inclusion of a theoretical chapter in what is essentially intended as a cookery book may be considered a little out of context. It is felt, however, that some knowledge of how foods are affected when preserved by freezing will be of great benefit when making practical use of this method of food preservation.

To this end each of the three stages in the 'life' of food preserved by freezing will be considered; the three clearly defined stages are (a) freezing, (b) frozen storage, and (c) thawing. The treatment received by the food at these three stages will be reflected in the final quality of the food. This quality will be indicated by the texture, colour and flavour, as well as by the nutritional value, which is considered separately in this chapter.

Consideration will first be given to the effects of freezing during which the food undergoes a change in its physical state.

THE EFFECTS OF FREEZING

When food is frozen the water it contains is changed into ice. It is this withdrawal or isolation of the water in the form of ice which in effect dehydrates the food and is the principle of freezing as a method of preservation.

Most foods are largely composed of water. The percentage of the total weight may be as high as 90 per cent, as in strawberries, and 70 per cent as in lean meat, although in some prepared foods the percentage may be much lower. But whatever the water content, all foods

25

are complicated substances. Dissolved in the water which normally makes up a large proportion of all foods, are salts and sugars, while complex protein molecules exist in the form of colloidal suspensions. The effect of the dissolved salts and sugars in the food is to depress the freezing-point of the water content of the food to a value below the freezing-point of pure water. The amount it is depressed is proportional to the concentration of salts present.

Meat, fish and most of the common fruits and vegetables which have a high water content have freezing-points which lie between 32°F and 25°F (0°C and −4°C). This is usually known as the *zone of maximum crystal formation*. In contrast, some prepared foods with a low water content and in which the salts and sugars are more highly concentrated have a much lower freezing-point. For example, the freezing-point of mince pies, a food with a high sugar content, is about 15°F (− 10°C).

Freezing curves

Although it is convenient to give a single freezing-point for a given food, the freezing of foods actually takes place over a temperature range and can most simply be shown in the form of a curve, known as a freezing curve, as shown in Fig. 1.

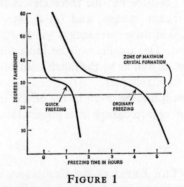

FIGURE 1

FREEZING CURVE FOR A TYPICAL FOOD

When food at room temperature is placed in a freezer the temperature of the food falls rapidly at first because of the wide difference in temperature between the food and the refrigerating medium. This is indicated by the steep descent at the beginning of the curve. The fall in temperature slows down as the food reaches freezing temperature, when the change in state occurs. The temperature of the food during this period is almost constant as indicated by the plateau section of the curve.

At this stage the latent heat of crystallization is released, and the rate at which this is removed determines the rate of freezing.

The conversion of some of the available water into ice as the food becomes frozen has the effect of increasing the concentration of the salts in the water remaining. The same quantity of soluble constituents must be dissolved by a smaller quantity of solvent. Such an increase in fluid concentration in a food results in a further lowering of the freezing-point. This is shown at the end of the plateau section by the final down-ward trend in the curve.

It is found, in fact, that even when the temperature reaches that of 0°F (-18°C) there may still be a very small percentage of water which remains unfrozen. When the temperature of the food is reduced still further a point is reached when the remaining water and soluble constituents solidify together. This is known as the *eutectic point*.

The eutectic point is defined as the maximum temperature at which maximum crystallization of the solvent and solute can occur. This point could be below the temperature at which commercial frozen food stores are normally operated. It must be appreciated that the presence of even small quantities of unfrozen water means that a region exists where micro-organisms could remain viable, even though they could not grow and multiply at this temperature.

Volume changes

A volume expansion of a little less than 9 per cent occurs when water is changed into ice. The common occurrence of burst pipes in severe frosts during winter-time is an unpleasant demonstration of this phenomenon. Foods with a high water content expand on freezing correspondingly more than foods with a low water content. Allowance must be made for this when packaging foods prior to freezing. For practical purposes this allowance is termed *headspace*, and thus greater headspace must be allowed for strawberries, apricots, and other foods with a high water content.

Rate of freezing

The quality of the frozen product is affected by the rate at which it is frozen, and today it is agreed that the best results are obtained with foods which have been 'quick frozen'. It has been common commercial practice to quick freeze foods since the method was first introduced by Clarence Birdseye more than thirty years ago. In a later chapter descriptions are given of various methods used commercially to quick freeze foodstuffs.

The quality effects of different freezing rates

When most foods of plant and animal origin are frozen water separates out in the form of ice crystals as it passes through the zone of maximum crystal formation. If the passage through this temperature range is slow, large crystals are formed, which are believed to cause damage to the cell structure. If the passage through this range is quick, then small crystals are formed and cell damage is slight. Quick freezing has been described as the process by which the temperature of the food passes through the zone of maximum crystal formation in thirty minutes or less.

Apparently the rate of freezing also affects the location of these ice crystals—a very rapid rate produced, for example, by direct immersion in liquid oxygen results in minute crystals being formed within the cells. As the freezing rate is reduced, even within the usually accepted bounds of quick freezing, so crystals are formed in the spaces outside the cells. It is not certain why there is this changeover from intra-cellular to extra-cellular crystal formation at a certain rate of freezing.

There are two explanations for the fact that a better-quality product results when food is frozen quickly. The first is usually known as the *cell rupture theory*. This assumes that when animal or plant tissue is frozen slowly, the large ice crystals which are formed puncture the cell walls so that when the food is thawed the cell fluids are free to drain away. This fluid which leaks away from a food when thawed is called *drip* and may contain valuable dissolved nutrients.

The improvement in quality claimed for foods which have been quick frozen is explained by the cell rupture theory as follows: The smaller ice crystals formed during quick freezing cause fewer cell walls to be punctured and consequently the amount of fluid or drip lost on thawing is reduced. The cell rupture theory, however, does not explain changes on freezing which occur in prepared foods which are non-cellular.

There is more support for the second theory of the quality improvement evident in quick frozen foods, as this theory not only accounts for changes in plant and animal tissue composed of cells, but also explains some of the changes occurring in prepared foods which are non-cellular.

Earlier in this section mention was made of complex protein

28

molecules existing in foods in the form of colloidal suspensions; in more general terms they might be described as being 'jelly-like'. When the food is frozen ice crystals form, and when the freezing rate is slow, the ice crystals grow, drawing water at first from the cells so that the cell contents becomes concentrated. Eventually water is drawn from the jelly-like proteins themselves so that they are dehydrated and denatured. This dehydration and denaturation may be irreversible, so that when the food is thawed and the ice crystals melt, the proteins are unable to reabsorb the water they lost during freezing. This water drains away and is recognized as drip.

On the other hand, when freezing takes place quickly, small ice crystals are formed uniformly throughout the tissue, and less time is available for the ice crystals to grow at the expense of the complex protein molecules. Thus denaturation of the proteins has not occurred to any great extent, and when thawing takes place there is a greater possibility of the smaller ice crystals being more completely reabsorbed by the protein molecules.

It is likely that some of the undesirable changes in prepared foods, such as the curdling and the separation of some sauces which cannot be rectified when the food is thawed, can be attributed to the irreversible denaturation of the protein content occurring during slow freezing. It is not possible to provide an explanation of these changes according to the cell rupture theory. However, it may be that the final quality of the thawed foods can only satisfactorily be explained by taking account of both theories under consideration.

Changes caused by freezing in fruit and vegetables

Better results are obtained for fruit and vegetables when they are quick frozen. Tissue damage may result when slow-freezing methods are used. This may be observed in some soft fruits which have a tendency to 'fall' and lose liquid when thawed, although it would be of little significance in vegetables which are cooked from the frozen state.

Changes caused by freezing in meat and fish

Meat is made up of a large proportion of connective tissue consisting of collagen and elastic fibres which give the meat resilience even after freezing. Thus the appearance of meat which has been frozen differs very little from the fresh equivalent. It seems that the amount of drip which leaks away from meat when it is thawed is greater if the freezing rate has been slow. But other factors which must be taken into consideration include the size and cut of the sample, the temperature at which it was stored, and the amount of 'ageing' it received before freezing.

There is less connective tissue binding together the muscular tissue in fish, and after freezing the flesh has a more open appearance. A slow rate of freezing allows more time for ice crystals to grow in the fluid surrounding the cells. As they increase in size water will be drawn from the cells and the fish protein become increasingly dehydrated and denatured. This denaturation of the protein is thought to lead to toughness in the fish when eaten, although it is doubtful if it would be noticed by the average consumer.

THE EFFECT OF FROZEN STORAGE

A temperature of $0°F$ ($-18°C$) has been proved an acceptable and safe temperature to store foods. No micro-organisms can grow at this temperature and enzymic action is negligible. Domestic freezers are designed to maintain a temperature of $0°F$ ($-18°C$), although it is usually possible to reduce the temperature by a further 5° or 10°F. Commercial frozen-food stores are usually maintained at $-20°F$ ($-29°C$).

Frozen foods stored under ideal conditions in a domestic freezer or commercial frozen-food store should not suffer any loss of quality. Ideal conditions consist of completely effective packaging for the frozen food and a steady non-fluctuating temperture of $0°F$ ($-18°C$) in domestic freezers. Although with the wide range of packaging materials available the first condition may be fulfilled, it is not possible in practice to prevent temperature fluctuations occurring within the freezer.

In fact a *temperature gradient* frequently exists in the freezer, with the result that heat will flow from the warmer to the cooler areas until a uniform temperature exists again in the cabinet.

Several factors contribute to the existence of a temperature gradient. Simply opening the freezer door or lid to remove a package of frozen food will introduce a little warm air into the cabinet, although of greater importance will be the rise in temperature produced by the addition of packs of unfrozen food. Even when these are thoroughly cooled and chilled first, they are still much warmer than the rest of the frozen food. They must be kept from physical contact with already frozen foods. But without opening or closing the door or lid, small rises in temperature are happening all the time in a freezer standing in a kitchen. However, these fluctuations are kept in check by the operation of a thermostat which brings the compressor into action when the upper temperature limit has been reached, and cyclically brings down the temperature.

Effect of a temperature gradient

The effect of a temperature gradient (or temperature difference between one part of the freezer and another) is for a circulation of air to be set up. As air becomes warmer, so it becomes unsaturated and it is able to take up moisture vapour from inadequately protected frozen foods. This moisture vapour is redeposited in the form of frost when the air is subsequently cooled by contact with a cold surface such as the evaporator coils or cooling surface.

In frozen foods the moisture content is present in the form of ice, and the removal of moisture in this form without first passing through a liquid phase is called *sublimation*.

Thus, when a temperature gradient exists there is a continuous loss of ice from unprotected frozen foods which is redeposited on the coldest surfaces of the freezer. The effect of this transference of ice away from the frozen food causes the food to be increasingly de-hydrated, with a resulting loss of quality.

Under certain circumstances it is possible for a temperature gradient to exist within a package of frozen food where pockets of air have been enclosed. This is more likely to be the case with irregular-shaped foods such as chicken, although it can occur with an in-adequately filled container. Thus it is very important that as much air as possible is excluded from all packs or containers before packaging. Where there is a movement of air within the package moisture vapour is removed from the food material and deposited on the interior walls of the pack as frost. Where this continues over a long period the food becomes dehydrated and the moisture which has been lost by the food exists on the inner walls of the pack as a layer of frost and ice crystals usually referred to as *cavity ice*.

Freezer burn

This is the term used to describe the severe desiccation of the surface tissue of a frozen food. The characteristic effects are to be observed in animal tissue. Freezer burn is caused by the sublimation of ice from the surface layers of the food. It can occur quite rapidly when foods are exposed to the effects of a temperature gradient within the

31

freezer, and when they are not protected by suitable packaging materials.

Freezer burn appears as patches of dark discoloration on the surface of animal tissue as the pigments in the surface layers are concentrated and oxidized. The surface of the tissue also takes on a greyish-white appearance caused by the presence of numerous cavities left behind after the sublimation of ice from the surface. As the tissue becomes increasingly affected by freezer burn the surface layers become more and more spongy and the layers immediately below are further condensed.

When a food is lightly affected by freezer burn the conditions may be reversible, and on exposure to moist conditions, such as are experienced on cooking, the effects of light freezer burn may rarely be detected.

When freezer burn is more pronounced the tissue becomes more open in texture and may also be affected by oxidation and chemical changes. Such a condition may not be remedied on cooking.

In the unlikely event of meat or fish being left unprotected in a freezer for two or more years, the surface would not only show these blemishes described, but the whole tissue would be completely desiccated. The weight of such a sample would represent only a fraction of the fresh weight.

Quality effects of a fluctuating or elevated storage temperature

Foods which have been properly packaged in suitable materials should not show the effects of desiccation or freezer burn, even when the temperature of the freezer is permitted to fluctuate or is maintained above $0°F$ ($-18°C$). However, in these circumstances, such foods may become progressively denatured, with the result that texture and colour changes are discernible in the thawed product.

When the temperature of a frozen product is allowed to rise and fall during frozen storage the ice crystals increase in size. Some of the water necessary for growth comes from that small percentage of available water which remains as a liquid when the food freezes, and some is taken from the complex protein molecules which as a result suffer dehydration and denaturation. It can be seen that poor storage conditions produce a similar effect on the quality of the frozen foods to those which result from a slow rate of freezing; in fact, they may be of greater significance. The extent to which the quality is affected depends upon the size of the temperature fluctuations and the frequency with which they occur.

1a. Microwave oven with automatic opening door and push button timer

1b. Small upright freezer, 3.2 cubic feet

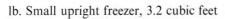

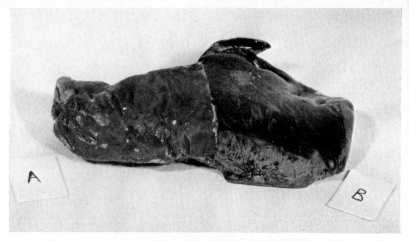

2a. Sample of liver showing freezer burn. Portion A was closely wrapped in moisture-vapour-proof film. Portion B, unprotected during frozen storage, shows dark discolouration from concentrated pigments in the surface and a number of greyish white cavities caused by sublimation of ice from the tissue.

2b. Apricots in syrup, showing effectiveness of ascorbic acid in the prevention of browning. Samples have been stored for six months at 0°F (−18°C), thawed and allowed to remain eight hours at room temperature. Sample A, treated with ascorbic acid before freezing, is unaffected; Sample B, untreated, shows considerable darkening.

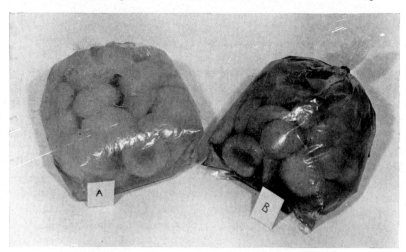

Changes during frozen storage in fruit and vegetables

The changes occurring in fruit and vegetables are less noticeable than those in meat and fish. Fruit is usually packed in syrup or sugar and for this reason is rarely affected by freezer burn. Although an increase in the size of the ice crystals and a slight deterioration of texture may be the result of a fluctuating temperature.

Inadequately packed vegetables, when severely desiccated, may show freezer burn, but such deterioration of the texture as may exist is not usually noticeable after cooking. The presence of cavity ice within the carton is usually indicative that the package has been exposed to a fluctuating temperature during frozen storage, and texture and colour changes may also be discernible.

Exposure to light which occurs in retail cabinets may also produce colour changes. For instance, peas are observed to turn slightly yellow and develop an 'off flavour'. This phenomenon only occurs if the peas are packed in transparent plastic bags, and not when packed in coloured bags or waxed cardboard cartons.

Changes during frozen storage in meat and fish

The effects of desiccation and freezer burn are more pronounced in animal tissue, and thus poorly packaged meat will show freezer burn on the surface, and where the loss of moisture from the tissue is considerable the texture may become tough. At the same time, the more open texture which is produced by severe desiccation and freezer burn permits oxygen to penetrate the tissue, with the result that the fat content becomes oxidized and rancid. It is generally agreed that the storage life of meats with a high fat content should be limited.

The effect of light on frozen meats is for the surface to develop a brown discoloration as a result of oxidation of the myoglobin pigments. This may occur in refrigerated display cabinets. Light also accelerates rancidity.

Frozen poultry is more likely to show evidence of freezer burn than most meats, because it is more difficult to exclude air from the irregularly shaped pack. The water lost from the poultry may be deposited on the inside of the pack in the form of cavity ice. Rancidity may also result due to oxidation of fats.

Loss of quality due to desiccation and freezer burn is also evident in stored fish unless the packaging is effective. Large fish are frequently glazed (given a coating of ice); this is very effective, although it must be repeated periodically if the fish is to be stored for many months.

Shellfish, such as prawns and shrimps, which because of their irregular wrinkled surface are more prone to desiccation, are usually supplied for the retail market in vacuum packs.

Denaturation of fish stored at a fluctuating or too high a storage temperature is more pronounced than in most meats. The poor quality which may be observed as toughness in the fish muscle even after cooking is more likely to have been caused by adverse storage conditions than a slow rate of freezing. For this reason a temperature of $-20°F$ ($-29°C$) is recommended as the maximum for storing fish commercially.

The storage life of fat fish is much shorter than that for lean fish. The fat content is unsaturated and may be readily oxidized, producing rancidity, and ultimately a brown discoloration of the muscle tissue.

THE EFFECT OF THAWING

Most of the changes manifested when a food is thawed are the results of experiences received during the freezing process and frozen storage. But the rate of thawing is also important.

Most fruits leak or lose fluid when thawed, although because they are usually contained in syrup or sugar, this loss of fluid may go unnoticed. Because most soft fruits have a tendency to 'fall' when all the ice crystals are melted, they are best eaten whilst still containing a little ice. Consumption immediately after thawing, or while still slightly frozen, is also recommended for peaches and apricots. Browning or discoloration will proceed rapidly after thawing if the fruits have not been treated with an antioxidant.

Vegetables are not thawed before cooking, with the possible exception of corn-on-the-cob. Even so, they do not have the same tendency to leak as do fruits, while starchy vegetables do not drip at all.

Meat, poultry and fish drip or lose fluid when thawed. The rate of freezing and conditions of frozen storage may influence the quantity of drip produced, but the amount of cut surface is also a significant factor. On the whole fish appears to drip more than meats, and it has been observed that fish with a fine texture drips less than varieties with a coarse one. Dipping fish fillets in a weak saline solution prior to freezing has been found to reduce the amount of drip on thawing (the effect is to increase the pH to a value of about 7 when proteins are apparently less prone to lose water). This process is not recommended for fat fish, as salt has the effect of accelerating rancidity—in any case the tendency of fat fish to lose fluid or drip is less marked.

Slow thawing in a domestic refrigerator is usually recommended for meat and fish to provide time for the reabsorption of the fluid into the tissue from the melting ice crystals. Once thawing is complete the meat or fish should be promptly cooked. It is possible to cook meat and fish straight from the frozen state, when any resultant drip is incorporated with the liquor or gravy.

NUTRITIONAL VALUE OF FROZEN FOODS

The earlier part of this section has been devoted to the effects of preservation by freezing on the physical qualities of foods. Consideration will now be given to the effects on the nutritional qualities of food when frozen.

It is well known that food is made up of several components called nutrients; these are carbohydrates, fats, proteins, vitamins, and inorganic elements; and they are present in varying proportions in different foods. To enable the body to function properly these nutrients must be available in sufficient quantities. This is most simply and satisfactorily achieved in the regular consumption of well-balanced meals. Unfortunately this is not the birthright of a large proportion of the world's population.

Frozen food was formerly dismissed by some critics as 'having lost all its goodness'—presumably meaning it was lacking in nourishment. Yet it is doubtful if a complete meal, lunch or dinner, prepared entirely from frozen foods, would vary substantially, from one prepared entirely from fresh ingredients. There may be some reduced values—Vitamin C is one notable example of a nutrient not found in frozen vegetables in the same proportion as the fresh equivalent—but a great majority of the world's population rarely gets the fresh equivalent. For those living in towns, several days can elapse after the vegetables have been picked before they are consumed at home; in this time the amount of Vitamin C may be much reduced. In fact, losses of this vitamin of up to 50 per cent have been reported in vegetables not marketed promptly.

The quality of frozen foods lies in the fact they are taken and preserved when in their prime condition. Frozen-food producers build their factories close to the source of supply in order to keep delays between harvesting and freezing to a minimum. Similarly the user of a domestic freezer should aim to freeze speedily only the best of the garden produce. Meat and poultry bred and reared for preservation by freezing should likewise be selected when at their best quality (i.e. with the right weight and fat content).

Nutrient losses

The original high nutritional quality of a food taken for freezing may not be maintained if it is not handled well during preparation and freezing, or if the storage temperature is allowed to fluctuate, or if the thawing conditions are not carefully controlled.

Nutrients are frequently lost from frozen foods in one of the following ways:

In solution   Inorganic salts, sugars, and soluble proteins may be lost from foods in the fluid which drains away from certain foods, particularly meat and fish, when they are thawed.

As a result of oxidation reactions   Fats and vitamins may be destroyed by oxidation reactions catalysed by certain enzymes occurring within the food.

As a result of denaturation   Irreversible denaturation of proteins may occur during the freezing process and during frozen storage.

The extent to which any of these nutrient losses occurs in frozen foods will vary according to the treatment received.

Nutritional value of frozen fruit and vegetables

Fruit, vegetables and cereals are the most important sources of carbohydrate in the diet. They also contain some proteins, but unlike animal proteins, they are not made up from the full complement of essential amino acids needed by the body; they are therefore frequently referred to as second-class proteins. Fruit and vegetables are important sources of Vitamins A, B, and C. Most fruits constitute a fair source, and most vegetables a good source, of inorganic elements. Only nuts contain fats in any quantity, but these are not usually frozen except in small amounts in pre-cooked foods.

The carbohydrate content may be in the form of starch, a polysaccharide, which forms the bulk of the food store in plant materials, or as sugar which is found in increasing quantities in fruits as they ripen. Sugars may be disaccharides or monosaccharides, the former being converted into the latter as a result of hydrolysis (i.e. the addition of water).

Preservation by freezing does not seem to affect the starch content of vegetables, but some sugars may be lost from vegetables in the normal course of washing and blanching prior to freezing. Also, during frozen storage some of the disaccharides are converted to monosaccharides, but this is not a detrimental change, as it occurs naturally in the body in the normal course of digestion. Fruits are frequently frozen in syrup or dry sugar packs, which, in fact, adds carbohydrate to the frozen product. Natural sugars which are dissolved from the fruit during preparation and storage are retained in the syrup in which the fruit is ultimately served.

Vitamin A is found in fruit and vegetables in the form of its precursor, carotene. Carotene is converted into the available form of this vitamin by the body. Those fruits which are of a yellow colour, such as apricots, melons, and peaches, and those vegetables with dark green leaves such as spinach and watercress, and also carrots, are good sources of carotene.

Vitamin B is made up of at least eleven different substances, but the principal ones found in fruit, and to a greater extent in vegetables, are thiamine, riboflavin, niacin, folic acid, and pyridoxine. All members of this group are water soluble and most are unstable at high temperatures.

Vitamin C is the most important vitamin found in fruit and vegetables, because whereas the other vitamins may be obtained in reasonable quantities from animal sources, only insignificant quantities of Vitamin C are found in foods other than fruit and vegetables.

Vitamin C is the least stable of the vitamins; it is water soluble, unstable at high temperatures, and easily destroyed as a result of oxidation. Thus the presence of Vitamin C in reasonable quantities in a food product preserved by freezing is usually regarded as an indication of good quality. It is for this reason that most of the work concerned with the nutritional quality of frozen foods has been based on the survival of Vitamin C, although some observations have also been made on the survival of some members of the B group of vitamins.

It has been found that losses of water-soluble vitamins from frozen fruits and vegetables occur not so much in the actual freezing process and during frozen storage—providing the temperature is not allowed to fluctuate above 0°F (−18°C)—but in the preparation which is necessary prior to freezing.

The preparation of all vegetables begins with a thorough washing, followed in most cases by cutting or slicing, which exposes a large number of cells at the cut surfaces. The vegetables are next blanched in boiling water or steam. This is to inactivate the enzymes which would otherwise cause discoloration and the development of 'off flavours', as well as reducing the vitamin content during frozen storage. After blanching, a thorough cooling is necessary before the vegetables can be packaged and frozen. It is during this period of preparation that the greatest losses of water-soluble vitamins occur, particularly during blanching, which supplies moisture and heat—the two conditions to which the vitamins are most vulnerable. The blanching time must be carefully controlled. The higher the temperature, the more quickly the enzymes are inactivated; however, in the home the blanching temperature may be only as high as that of boiling water. Thus it is important when adding batches of prepared vegetables to the blanching kettle not to overload it, or it will be several minutes before the temperature of the water in the kettle returns to boiling-point.

Vitamin losses will continue during cooling. Cooling should thus be as rapid as possible, so that the vegetable can be quickly transferred to the freezer.

Even under good conditions of preparation, losses of 30 per cent for Vitamin C and 20 per cent for the Vitamin B group, are not unusual, and under poor conditions the losses may be much higher. Small quantities of inorganic elements may also be lost in the blanching and cooling waters.

Fruits do not suffer the same losses of vitamins as vegetables, as they are rarely blanched prior to freezing. In fact, Vitamin C, as ascorbic acid, is added to some fruits to inactivate those enzymes responsible for the browning which occurs in certain fruits. Thus the final result is more likely to be an increase than a decrease in this vitamin. Blanching may be used occasionally with apples or pears, but where it is likely to produce a loss of flavour or texture in dessert fruits, ascorbic acid is added instead.

During frozen storage at $0°F$ ($-18°C$) there should be no losses of nutrients from frozen fruit or vegetables. However, when the temperature has been allowed to fluctuate above this level losses of Vitamin C have been recorded in fruit and vegetables. The losses of this vitamin can be shown to increase twofold with each $5°$ rise in temperature.

However, when frozen vegetables are cooked, the losses of vitamins will be less than from the fresh equivalent. This is because less cooking time is required, and because the enzymes responsible for the destruction of vitamins have been largely destroyed by the blanching received by the frozen vegetable. These enzymes will of course still be present in the fresh equivalent.

The retention of Vitamin C in some soft fruits after thawing has been shown to be as high as 80 per cent, particularly if packed in sugar. High Vitamin C contents have also been obtained with fruit juices.

Nutritional value of frozen meat and fish

Meat, poultry, and fish contain little or no carbohydrate, but they are important sources of proteins, fats, and the fat-soluble vitamins A and D. They are also valuable sources of thiamine, riboflavin, and niacin—members of the B group of vitamins. Meat is a major source of iron, and an important source of sodium and potassium.

Fats are composed of fatty acids which, if they are fully saturated, result in a solid fat at room temperature—the form largely found in meats. If the fatty acids are unsaturated, the result is a liquid or oil at room temperature—this is the state in which fats occur in most fat fish. Unsaturated fats will tend to become oxidized, which results in a loss of sweetness, a tendency towards rancidity, and an accompanying loss in nutritive value. It has been shown that rancidity is accelerated when the temperature at which meat or fish is stored is allowed to rise above $0°F$ ($-18°C$). Destruction of the fat-soluble vitamins is seen to accompany the development of rancidity in meat and fish. It is therefore not advisable to extend the storage life of meat with a high fat content such as pork; and where possible all surplus fat should be trimmed from joints which are to be frozen.

A short storage life is also recommended for all fat fish if they are to retain their characteristic sweetness.

Meat, poultry, and fish are important sources of proteins, not only because of the quantities in which they occur, but because they contain all the essential amino acids which cannot be manufactured by the body and which cannot be supplied from any other single food source.

Slow rates of freezing, and fluctuating storage temperatures cause the proteins to be dehydrated and denatured, sometimes irreversibly. Water withdrawn from the complex protein molecules by the growing ice crystals is not always reabsorbed when the tissue is thawed. This unabsorbed water which drains away may carry with it some soluble proteins, inorganic elements and small quantities of the Vitamin B group of nutrients. However, when the drip is contained in stock or gravy, these nutrients are not lost.

Where denaturation occurs, the denatured proteins may appear tough and result in a small decline in the quality of the muscle.

39

However, it is unlikely that this change will be discernible after cooking, and there is little possibility of the protein content being substantially reduced or rendered unavailable to the body after consumption.

Time-Temperature-Tolerance

In conclusion, a brief mention may be made here of an extensive survey carried out in America by the Western Utilization Research Branch of the United States Department of Agriculture, Albany. This sought to investigate the behaviour of frozen foods in conditions which are equivalent to the type of experience received during commercial distribution. It is usually referred to as *Time-Temperature-Tolerance*. The aims of the survey were to discover what degree of tolerable deviation from ideal conditions was permissible before a discernible change and deterioration occurred in frozen foods, and if possible— to discover tests of quality control that could be applied to frozen foods at any point in the distribution system.

Surveys were made of many food products stored for specified times at specified temperatures. Various factors were studied as indicative of a change in quality. For instance, the quantity of ascorbic acid (Vitamin C) present, the degree of chlorophyll conversion, colour, and flavour changes. The following observations are of significance to both commercial and domestic frozen food users:

(a) The rate of deterioration doubles with each rise in temperature of 5°F (above 0°F) during storage.
(b) Most changes were cumulative. Thus where a product was exposed to adverse temperature conditions on several occasions the degree of change was the sum of all the time temperature experiences; thus it may be said that frozen foods have a 'memory'.

REFERENCES

Food Texture. S.A. Matz. A.V.I. Publishing Co. Inc. 1962.
The Freezing Preservation of Foods. Tressler and Evers. A.V.I. Publishing Co. Inc. 1965.
'Time-Temperature-Tolerance.' Survey carried out at Western Utilization Research Branch, U.S. Dept. Agriculture, Albany—reported in *Food Technology* beginning January 1957.

'Fundamentals of Low Temperature Food Preservation.' O Fennema, W. D. Powrie. *Advances in Food Research* 13. 1964.

'The Effect of Freezing on Fish Muscle.' R. M. Love. *Recent Advances in Food Science*, Vol. I. Butterworth. 1962.

'Ice and its Role in the Ecology of Frozen Foods.' Report by Technical Staff of Quick Frozen Foods. *Quick Frozen Foods*. Nov. 1961.

'Control of Freezer Burn.' G. Kaess, J. F. Weiderman. *Food Preservation Quarterly*, Vol. 22, No. 2. 1962.

Manual of Nutrition. H.M.S.O. 1961.

3

The Importance of Enzymes and Micro-Organisms in Freezing

THE reason for the improved palatability and nutritional value of preserved foods today is mainly due to an increase in the knowledge of those agents which can cause it to decay—namely enzymes and micro-organisms.

ENZYMES

Enzymes exist in all plant and animal tissue and are responsible for the normal processes of metabolism which go on within living tissue. These enzymes assume a new role after the plant is harvested or the animal is killed. Whereas during the life of the tissue they were concerned with its growth and nourishment, after its death their function becomes largely one of destruction. Examples of this may be seen in the browning which occurs at the cut surface of a fallen apple and which spreads to the whole fruit, and the process of ageing which goes on in a carcass of meat when it is 'hung'. Frequently the activities of the micro-organisms are associated with those of enzymes and may, in fact, occur at the same time. However, the activities of the micro-organisms are considered in more detail later.

The type of reaction catalysed by the enzymes in the tissue is largely that of oxidation—oxygen from the air, or sometimes from within a carton of a food prepared for freezing, combines with the food and gives rise to discolorations, 'off flavours', a loss in nutrient value, and eventually to spoilage.

Although enzymes can be permanently inactivated by exposure to high temperatures, their activities are merely inhibited when retained at low temperatures such as are experienced in the freezer. Enzymes will continue to function during the actual process of freezing and although they will remain inactive during frozen storage at 0°F (−18°C), when thawing occurs they will resume their activities.

Enzyme action in vegetables

It was found that vegetables developed hay-like off flavours and discoloured during frozen storage. The rate at which these changes took place was accelerated when the temperature of frozen storage was allowed to rise above 0°F (−18°C). Peroxidase and catalase were discovered to be the two enzymes principally responsible for these changes.

Blanching the vegetables before freezing has been found effective in destroying most of the enzymes present, although peroxidase may survive. This enzyme has been shown to survive temperatures of 250°F (120°C) for several minutes, although the activities of most other enzymes are negligible after treatment at 175°F (80°C).

It is possible to test for the presence of peroxidase by the following method:

A freshly prepared mixture of 1 per cent quaicol and 1.5 per cent solution of hydrogen peroxide (5 volumes) in equal volumes is 'spotted' on to the blanched vegetables—a red-brown colour develops in under a minute if peroxidase is still present.

However, extending the blanching times beyond those recommended in the section on Freezing Fresh Vegetables is not advisable, even in the event of getting positive peroxidase tests; this is because the texture of the vegetables, which are almost cooked as a result of this longer time, is inferior to those frozen after blanching only. Maintaining a non-fluctuating temperature of 0°F (−18°C) during frozen storage is more effective in ensuring high quality.

Blanching before freezing also helps to preserve the green colour in vegetables. This is due to the inactivation of the enzyme lipoxidase which can cause the decomposition of chlorophyll. During blanching, air and the other volatile substances are removed from the large parenchyma cells, which make up a large percentage of the plant tissue, and are replaced with water; the translucency of the tissue gives a ready indication that the enzymes have been inactivated.

Enzyme action in fruit

The enzymes contained in some fruits such as apricots and peaches cause undesirable discoloration. This discoloration occurs most rapidly during the preparation for freezing and on subsequent thawing. It is the result of the enzymic oxidation of the catechol tannins in the fruit by the enzyme polyphenol oxidase. The darkening of fruits can be prevented by blanching as with vegetables. This method is not suitable for those fruits usually eaten raw as some of the subtle flavours and aromas may be lost. However, fruits which are stewed or served in syrup, may be blanched for a short time in hot syrup before being transferred to cold syrup for packaging.

Lowering the pH by the addition of acids is another means of limiting darkening. Citric acid may be added to the water used in preparation or be added directly to the fruit or syrup where the resulting tartness is acceptable.

Ascorbic acid (Vitamin C) is widely used to control the browning action in fruits. It slightly lowers the pH value of solutions to which it is added, but its principal function is that of a reducing agent.

The action of ascorbic acid is indirect; it inactivates the poly-phenol oxidase by acting upon the oxidation product produced by it, and not on the enzyme itself. (This is in contrast to the action of sulphur dioxide sometimes used in fruit bottling which acts directly upon the polyphenol oxidase.)

It is important that sufficient ascorbic acid is added so that it is not spent before all the polyphenol oxidase has been inactivated. When freezing fruit juices, discoloration can be permanently prevented by adding ascorbic acid a little in excess of that required to inactivate the enzyme. But fruits frozen in thick sections cannot be completely penetrated by the ascorbic acid, and the discoloration may only be delayed and not completely prevented. However, if the fruit is treated with the ascorbic acid immediately prior to freezing, the discolora-tion may be delayed for several hours after the fruit is thawed. Another contributory factor to the maintenance of a good colour is the extent to which the container is filled. It is found that fruit in a well-filled container keeps the colour better during freezing and thawing than fruit in a partly filled container, as the amount of entrapped oxygen is reduced.

The enzymes pectinesterase and polygalacturonase (pectinase) are important in the production of frozen concentrated juices. Their effects are to hydrolyse the pectins which are responsible for keeping in suspension many other constituents. Pectinesterase and poly-galacturonase continue to be effective when the juices are frozen, with the result that sedimentation occurs; this is normally referred to as 'loss of cloud'. Heat treatment to 190°F (90°C) before freezing is effective in inactivating these enzymes and ensuring a stable product.

Enzyme action in fish and meat

Although fish and meats also contain enzymes, their activities during frozen storage are not considered sufficiently troublesome to warrant any special treatment prior to freezing.

It might be expected that enzymes present in fish would continue to function during frozen storage since they normally function at the lower temperatures found in the sea. It may be that the storage temperature of $-20°F$ ($-29°C$) used commercially to obtain a high-quality product serves to inhibit enzymic activity.

Enzyme action in pre-cooked and prepared foods

The cooking given to most dishes cooked before freezing is sufficient to destroy the enzymes present. However, those dishes prepared but not cooked before freezing may require some treatment when they incorporate raw fruits or vegetables. For example, it is recommended that an apple tart be treated with an antioxidant when frozen raw.

Because most treatments aimed at inactivating the enzymes in fruit and vegetables are rarely completely effective, it is important that they are always combined with good practice during freezing, frozen storage, and ultimate use.

Preparation and freezing should always be carried out with the minimum delay possible; it is important that the containers be well filled and properly sealed and the temperature reduced to $0°F$ ($-18°C$) as quickly as possible.

Frozen storage should be maintained at $0°F$ ($-18°C$). When a higher temperature is maintained, or fluctuations permitted, the enzyme activities may recommence.

Finally, foods should be used as soon as they are taken from the freezer. Transferring frozen vegetables to fast boiling water is the acknowledged recommended practice, as this provides little opportunity for the recommencement of enzymic activity. However, some dessert fruits which are not cooked may start to show signs of discoloration soon after thawing is completed, particularly if they have not been treated with an antioxidant. Dessert fruits will be found to keep their appearance better if they are served while still containing a few ice crystals.

MICRO-ORGANISMS

Micro-organisms are present almost everywhere, for example in the soil, in the air, in water, on the skin of human beings in the intestines of animals and on foods. They are able to grow and multiply rapidly when conditions are favourable. The vast majority of micro-organisms are harmless, in fact some are actively beneficial, but there are some micro-organisms capable of causing spoilage in foods and others whose presence may cause food poisoning and infections. Both spoilage and food-poisoning organisms can occur in foods which are frozen; they can survive many months of frozen storage and may start to multiply when the food is ultimately thawed. However, food spoilage and food poisoning only occur when the number of such organisms reaches a high level.

In order that frozen foods shall be wholesome and safe it is important to appreciate the distribution and nature of the activities of the micro-organisms, and the factors which affect their growth.

Factors effecting the growth of micro-organisms

(a) All micro-organisms require moisture in order to grow. The absence of available moisture in frozen foods is a contributory factor to the absence of growth of micro-organisms in frozen foods stored at 0°F (−18°C).

(b) Yeasts, moulds, and bacteria, differ in their requirement for oxygen. Yeasts grow in the presence or absence of oxygen, i.e. they grow aerobically or anaerobically. Most moulds and many bacteria are aerobic, but certain bacteria, including some food-poisoning strains, are anaerobic.

(c) Both yeasts and moulds grow well in acid conditions. With some exceptions (notably the lactic acid bacteria), bacteria flourish in a more alkaline medium; thus a pH value of 6–7 is very common for bacterial growth. Therefore yeasts and moulds can be expected to grow and cause spoilage on fruits which are acid, while bacteria can grow on vegetables which are more alkaline. However, it is not unusual for one organism to raise or lower the pH value of the medium and so make conditions suitable for a second organism.

(d) All micro-organisms need certain nutrients for growth, but some organisms are more exacting in their requirements than others. The drip which leaks away from certain food materials when they are allowed to thaw is rich in dissolved nutrients and provides a favourable medium for the growth and multiplication of micro-organisms in thawed foods.

(e) Microbial growth can occur between 15° and 165°F (−10° and 75°C), but there is for each organism a temperature range within which growth is most rapid.

46

Micro-organisms may be classified according to their temperature requirements, as follows:

	Temperature range for growth	Optimum temperature for growth
Psychrophilic organisms	15 to 70°F (—10 to 20°C)	45 to 55°F (7 to 13°C)
Mesophilic organisms	70 to 115°F (20 to 45°C)	85 to 105°F (30 to 40°C)
Thermophilic organisms	115 to 165°F (45 to 75°C)	130 to 140°F (55 to 60°C)

It is the slow growth of the psychrophilic organisms which can cause spoilage in frozen foods should large temperature fluctuations occur during frozen storage, or when too high a storage temperature is maintained. Moreover, holding foods at unsuitable temperatures during the preparation prior to freezing can result in multiplication of mesophilic and thermophilic spoilage and food-poisoning organisms.

It is important to appreciate that when environmental conditions are favourable to the micro-organisms present, they are able to multiply extremely rapidly. The organisms may divide several times an hour, with the population doubling at each generation. For example, 1,000 bacteria dividing every fifteen minutes reach a total of 65,536,000 in four hours!

Food-poisoning bacteria

Bacterial food poisoning is brought about either by eating foods which contain toxins or poisons already produced in the food by certain bacteria, or alternatively by eating foods containing bacteria which are capable of multiplying and producing their toxins within the body.

The table on page 48 sets out the principal food-poisoning bacteria and the type of food poisoning for which they are responsible; the table also gives the sources of organisms which may be man, animal, animal products, soil, dirt, and also inanimate objects such as food mixers.

The table also shows the nature and distribution of the food-poisoning organisms; they are already present in certain foods and may be conveyed to other foods by human and other sources.

FIGURE 2

PRINCIPAL FOOD POISONING BACTERIA

Organism	Type of food poisoning	Source of organisms	Foods most usually implicated	Resistance of organism and toxin to heat
Staphylococcus (tolerant of high sugar and salt concentrations)	Toxin, i.e. toxin produced in food before consumption	Chiefly human beings in nose, boils and lesions (on the hands and elsewhere)	Sweet foods, e.g. custards, synthetic creams; cooked meat, especially ham and boiled bacon	Toxin may withstand boiling for ¼ hour, non-sporing vegetative organisms are heat sensitive
Clostridium botulinum	Toxin	Soil, dirt, sewage	Improperly processed, canned, and bottled vegetables, and less acid fruits, e.g. pears, and meats	Toxin destroyed by boiling, but spores resist boiling for some hours
Clostridium welchii	Toxin or infection (not determined)	Human and animal faeces, carcass meat	Cooked meats, reheated meat, and meat pies	Spores resist boiling for some hours
Salmonella	Infection, i.e. bacteria multiply in the intestine	Pigs, poultry, and hence meats. Animal produce, e.g. eggs (usually duck eggs), bulk frozen eggs (since 1964 required to be pasteurized), human beings, dogs, cats, mice, rats, flies and insects	Cooked meats and poultry, reheated meat and meat pies, sweet foods such as custards, eggs and egg products, cream cakes	Non-sporing vegetative organisms are heat sensitive

48

3a. Small chest freezer,
4.0 cubic feet

3b. Large upright freezer,
9.5 cubic feet

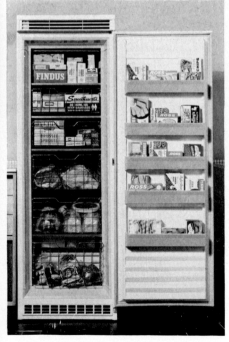

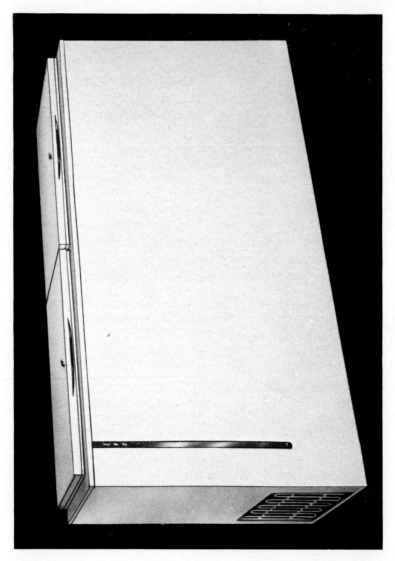

4. Large chest freezer, 18.0 cubic feet

Preventing the organisms from reaching the final food product in significant numbers must form the basis of all methods of food preparation. These preventative measures are discussed more fully in a later section giving special emphasis to the preparation of foods for freezing.

Food-poisoning organisms can occur in frozen foods. The process of freezing, although substantially reducing the numbers of organisms, does not destroy all those present, and if food-poisoning organisms were present before freezing, they could number among the survivors. However, at 0°F (-18°C), the organisms are dormant, vegetative cells cannot grow, and spores cannot germinate; there can be no multiplication of organisms or production of toxins within the food at this temperature. However, once food is thawed and partially warmed up, any food-poisoning organisms present could become active again.

Food-spoilage organisms

Spoilage organisms, which include species of yeasts, moulds, and bacteria do not represent the same danger to health as the food-poisoning organisms, but may be responsible for a great deal of food wastage.

Many species are found as the natural flora of foods: for example, yeasts and moulds are found on the skins of ripening fruits; certain bacteria are found in milk. Normally their presence is not significant, but under certain conditions they can multiply rapidly and cause spoilage. Many species are airborne so that it is impossible to prevent contamination under normal conditions of preparation, although keeping foods covered as much as possible greatly reduces this form of contamination.

Food spoilage manifests itself in a number of ways. The presence of small numbers of spoilage organisms in food cannot be detected by the naked eye, and is not significant unless multiplication occurs; but if conditions are favourable for growth they may form colonies which are clearly visible. The colonies formed by yeasts and bacteria are often pigmented and brightly coloured; in some cases they are slimy. Moulds, which are essentially aerobic, form a fine network of threads (mycelium) resembling cotton wool over the surface of the foods.

Most spoilt foods become discoloured and show texture changes, for example spoilt food may appear 'mushy' and collapse when lifted from the container. Frequently an unpleasant and sour smell

49

accompanies the putrefaction of certain foods, leaving the potential consumer in no doubt as to its condition.

Many spoilage organisms are psychrophilic and therefore of particular interest when using freezing as a means of food preservation. Some species have been known to grow at temperatures below freezing-point; for example, the airborne mould *Cladasporium* has been found to cause spoilage within cold stores at temperatures little above 20°F (−7°C).

Spoilage organisms can occur in frozen foods. The preparation given to most foods would not destroy all the vegetative organisms present, although all practical measures should be taken to minimize all further contamination. Many of the organisms present will be killed by the freezing process, although some vegetative cells may survive. Providing a frozen storage temperature of 0°F (−18°C) is maintained, the surviving organisms will remain dormant. Large temperature fluctuations or the maintenance of too high a storage temperature within the freezer could result in renewed activities by the spoilage organisms, with resulting deterioration of the stored product.

When a food is removed from the freezer and thawed, it should be used promptly; transferring it to a domestic refrigerator, although preferable to allowing it to remain at room temperature, will not prohibit the growth of psychrophilic spoilage organisms.

The aims of food preservation

At this point it is appropriate to summarize the aims of food preservation, since these principles underly all procedures and methods adopted in the following sections.

These aims are:

(a) To reduce the number of spoilage organisms present in the food.
(b) To destroy any pathogenic organisms present in the food.
(c) To prevent further contamination of food, either by large numbers of spoilage organisms or by any pathogens.
(d) To prevent the multiplication during storage of any organisms still present.

Preparation and freezing of foods

Most foods contain some micro-organisms, the number and varieties of which vary with the origin and history of the food. For example, together with those organisms which are present as their

50

natural flora, vegetables contain organisms normally found in the soil, and fish contain organisms which are water-borne. All foods may contain organisms which are transmitted to them by human agencies.

Under good conditions of preparation the number of micro-organisms initially present in a food can frequently be substantially reduced. This may simply be as a result of washing as for fruits, or by washing and blanching as for most vegetables. Normally a number of micro-organisms remain after preparation or after completion of preservation—and this number may include some pathogenic organisms, although the elimination of these organisms should be the aim of hygienic methods of preparation and production.

In all methods of food preservation, and in freezing in particular, where there is no final heat sterilization of the food to destroy pathogenic organisms or toxins produced by them, the need for a high standard of hygiene cannot be over-emphasized.

Whether dealing with a large-scale operation, or in a small domestic kitchen, the essential points are unchanged and can be listed as follows:

(a) A high standard of hygiene on the part of the food handler is the first consideration—for example clean overalls, suitable hair covering, clean hands and nails, and not least, the observance of a strict code of personal hygiene.

(b) Animals, including domestic pets, should be excluded from areas where food is prepared or stored, and food protected from other carriers such as insects and flies.

(c) All possible precautions should be taken to prevent the food from becoming further contaminated during the period of preparation, e.g. slicing and mincing machines, chopping boards and other equipment should be thoroughly cleaned each time they are used. The fluid or juices extruded from foods provide an ideal medium for the growth and multiplication of micro-organisms, and when left on pieces of equipment can easily cause contamination of subsequent batches of food with large numbers of bacteria. The selection of suitable equipment which can be properly maintained and cleaned minimises the risk of this type of infection.

(d) Frequent changes of clean water should be used for all washing purposes. Where the same washing water is used for successive batches of food it may acquire a high bacterial load. This is because organisms may multiply rapidly in wash water containing organic matter. They can then be deposited in large numbers on successive batches of food.

(e) Certain foods are regarded as 'suspect foods' in that they are known from past experience to be more likely than other foods to contain pathogenic organisms: for example poultry may carry salmonellae, and reheated meat may carry *Clostridium welchii*. Such foods therefore warrant special care.

(f) It is important when preparing food for the freezer (as in all forms of food preservation) to avoid holding prepared and partially prepared foods at temperatures at which pathogenic organisms are able to grow. (The Food Hygiene Regulations 1960, stipulate that food must not be held at temperatures between 50° and 145°F (10° and 63°C)). Thus blanched vegetables, cooked meat pies, etc., must be cooled rapidly, and if they cannot be immediately accommodated in the freezer, they should be stored temporarily in the refrigerator. Where possible the programme of food preparation and freezing should be planned ahead to avoid delays of this nature.

Frozen storage

The freezing process results in a considerable reduction in the number of micro-organisms present in a food, although a substantial number may survive.

The number of surviving micro-organisms will depend to some extent on the type of organisms initially present and the nature of the food; for example, fatty substances offer protection for some micro-organisms, but a low pH value hastens the rate at which organisms are killed.

It has been found that many micro-organisms die off more quickly at 30° to 25°F (−1° to −4°C) than at lower temperatures, but a significant number may remain. Some organisms, including pathogenic ones, have been found to remain viable for long periods at temperatures considerably below 0°F (−18°C). The spores of spore-forming organisms are capable of surviving frozen storage.

As long as the food material remains frozen at 0°F (−18°C) or below, there will be no danger of an increase in the surviving population of micro-organisms, even though a very small percentage of available water remains unfrozen. As seen in Fig. 3 the lowest temperature at which growth of psychrophilic organisms has been recorded is at 15°F (−10°C).

Fluctuating temperatures, or a storage temperature of above 0°F (−18°C) have an undesirable effect on the quality of the food; moreover fluctuations of sufficient magnitude can produce conditions allowing growth of spoilage organisms. Only in the event of gross mishandling when frozen foods are allowed to thaw out and partially warm up, do conditions exist which permit pathogenic organisms to multiply and spores of the spore-forming organisms to germinate.

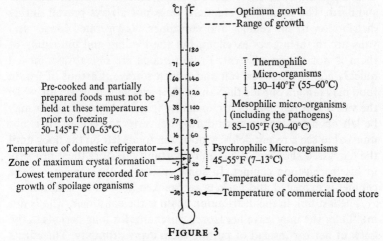

FIGURE 3

SIGNIFICANT TEMPERATURES FOR FROZEN STORAGE

Thawing

When a food is taken from frozen storage and thawed, the combined effect of raising the temperature and increasing the available water content as a result of melting ice crystals, can produce conditions allowing the growth of micro-organisms present. The water contains dissolved nutrients withdrawn from the cells during freezing, and this may constitute a suitable medium for the growth of micro-organisms. Reference to Fig. 3 will show that no pathogenic organisms can grow at defrost temperatures. However, once a food is thawed, it should not be left at room temperature, but should be put into a refrigerator or consumed immediately.

Opinion is divided as to whether some foods are more perishable as a result of frozen storage. It is possible that the more delicate plant tissue is affected by the combined blanching and freezing, and thus rendered more susceptible to attack by micro-organisms. However, it would seem more probable that as a result of metabolic injury the growth of most micro-organisms is retarded rather than accelerated by frozen storage.

Handling and distribution problems

The frozen food industry has an excellent history in that no cases of food poisoning have been attributed to frozen foods. This would indicate that conditions under which frozen foods are produced are of a high standard; however, it is important to maintain this high standard. The same high standard does not always prevail during distribution to the retailer and consumer. Refrigerated lorries and vans are in themselves excellent, but the loading and unloading of them is not always efficient. Frozen goods are not always moved quickly enough between van and frozen storage. Cartons of frozen food may remain outside the van on the pavement to 'warm up' while the van driver reorganises his load, and frequently the van door may be left open for long periods during deliveries to retailers. It is not unusual to see a recent delivery of frozen foods left on the floor until the harrassed shopkeeper has an opportunity to store them properly in his frozen-food cabinet. In some cases also the capacity of the cabinet is exceeded and packets are stacked above the 'load line'.

The last link in the distribution chain is the consumer, who is not infallible; she may leave her frozen purchases for long periods in the back of her car instead of putting them away promptly. The effects of such mishandling are cumulative—several fluctuations in the temperature occurring to the same packet will not improve the quality and where the temperature rises are of sufficient magnitude the result may be an increase in the number of micro-organisms.

Since the mishandling of frozen foods can constitute a danger to health, there have been requests from some authorities for legislation and minimum standards to be laid down for these foods. There is much to be said in favour of such a proposal, but to produce such a set of standards to embrace all types of organisms in all types of food is a formidable if not impossible task. Meanwhile it is essential that emphasis be placed on high standards of hygiene and a good code of practice in the production of frozen foods, and on maintenance of correct and unfluctuating temperatures throughout the distribution and storage system.

REFERENCES

Enzymes. H. W. Schultz. A.V.I. Publishing Co. Inc. 1960.
Food Microbiology. W. C. Frazier. McGraw Hill. 1958.
Practical Food Microbiology. H. C. Weiser. A.V.I. Publishing Co. Inc.

THE IMPORTANCE OF ENZYMES

The Technology of Food Preservation. N. W. Derosier. A.V.I. Publishing Co. Inc. 1959.
'Microbiological Problems of Frozen Foods.' G. Borgström. *Advances in Food Research,* 6. 1955.
'The Physiology of the Microbial Spoilage of Foods'. D. A. A. Mossell and M. Ingram. *Journal Applied Bacteriology,* 18. 1955.
'The Survival of Food Poisoning Bacteria in Frozen Foods.' D. L. Georgala and A. Hurst. *Journal Applied Bacteriology,* 26. 1963.
'Microbiological Standards and Handling Codes for Chilled and Frozen Foods.' R. P. Elliot and H. D. Michener. *Applied Microbiology,* 9. 1961.
Food Hygiene Regulations, 1960. Part IV Relating to Food Premises.

4

The Freezer

THE increasing choice of freezers on the domestic market today may seem confusing to anyone who is contemplating ownership of one of these pieces of household equipment. Their range of capacity is from 2 to 25 cubic feet, or more. They may be tall and slim, or long and low; they may be well endowed with shelves and baskets, or fitted out sparsely.

WHAT SERVICE DOES A DOMESTIC FREEZER PROVIDE?

To store fruit and vegetables

The most obvious function of a freezer is as an alternative method of storing surplus fruit and vegetables. The freezing of garden produce is less complicated and time consuming than either canning or bottling. At the same time, produce placed in a freezer more closely resembles the fresh product than that preserved by other methods. There is the added advantage that while it is hardly worthwhile getting out canning equipment to preserve the odd pound of fruit or vegetables, small quantities of produce may sensibly be preserved by freezing with very little interruption of the daily schedule.

Many families think first about buying a freezer when confronted with bumper crops in their gardens. However, just as important are some of the other functions of a freezer, which are listed below, and which also provide sound reasons for such a purchase.

To store meat and fish

Deep freezing provides a means of preserving various quantities of meat and fish for which alternative domestic methods do not readily exist.

THE FREEZER

For convenience foods

The freezer may be used to store made-up and pre-cooked foods, whether purchased or made at home. It is good practice when preparing a main course such as a goulash or steak and kidney pie, to double or treble the quantities prepared. Sufficient for one or two meals may then be stored in the freezer. Prepared or baked pies, cakes, and bread, may also be stored satisfactorily, so that unexpected visitors can be catered for at any time with a minimum of effort.

In emergencies

The freezer may function as an invaluable service in an emergency. It is always worthwhile to keep in a freezer a small amount of some of the usual daily requirements—a loaf of bread, a carton of milk, butter, some eggs (broken individually into waxed cartons). These may be brought out when necessary. They should, however, not be allowed to remain indefinitely in the freezer, but used and replenished at regular intervals.

Entertaining

A freezer may be put to useful service when a party is planned. A great deal of preparation may be carried out ahead of time, and a variety of foods stored at zero Fahrenheit, thus eliminating much of the strain of final preparation. Suggestions will be made for party preparations in a later section of this book.

When week-end guests are expected menus can be planned, and dishes prepared in advance, and frozen. Thus instead of spending much of her time in the all too familiar surroundings of her kitchen, the hostess has more time to spend in the company of her guests.

CHOOSING A FREEZER

The selection of the freezer will be determined largely by the services expected of it.

How many cubic feet?

The type of produce to be stored and the size of the family help to supply the answer to this question. Consider, for example, the following three families:

The first family is that of three adults living on a smallholding where fruit and vegetables are grown for sale and a few dozen poultry are kept.

If the birds are plump but the market price below average, it may well benefit the smallholder to kill and prepare his birds and to hold them in the freezer until the price is more favourable. Or it may be that unsettled weather prevents his outdoor work, but permits time for plucking and preparing poultry. Again, these birds may be held at zero Fahrenheit until they are required for sale. This semi-commercial use of the freezer combined with an abundance of fruit and vegetables necessitates a unit of good capacity—say 24 cubic feet.

The second family consists of two adults, and three children, living in a suburban home with an average-sized garden. Their garden yields more soft fruit and vegetables than they can eat, so that any surplus may conveniently be frozen. At the same time, the family like pies and other baked produce which may be prepared in quantity, and frozen until required—this is particularly useful when the children are on holiday. A unit of, say, 8–12 cubic feet would be satisfactory for their needs.

The third example is of a couple living in a city flat. They have no garden, but are able to buy frozen produce in bulk, for example peas in 5 lb. packs, and chickens when cheap from the supermarkets. Both members of the household work during the day, but they entertain a great deal, and therefore find it useful if dishes for a dinner party may be prepared in advance. A freezer of 4 cubic feet capacity would probably be adequate for their needs.

It is not possible to generalize on the number of cubic feet to allow for each member of the family. Estimates would range from 2 to 5 cubic feet, depending largely on the availability of garden produce and home-killed meat and poultry.

It is worth noting that the life of a freezer could exceed fifteen years, and allowance should be made for the increase in children's appetites as they grow older. It would be false economy to outgrow the freezer before the end of its useful life.

A would-be purchaser of a freezer who has made a calculation of his requirements is usually advised to increase his estimate and buy as big a freezer as he can accommodate or afford, there being no limit to the uses to which the freezer may be put!

Turnover

It is usually assumed that each cubic foot of freezer space will accommodate between 25 and 30 lb. of frozen produce. Baked foods are lighter and the quantity stored in terms of weight will be less. Bulky packaging also reduces capacity. It is worth while to make an estimate of the quantity of fruit and vegetables which may be expected annually, and at what period they become available—early or late summer, or autumn crops.

It must be stressed that a freezer should not be used for hoarding foods. Although it is quite possible to be eating the previous year's French beans just two weeks before the new season's crop is ready, there is no great virtue in doing this. As with any piece of equipment, the more it is used, the greater will be the economy, and this applies particularly to freezers whose running costs remain the same whether full or empty.

By careful planning and management the quantity of food stored in a freezer during the year may be double or treble the actual capacity of the unit. For example, a freezer of 250 lb capacity may accommodate during the year:

300 lb. fruit and vegetables
200 lb. meat and poultry
100 lb. baked and prepared foods
50 lb. commercial products.

Location

The position of the freezer is of utmost importance, and must be decided before the relative merits of the different units can be considered.

It may be sited in the kitchen, providing a position can be found not adjacent to sources of heat. If the unit is in frequent use the kitchen would be a particularly convenient location. Often space is more readily available in the garage, vestibule, or utility room. Wherever the freezer is to be installed there are some important points to remember.

If the surrounding air is very warm, as in a utility room which also houses the central heating boiler, the compressor is forced to operate for many more hours than is desirable each day, and consequently the life of the unit is reduced. In contrast, the compressor will not function correctly if the ambient temperature is very low, as might occur in an exposed outbuilding during frosty conditions.

It is important that there is a good circulation of air around the freezer. An alcove or disused pantry may seem a good place to put the freezer, but not when it fits so snugly that the air cannot circulate around and behind it. This is because the condensers of most freezers lie between the inner and outer linings of the cabinet, and the heat removed from the foods is dispersed through the walls of the cabinet into the surrounding air. This type of condenser is usually described as the 'skin' type.

Some manufacturers use a fan to get rid of this heat extracted from the storage chamber, and they claim that no allowance need be made for circulation of air around the unit. This type of unit with a fan-cooled condenser will be recognized by the presence of a grille on the front of the cabinet, usually to be found in a lower corner. When a freezer with a fan-cooled condenser is running no heat is transmitted through its walls and the cabinet feels cold to the touch. If a freezer with a fan-cooled condenser is situated in a damp atmosphere, it will become covered with condensation, which results in deterioration of the enamel and fittings. The exterior of a freezer with the skin type of condenser is warm to the touch and therefore is not liable to the effects of condensation.

Two rather elementary points should be remembered before purchasing a freezer. The first is the weight of the unit. This is usually given on the specification of the freezer, but to this must be added the capacity of the unit in terms of pounds of frozen food. Obviously a floor must be strong to support a fully loaded freezer.

Secondly, consideration must be given to the doorways and staircases through which the freezer must be taken in order to install it. This must be done before the freezer is unloaded from the delivery van!

Finally, of course, there must be a suitable power point available. Wherever possible this should be used exclusively for the freezer; where a twin socket or an adaptor is used there is always the possibility of the unit being switched off by mistake. Remember that if there are any young children in the household this plug should be inaccessible to them.

Chest or upright

Freezers are available in two types as shown in Figs. 4 and 5. The chest type is available in capacities from about 4 to 24 cubic feet. With larger models the lid is counterbalanced. Because cold air is heavier than warm air there is little loss of cold when the freezer is opened. The depth of the storage compartment may be 30 inches, which represents a considerable reach for the average user to recover a package from the front, and an even greater reach to the back. In the large chest freezers there is the danger of some packages being

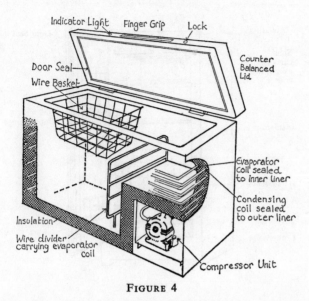

Indicator Light   Finger Grip   Lock

Counter Balanced Lid

Door Seal

Wire Basket

Evaporator coil sealed to inner liner

Condensing coil sealed to outer liner

Insulation

Wire divider carrying evaporator coil

Compressor Unit

FIGURE 4

DIAGRAM SHOWING MAIN FEATURES OF CHEST
FREEZER
(skin type evaporator)

overlooked or temporarily lost in the bottom corners of the storage compartment. However with the use of wire baskets which slide across the top, and at the same time the employment of a methodical storage system, these disadvantages may be overcome.

Problems also exist with the small chest freezers. The location of a particular package of frozen food in a well-filled cabinet is very difficult. It is necessary to take out most of the packages of frozen food before a packet can be removed from the bottom. A considerable amount of rearrangement of a well-stocked cabinet is also necessary in order to freeze new produce, if it is to be placed in contact with a freezing surface without touching any of the already frozen food. Some manufacturers supply a wire basket to fit into the top of the cabinet to overcome these problems, but when fully loaded this becomes a heavy item to lift out.

Upright freezers can now be obtained with capacities of up to 20 cubic feet. Because of the loss of cold air from the freezer each time the door is opened, they are usually considered to be less efficient and at the same time less economical to operate than the chest type. However, the temperature recovery of modern upright freezers is

61

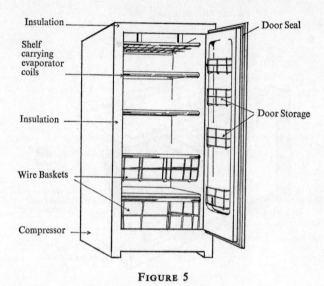

Insulation

Shelf
carrying
evaporator
coils

Insulation

Wire Baskets

Compressor

Door Seal

Door Storage

FIGURE 5

DIAGRAM SHOWING MAIN FEATURES OF UPRIGHT
FREEZER

quite rapid, so that this is not a serious drawback. Upright freezers which have evaporator tubes running in some of the shelves have the advantage of providing more space for faster freezing. To reach to the back of the shelves may be as much as 24 inches, and in order to recover an item from the back the whole contents of the shelf must be unloaded. This has the combined disadvantage of keeping the door open for several minutes and a possible warming up for those packages being rearranged. The speedier location of frozen produce has been achieved in some upright models by pull-out and swivel baskets. However, where the door is opened several times daily, as for example by small children in search of ice creams and water ices, a layer of ice builds up on the shelves and baskets, making their withdrawal very difficult.

Upright freezers are usually more expensive to buy, and at the same time involve slightly higher running costs. Chest models may take up twice as much floor space as upright models of the same capacity, although an adequate allowance has to be made for the opening of the door on an upright freezer, particularly if there are pull-out baskets for easier access as shown in Fig. 6. The tops of chest models may provide good work-top space if there is a generous toe allowance at the front.

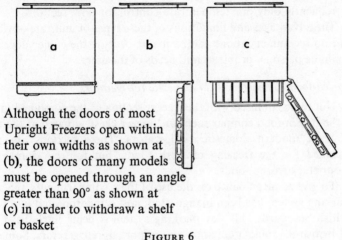

Although the doors of most Upright Freezers open within their own widths as shown at (b), the doors of many models must be opened through an angle greater than 90° as shown at (c) in order to withdraw a shelf or basket

FIGURE 6

FREEZER DOOR OPENING

The ultimate choice of a chest or upright freezer thus depends upon the personal problems and requirements of the user.

A second freezer

In these days of second cars, second bathrooms, and second television sets, it is not difficult to produce a very good argument for the second freezer. The need for this might quite simply be due to the first freezer selected having proved too small. Or alternatively, whereas space was not available to install one big unit, space could be found for two smaller ones. For example, a chest type in the garage and an upright one in the kitchen. Two freezers might prove a better proposition where little turnover of frozen produce is expected, so that after one freezer was emptied it might be turned off until next required. A second freezer also provides a type of personal insurance—should one freezer break down it may be possible to transfer much of the contents into the one which is working.

Refrigerator and freezer

Several manufacturers produce what are described as 'twin units', when refrigerators and freezers are similarly styled and intended to stand side by side in the kitchen. In this case each unit would have

have its own temperature control. Some manufacturers produce a combined refrigerator and freezer. The two units are housed together, either side by side or one above the other, in the same cabinet. There is frequently only one temperature control for both sections.

The advantages and limitations of these types of units are obvious and no recommendations can be made, because the choice depends upon the personal problems and needs of the user.

Frozen-food compartments in domestic refrigerators

Not to be confused with the freezer section of the combined units is the frozen-food compartment or 'freezer chest' situated at the top of most modern domestic refrigerators. These sections are not intended for the freezing of fresh foods, but for the storage of proprietary frozen foods.

To give some guidance on the use of these compartments, the star marking system has been inaugurated, and is incorporated in a new British Standard. This star marking system is being introduced on all British-manufactured, and some imported, refrigerators. Some of the frozen-food manufacturers use the same star marking on their products so that the consumer has a clear and simple directive on how long a specific product may be kept in the frozen-food compartment of a particular refrigerator.

MAXIMUM TEMPERATURE IN THE FROZEN-FOOD COMPARTMENT		LENGTH OF TIME FROZEN FOODS MAY BE STORED
*	21°F (− 6°C)	1 week
**	10°F (−12°C)	1 month
***	0°F (−18°C)	3 months

It is possible to use a three-star frozen-food compartment which has a maximum temperature of 0°F (−18°C) for freezing small quantities of fresh produce. However, it is important not to overload such a small unit. The capacity may be as little as 1 cubic foot or less, so that the amount of food which can be added per twenty-four hours is probably less than 3 lb., and this should not be allowed to contact food already frozen.

ECONOMICS OF A FREEZER

This section is of importance to the serious freezer user, particularly where home-grown produce will be frozen. With the continually improving range of products and improved distribution methods, the commercial foodstuffs—particularly when purchased in bulk—can prove a more economical proposition than those frozen domestically. Thus the economics of home freezing must be appreciated.

The cost of keeping a freezer is more than just the cost of electricity to operate the unit. In strict accounting language there are the fixed and the running charges.

Fixed charges

The fixed charges comprise the following two items:

(a) Interest on capital used to purchase the unit

The theory here is that the money, if not used to purchase a freezer, could be employed in earning money in other ways, such as interest in stocks, savings certificates, etc. Thus instead of spending £100 on a freezer, it could be earning £5 per annum if used to purchase 5 per cent Loan Stocks. If the freezer is on hire purchase, there is still the capital outlay, usually larger than when initially purchased outright, but spread over some years; there is still, however, the loss of interest on such capital.

(b) Depreciation of capital used to purchase the unit

Nearly all equipment needs replacing after so many years, and in order to purchase a replacement freezer money should be set aside as a depreciation charge to accumulate to the required purchase cost. For a freezer, fifteen years is a realistic life, and thus nearly 7 per cent of the purchase price needs to be set aside each year to accumulate to the original purchase cost after fifteen years. With the inflationary spiral which appears unavoidable in our society, something more, say 10 per cent, needs to be put aside to purchase an equivalent size of freezer after fifteen years. However, to possess the original capital sum after fifteen years, the 7 per cent depreciation charge is applicable.

Running charges

The running charges involve the following three items:

(a) Cost of electricity

According to the size of the freezer, so a large or small compressor is required to keep the freezer at the required temperature. A small unit uses a $\frac{1}{12}$ h.p. compressor, while a large unit could use a $\frac{1}{2}$ h.p. compressor. Take as an example a small 4 cubic foot freezer with a

$\frac{1}{12}$ h.p. compressor, the cost of power is calculated as follows:

$$\frac{1}{12} \text{ h.p.} = \frac{1}{12} \times 0.746 \text{ kilowatts.}$$

Daily power consumption $= \frac{1}{12} \times 0.746 \times 1.75$ (per Kw) $\times 8$ hours (assuming electricity is 1.75d. per unit and the motor is running 8 hours per day on average throughout the year. Obviously the motor will function more in the summer than in the winter months, and also when additions of food are being made).

$= \frac{1}{8}d.$ for 4 cubic feet.

There is a slight decrease in the ratio of current consumed per cubic foot as the size of the freezer increases, because the losses depend upon the surface area (a square law), while the initial power required depends upon the cubic capacity (a cube law). However, as an approximation, the power requirement is roughly *pro rata* with the cubic capacity; thus the running costs are slightly less than $\frac{1}{4}d.$ per day per cubic foot, or $1\frac{1}{2}d.$ per week per cubic foot.

(b) Repair costs

With modern freezers the repair costs are low. Nearly all compressors have a five-year warranty and few give trouble within this period; many will run for ten or more years without any trouble. However, where a repair is required it usually involves the compressor, and the costs can be considerable. An average allowance of 2 per cent per annum would accumulate 30 per cent of the original cost over fifteen years, and would be realistic to set aside for repair bills.

(c) Packaging costs

Foods to be stored in a freezer need to be packed in special packaging materials. There is a wide range of suitable materials— polythene bags, waxed cartons, aluminium containers. Many of these, with careful treatment, may be used a second or third time. Taking into account the purchase of some expensive and some cheap materials, combined with the re-use of others, an estimate of 2d. for each lb. of frozen food stored seems to be fairly realistic.

Summary of costs

The costs are summarized in the following two typical examples:

(a) A £50 freezer containing 4 cubic foot, or 120 lb. full capacity load.

(b) A £100 freezer containing 10 cubic foot, or 300 lb. full capacity load.

Assume the contents are turned over every four months:

	(a) £ per annum 4 cu. ft. unit		(b) £ per annum 10 cu. ft. unit	
Interest on capital	(5%)	2·5	(5)	5·0
Depreciation	(7%)	3·5	(7%)	7·0
Electricity costs	(6d. per week)	1·3	(12d. per week)	2·6
Repair costs	(2%)	1·0	(2%)	2·0
Packaging costs	(2d. per lb. × 120 lb. × 3 loads per annum)	3·0	(2d. per lb. × 300 lb. × 3 loads per annum)	7·5
		£11·3		£24·1

Cost of freezing per lb.

$$\frac{£11\cdot3 \times 240d.}{3 \times 120\,lb.} \qquad \frac{£24\cdot1 \times 240d.}{3 \times 300\,lb.}$$

$$= 7\tfrac{1}{2}d. \text{ (approx.)} \qquad = 6\tfrac{1}{2}d. \text{ (approx.)}$$

Turning over the contents every three months:
Cost of freezing per lb.

$$\frac{£12\cdot3 \times 240d.}{4 \times 120\,lb.} \qquad \frac{£26\cdot6 \times 240d.}{4 \times 300\,lb.}$$

$$= 6d. \text{ (approx.)} \qquad = 5\tfrac{1}{4}d. \text{ (approx.)}$$

Turning over the contents every six months:
Cost of freezing per lb.

$$\frac{£10\cdot3 \times 240d.}{2 \times 120\,lb.} \qquad \frac{£21\cdot6 \times 240d.}{2 \times 300\,lb.}$$

$$= 10\tfrac{1}{4}d. \text{ (approx.)} \qquad = 8\tfrac{3}{4}d. \text{ (approx.)}$$

Thus, providing full capacity can be assured, it is cheaper to use the largest possible freezer and to turn the contents around as often as possible.

The above is strict accounting theory. If, however, one ignores the interest and depreciation charges, which is fair if a freezer is received as a gift and one does not plan to replace it in fifteen years' time, say when the family is grown up and married, then the costs per cubic foot are approximately halved.

Labour costs have also been ignored throughout. It is assumed that no savings would arise if the deep-freezing work was not undertaken; some other chores or garden work would doubtless be completed, but no financial advantage would arise.

CONSTRUCTION OF THE FREEZER

The operating cycle

The following three components combine to produce a low temperature within the storage compartment of a freezer: the compressor, the condenser, and the evaporator. The cycle is easily understood with reference to Fig. 7.

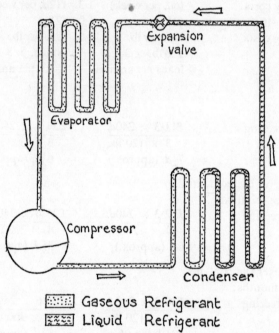

FIGURE 7
THE OPERATING CYCLE OF A FREEZER

The compressor receives the refrigerant gas—usually Freon 12, from the evaporator. It forces it into the condenser, where the effect of temperature and pressure cause it to condense as a liquid. An expansion valve restricts the flow of the liquid into the evaporator, where at a lower pressure it changes into a gas. The gas returns to the compressor and the cycle is repeated.

The effect of the gas condensing into a liquid which it does during its passage through the condenser results in the production of heat which is given off by the condenser into the room. When the liquid changes into a gas in the evaporator, heat is absorbed from the storage compartment and thus from the food contained within.

Compressor

In most freezers the compressor is a hermetically sealed unit, the motor and all the parts are assembled in a dust-free environment and then sealed in a gas-tight compartment. Adequate lubricant is sealed in, and the unit needs no attention. In fact, most manufacturers guarantee the compressor for five years. Some of the larger freezers have what is termed an open type of unit. This needs to be oiled regularly and maintained, but should a fault develop it may be repaired by a local service man and need not be returned to the factory. Thus a freezer of the open-unit type may be put back into service more quickly should a fault occur.

The size of the motor selected in terms of horsepower will mainly depend upon the cubic capacity of the freezer, and on the nature and thickness of the insulating material; a freezer of 4 cubic feet being fitted with a motor of $\frac{1}{12}$ h.p. rating, and a freezer of 24 cubic feet with one of $\frac{1}{2}$ h.p. rating. However, the size of the motor and cubic capacity of the freezer are not always directly related. The precise size of the motor installed by the manufacturer will be modified by his judgement of how he expects the freezer to be used. For instance, upright and chest freezers of the same capacity may have motors of different horsepowers. This is because the manufacturer may consider that the upright model will be installed in a warm kitchen, but the chest model in a cooler place. Also he may presuppose that the upright freezer is the more likely choice of an enthusiastic cook who uses her freezer constantly for advance meal preparations; whereas the chest freezer is the more likely choice of the enthusiastic gardener with an abundance of fruit and vegetables to preserve, when the primary function of this freezer will be one of storage. Thus the manufacturer may install a slightly larger motor in the upright freezer than in the chest type.

Whatever their size, most motors are supplied with an automatic cutout which functions should the motor become overheated.

Condenser

The condenser is basically a system of tubes presenting a large surface area for the transference of heat to the outside air.

The condenser fitted in a chest freezer may be one of two types. The first one is fan cooled; a fan forces air over a system of tubes through which the freon gas is circulating. The air is heated up as a result of the gas condensing to a liquid in the tubes, and the heated air passes through a grille on the front of the cabinet into the room beyond.

The second type is known as the skin condenser. A system of tubes, of copper or aluminium, is welded to the inside of the outer wall of the cabinet and the heat resulting from the condensing gas is dissipated by convection and radiation into the room directly through the wall of the cabinet. Freezers with skin condensers must be allowed an adequate circulation of air around them.

The condenser of an upright freezer resembles that of a domestic refrigerator; on the back of the cabinet is a system of tubes which gives off its heat into the room by convection and radiation, as the room is invariably at a lower temperature.

Evaporator

The evaporator is also a system of coils, with the function of absorbing heat from the food being frozen.

In the chest freezer the coils carrying the liquid refrigerant are welded on to the inner lining of the cabinet. The more closely they are in contact with the lining, the more efficient will be the cooling system. Sometimes the larger chest freezers have evaporator coils within the vertical wire-frame dividers, thus increasing the area of the cooling surface. It usually follows that a greater area of cooling surface produces a more efficient system. Where the cooling surface is small, the temperature at which it operates must be lower if an equivalent volume of food is to be reduced to $0°F$ ($-18°C$). In such cases the compressor is required to function for longer periods.

Upright freezers may have evaporator coils attached to the inner lining of the food compartment, but more frequently there is a refrigerated grid situated in the top of the cabinet, and two or three shelves are refrigerated also. Then the tubes carrying the refrigerant are welded to the base of a solid shelf or within the wire structure of an open one. This method of welding the tube containing refrigerant to the shelf is being superseded by the development of evaporator plates produced by the process of 'roll bonding'. For this process two aluminium plates, one of which has been 'printed' with the desired path for the refrigerant, are sandwiched together. The passageway between them is opened up as a result of hydraulic pressure, and the refrigerant caused to follow the route between the two fused plates. These plates are very efficient and may be used to divide up the storage space in a large chest freezer.

THE FREEZER

General construction

In order that a freezer shall give a satisfactory performance over a number of years, it must be soundly constructed.

The basic structure of the cabinet is usually of steel, which has received an anti-corrosive treatment; sometimes it is made of aluminium. The interior is sometimes stove-enamelled steel, although aluminium and polystyrene are both frequently used for the inner lining. The exterior is usually stove-enamelled steel.

Insulation

The compromise of adequate insulation without excessive bulk or cost is one of the most important design problems. Various materials have been used over the years; asbestos, cork, fibre glass, and more recently, polyurethane. The improvement in the quality of insulating materials may clearly be shown by the reduction in the bulk of all domestic refrigerators and freezers with no reduction in capacity. Polyurethane is used fairly extensively for insulation at the present time. It is pumped in liquid form between the inner and outer linings of the unit, where it expands to a foam which completely fills the space between.

Door and lid seal

In an upright freezer the effectiveness of the door seal is of great importance. Most models are fitted with rubber or, more recently, with plastic gaskets and a simple latch or magnetic method of closure. The large chest models usually employ the same system, although the small top-opening models rely solely on a plastic lid resting on top of the cabinet. With some of these models it is necessary to remove frost and ice which may collect around the top of the cabinet, which would otherwise prevent the lid fitting properly into place.

SPECIAL FEATURES

Accessibility of controls

Many of the small top-opening domestic models have their temperatures pre-set at the factory and no adjustment is necessary even when fresh produce is added. Should an adjustment be necessary, for example when the ambient temperature is higher than normal, this can be achieved by a small knob (frequently coin operated) at the

71

base of the machine. Most of the large chest models have a control situated in the top of the food compartment and adjustments can be made when food is added.

Upright cabinets have a temperature control situated inside the cabinet, similar to domestic refrigerators, and it is usual to increase the setting (and so lower the temperature) when making additions of unfrozen food.

It is worth remembering that when it is not clearly indicated on the control knob, turning it in a clockwise direction, or where numbers are given, going from a lower to a higher number, reduces the temperature within the cabinet.

Special freezing compartment

Special freezing compartments, although not very common in the past are being featured in more of the models produced today. The temperature of the compartment may be lowered by a special switch without altering the thermostat setting for the freezer as a whole. Once frozen the food should be removed from the special compartment and stored elsewhere in the freezer. Where a special freezing compartment does not exist food should be placed against the coldest surfaces—in a chest unit this is against the base or sides. As previously mentioned this may impose difficulties in a small chest unit when fully loaded.

Many of the upright models, even the smaller ones, carry evaporator coils in many of their shelves, thus providing ample surfaces for freezing without the necessity of rearranging previously frozen stock.

Baskets for easy access

These are essential in large chest freezers, and are very helpful in the small domestic top-opening models. In some cases they are featured as an extra to the purchase price. In the larger freezers the baskets can slide across the top of the cabinet and do not need to be lifted out in order to reach food stored lower down. Some upright models have wire baskets in the form of drawers, which can slide in and out above the evaporator coils, or alternatively baskets which can pivot to one side.

These aids have been criticized on the grounds that they take up too much of the storage space, but if they contribute to the efficient use of the unit, then their presence is justified. The doors of upright models may also be fitted to accommodate packets, bottles, and tins.

THE FREEZER

Ease of servicing

Some freezers have a removable panel which gives easy access to the working parts of the unit. This is essential where the condenser is fan cooled and regular oiling and dust removal is necessary.

Lids and locks

Small top-opening freezers escape purchase tax if the lids are not hinged. Lids of the large chest models (which escape purchase tax if over 12 cubic feet capacity) are not only hinged but usually also counterbalanced. Several top-opening models are provided with a special working surface which proves a useful asset when the unit is situated in the kitchen or utility room.

Upright and most large chest models are provided with a lock, and where the freezer is kept in a porch or open garage this is a useful feature.

Interior light

This is obviously an advantage, particularly when the freezer is situated in a dark corner.

Ease of movement

When a freezer is situated in a kitchen the presence of castors or a 'skid rail' will protect the floor from undue denting and scratching, and at the same time permit easy movement when necessary.

Frost-free device

Some large upright freezers are supplied with an automatic defrost mechanism. In such a freezer the evaporator coils are outside the food compartment and a fan blows air across the evaporator coils into the frozen-food compartment, thus cooling it. Once every twenty-four hours the compressor is automatically switched off for a short period, and usually a heating element is turned on to melt the ice which has formed on the coils. Water resulting from the melting ice is collected in a drip tray and usually evaporated off by the heat produced by the normal working of the compressor.

The advantage of this device is that there is never any need to empty out the contents of the food compartments in order to defrost the unit; and ice does not build up on removable shelves and baskets, making them difficult to withdraw. The absence of frost or ice from the food compartment ensures that food packages do not become stuck together, and their labels are always easily discernible.

73

The disadvantage of this system, which relies on a forced-air circulation for its cooling effect, is that unless efficient packaging materials and sealing methods are employed the stored foods may become desiccated. This is particularly so during the defrost cycle, when the air temperature within the food compartment is liable to fluctuations. Another disadvantage is quite obviously one of cost—such a system is more costly, both to produce and to operate. Freezers incorporating the frost-free cycle are usually imported models.

Guarantees and insurances

Most manufacturers guarantee the complete freezer for at least one year, and the refrigerating system for five years. A few manufacturers give a food-spoilage warranty, which covers food spoilage caused by the failure of their product within a specified number of years.

It is worth making inquiries as to the servicing facilities offered by the retailer supplying the unit before making a final decision as to the model to be purchased.

Alarm system

An efficient alarm system is required to indicate a rise in temperature within the food compartment to 5°F (-15°C), as well as giving a warning when there is a break in the electrical circuit. Many freezers claim a warning system, but this is frequently only a red-coloured bulb which glows when the power is flowing to the unit. It does not indicate a rise in the temperature of the food compartment, but is a useful warning.

Proprietary alarm systems are available which can be easily fitted to a freezer. Alternatively a thermocouple can be fitted to the lining at the back of the unit, level with the top layer of food when the compartment is full. It is important to ensure that the connecting wires to the thermocouple affect the efficiency of the lid seal as little as possible. An audible signal is better than a visible one, particularly if the freezer is housed in the garage or outbuildings.

MANAGEMENT OF A FREEZER

Installation

After the freezer has been unpacked, it is advisable to wipe out the food compartment before using it for food storage. A solution of bicarbonate of soda—1 tablespoon to 1 quart of warm water—should be used.

The freezer may then be connected up to the power socket, taking care to earth it correctly. The wisdom of not using the socket for other electrical appliances has been pointed out earlier in this chapter. It is advisable to let the unit run for several hours before using it for the first time, in case there should be any fault.

Adding unfrozen food

The manufacturer's instructions should be followed when adding unfrozen food to the freezer. Some advise that the temperature setting should be lowered when food is being frozen, thus ensuring that the compressor will function continuously. Other manufacturers advise that in order to combat the rise in temperature brought about by adding unfrozen food to the freezer, the temperature setting is lowered some hours beforehand. In no case should additions of fresh produce exceed one-tenth of the total capacity of the food compartment per twenty-four hours. Thus up to 60 lb. of unfrozen food may be added to a unit of 600 lb. capacity, and up to 12 lb. to one of 120 lb. capacity. No domestic or semi-commercial freezer is able to produce quick-frozen produce according to the strictest definition. Some foods become frozen more slowly than others, notably those with a high sugar content.

The rate of freezing may be improved by:

(a) packing the food to be frozen in flattish packs to ensure quick and even penetration;
(b) placing the packs in contact with the sides of the unit, or plates containing evaporator coils;
(c) allowing a little air space between the packs, and never stacking them together.

The placing of unfrozen food in contact with food previously frozen should be avoided, as the temperature fluctuation produced in the latter may result in a deterioration in quality.

Once the food has become frozen it may be stacked in the food compartments, ensuring where possible a rotation of stored food so that it can be used in correct sequence.

Defrosting

Frost builds up on the walls and refrigerated surfaces of a freezer when it is in use. The presence of frost does not lessen the efficiency of the freezer except to reduce its capacity, prevent the easy withdrawal of shelves and baskets, and to cause the packages of food to stick together. When the deposit is of frost only and not hard frozen ice, it may be removed without putting the freezer out of commission. A plastic spatula or wooden ruler may be used to scrape frost from the sides of the food compartment. A sharp bladed tool must not be used, as permanent damage may be done to the inner lining. A most convenient receptacle in which to catch the dislodged frost is a clean dustpan. It is not necessary to take out the food from the food compartment during this operation; it may be simply moved to another part of the freezer.

The freezer should be switched off and completely defrosted when the layer of frost is of the order of $\frac{1}{4}$ to $\frac{1}{2}$ inch thick, and particularly if it also contains hard frozen ice. This will probably not be necessary more than once or twice a year. It always provides a good opportunity for stock-taking, particularly if the system of record-keeping has been allowed to lapse.

The amount of work is reduced if defrosting is carried out when the stocks of frozen food are low. After switching off the current to the unit, the frozen food can be lifted out and stacked in boxes or cartons. These should be placed in a cool place on a layer of newspapers, and given as much protection as possible by rugs and blankets to minimize the circulation of air. The more compactly they are stacked together, the colder they will remain. Frozen food may be stored temporarily in the domestic refrigerator if accommodation exists there.

The process of defrosting can be accelerated if as much of the frost as possible can be removed first with a spatula. Mopping up large quantities of water in chest freezers may prove a messy and time-consuming task unless there is a drain at the base of the food compartment. If a drain exists, warm (not hot) water, may be applied to the refrigerated surfaces to hasten the melting of the ice layers. Hot water applied to the refrigerated surfaces may result in pressure being built up in the evaporator coils and subsequent difficulty in restarting the compressor.

A hand-held electric hair drier can be used to direct warm air over the ice-covered walls, providing it is moved constantly to and fro, and never directed continuously on one area.

After all the ice has been removed from the food compartment the surface and surrounding gasket may be wiped first with a solution of bicarbonate of soda and then with clean water. Should a musty odour persist, for example as a result of an unnoticed leak from a package of frozen food, this may be removed by rinsing the surface with a solution of vinegar and water—2 tablespoons vinegar to 1 pint water.

After thoroughly drying the inside of the cabinet, the power may be switched on while the outer cabinet is washed with soapy water and afterwards dried. After the freezer has been running for about thirty minutes the food may be replaced inside. The whole defrosting operation should be completed in one to two hours, although this will depend on the size and condition of the unit.

Likely causes of trouble in a freezer with a sealed unit

Symptom	Diagnosis	Remedy
Motor running normally, but temperature more than 5°F (—15°C)	(a) Too great a load of unfrozen food has been added	Increase setting, thereby lowering the temperature, or remove part of the new load to a domestic refrigerator until the temperature has dropped
	(b) If (a) is not the cause, setting may be too low for ambient temperature	Increase setting, thereby lowering the temperature in the food compartment.
	(c) If (a) and (b) are not the causes, the fault is in the refrigerating system	Call service man
Motor running continuously	(a) Door or lid prevented from fitting properly by the presence of ice or frost	Remove ice with scraper
	(b) If (a) is not the cause, the fault is probably in the refrigerating system	Call service man
Motor not running	(a) Setting inadvertently altered—by a child? (b) Plug loose or removed from socket (c) Fuse blown in plug (d) Fuse blown in fuse box	Replace or repair
	(e) Local power failure	Find out how long power likely to be off, and if necessary take emergency action as given later.
	(f) If (a) to (e) is not the cause, fault is in refrigerating system	Call service man

Freezer breakdown

Freezers and refrigerators have a good reputation as far as breakdowns are concerned, and most of them give many years of trouble-free service.

However, the replacement value of a fully loaded freezer is considerable, both in terms of cash, and time spent in preparation; therefore it is necessary to determine a course of action should a breakdown occur.

If the freezer is not fitted with an audible alarm system, and few domestic freezers are, it is advisable to insert a thermometer in the top of the freezer at the level of the maximum load line. Where a freezer is not in daily use, and where it is situated in a garage or outbuilding, a daily check of the temperature is recommended to prevent the freezer being out of order for some days before this is discovered.

Faults with open type of compressor

The faults and diagnosis given on page 77 also apply to freezers with open-type compressors, but there is an additional problem.

When the motor is running and the temperature in the food compartment is too high, the fan belt may be the cause of the trouble. It may have become stretched in use and therefore be slipping, or it may have broken, so that the refrigerant in the condenser is not being cooled. It is obviously a wise precaution to keep a spare fan belt in reserve. Should the fan belt not be the cause of the trouble, it will be necessary to call a service man.

Emergency action should a freezer be out of commission for several days

Power failures seldom occur these days; and if they do happen it is more likely to be in a cold spell in the winter when demand for electrical power is at a maximum, rather than in a hot spell in the summer. A power cut rarely lasts more than a few hours, which would have little effect on the temperature of the frozen food, particularly if the external temperature is low. The freezer must be kept closed while the power is off.

Should a power cut occur, it is advisable to find out how long the power is likely to be off. In the event of this being more than twelve hours, emergency action may be necessary. Two possibilities exist. The transference of the frozen food to a low temperature store near by, or the addition of dry ice (solid carbon dioxide) to the top of the freezer. The work involved in transferring the food to a cold store, and the possibility of the power being restored shortly afterwards, makes this a less desirable alternative.

Dry ice may be purchased in large pieces, about 25 lb. in weight. They may be chopped into small pieces and put into the top of the freezer. It is necessary to separate the ice from the food packages by layers of cardboard; to avoid burns when handling the dry ice, thick gloves should be worn. Depending on the external temperature, new additions of dry ice will be necessary every one to two days.

It may be necessary to take emergency action when a freezer breakdown occurs, and in the event of the service man being unable to attend promptly. Most service depots are either able to lend a replacement freezer while repairs are carried out on the original one, or alternatively they have frozen storage available to which the food may be transferred temporarily.

Keep by the telephone a note of the telephone number of the local service man, and a supplier of dry ice.

The procedure to prevent spoilage and wastage of frozen foods in the event of a freezer breakdown and power failure has been outlined. There still remains the human failure. It is not unknown for a family about to depart on a holiday of one or two weeks to switch off the power at the mains before leaving home. To return home to a freezer containing spoiled and unpleasant-smelling foods rather than one which offers a selection of pre-cooked frozen dishes ready for a quickly prepared meal is a most unpleasant experience.

This unfortunate occurrence can be prevented by simply tying a label to the master switch with the warning—'Remember the freezer'. This should be done the day the freezer is installed.

When food has begun to thaw

If a little time has elapsed before it is discovered that the freezer has not been operating correctly, a decision has to be taken on what to do with the food it contains. The advantage of keeping a thermo-meter in the top of the freezer is obvious, as it gives an immediate indication of how much the food has warmed up. If it is not known how long the freezer has been out of commission, the course of action must be determined by the state of the food when examined. Usually the food in the bottom is in better condition that that at the top.

If the packages of food still appear to be hard frozen, no problem exists, as thawing can scarcely have begun. However, the recom-mendation of the U.S. Department of Agriculture for packages which have begun to soften is as follows: 'Partially thawed foods may be

THE FREEZER

refrozen as long as ice crystals are still present in the foods. Packets whose temperature has risen about 40°F (5°C) should not be refrozen.'

As thawing continues a stage is reached when finally no ice crystals remain in the food, although the food itself may still feel cold. Only fruit, bread, and some baked goods which have reached this condition may be refrozen. Raw meat, fish, and poultry may not be refrozen in a raw state, but providing they can be promptly and thoroughly cooked, they may be returned to the freezer in this form, so that losses of the more costly foods may be prevented. Because the alkaline media of vegetables renders them more susceptible to invasion by undesirable bacteria, they should not be refrozen once completely thawed. However, these vegetables may still be safely eaten if they are thoroughly cooked without delay.

Pre-cooked meat, poultry, and fish dishes which, although no longer containing ice crystals are still very cold, may not be refrozen, however, they may be eaten after a thorough reheating. Sufficient time must be allowed for the heat to penetrate to the centre of the dish; merely warming up is not adequate.

Food which is completely thawed and which no longer feels cold may have been well above freezing temperature for some time. Vegetables, meat, fish and poultry, and pre-cooked dishes containing them, should be destroyed without sampling them, as they represent a potential danger to health. It may be that an unpleasant appearance and smell, the evidence of spoilage, is already present. Fruits, being acid, do not spoil in the same way—they frequently show signs of fermentation in these conditions. Fruit need not necessarily be condemned with the other foods. The package should be opened and if the appearance and taste is satisfactory, then the fruit may be used for purées, sauces, or jams.

Foods which have been condemned as unsafe for consumption are best disposed of by burning, or by burying.

Should a quantity of partially thawed food be pronounced fit for refreezing, prompt action is necessary. In such emergencies it may be possible to obtain space in the freezers of friends, tradesmen, or at the service depot, in order to refreeze this produce as quickly as possible.

Locker plants

There are a number of locker plants now operating in Great Britain. These are centres where it is possible to rent frozen storage—usually annually. They are supervised by an attendant who will also carry out a certain amount of preparation—for example, the butchering and packaging of a carcass of meat.

Locker plants are more frequently found in farming areas where there are farmers and smallholders who do not have sufficient space in their domestic freezers for all their own produce. Compared with the United States, the impact of locker plants has been slight in this country—no doubt because only a small proportion of the population produces meat, vegetables, and fruit in sufficiently large quantities to warrant additional frozen-storage space.

REFERENCES

Home freezers, their selection and use. Home and Garden Bulletin No. 48. United States Department of Agriculture. 1964.

5

Packaging Materials

THIS chapter is in two parts; the first part lists the functions and properties desirable in a packaging material intended for use in a freezer, and the second part reviews the materials suitable for this purpose which are available at the present time.

THE FUNCTIONS OF PACKAGING MATERIALS

Primarily the function of any packaging material is to maintain the food contained in as perfect a condition as possible during frozen storage. Other factors which affect the quality of the food include the conditions of frozen storage and the initial quality of the food. Storing foods for long periods at 0°F (-18°C) cannot be expected to improve them; however, the use of appropriate packaging materials and packaging methods plays an important part in preventing any deterioration.

The main requirements of effective packaging materials will be considered in turn.

Low permeability

The permeability of a material is the extent to which gases and fluids are able to pass through its surface; the greater the diffusion the higher the permeability. High permeability of a material may be due to the presence of a great many minute holes, through which the gas or vapour is able to pass; this is the type of permeability found in aluminium foil, and is more correctly termed porosity, with each hole constituting a tiny pore. True permeability occurs when gases are able to diffuse directly through the material as in paper and untreated cellophane etc.

82

The first physical requirement of a packaging material is that it shall have low permeability at 0°F (−18°C) to

(*a*) moisture vapour
(*b*) oxygen and
(*c*) volatile odours.

Moisture vapour

Frozen foods become dehydrated within the freezer as a result of *sublimation*. Under certain conditions some of the ice sublimes away from the frozen food as water vapour (without passing through the intermediary liquid phase); this water vapour is redeposited as frost on the coldest part of the freezer. This may occur whenever a temperature gradient exists within the freezer, in other words whenever one part of the freezer is at a different temperature to the rest of it.

Conditions leading to the existence of a temperature gradient, and the effect that dehydration makes on the quality of the food, are discussed more fully in an earlier section on frozen storage (page 31). However, selecting packaging materials which have a low permeability to moisture vapour is a primary factor in maintaining a high quality in foods during frozen storage.

Oxygen

Some of the undesirable changes which may occur in frozen foods are the results of oxidation reactions. Oxidation of fats in meat leads to rancidity, lack of sweetness, and a loss of nutrient value; oxidation of some of the pigments in fruits results in discoloration. All possible care should be taken when packaging foods to exclude as much air as possible from the pack, and to select a material which will prevent the passage of any further oxygen during low-temperature storage.

Volatile odours

Low permeability to odours is necessary in order that the food may neither lose volatile odours which are characteristic of it, nor be contaminated with other flavours or taints from an adjacent food. It also follows that any material chosen as a frozen-food container must not itself have an undesirable odour, as a result of plasticizers or other additions included in its manufacture, and which may be transmitted to the frozen product contained in it. Those foods which have a mild flavour or odour, such as most dairy produce, may easily be contaminated by a stronger one.

83

Mechanical strength

The material selected must be strong and stand up to the type of handling received during preparation and in the freezer. Folding and creasing some materials results in increased permeability—particularly along folds and seams, and a certain loss of strength. To the frozen-food producer, who relies considerably on mechanical aids for the handling of his products, the selection of a sufficiently strong material is of great importance.

Easily sealed

An important property of a packaging material is that it is capable of being effectively sealed. Some materials may be heat-sealed, others must be used with an adhesive, but whatever method is applied, the seal must be air-tight. The quality of the product is threatened as much by improper sealing as by any imperfection in the material selected for packaging.

Resistance to water, weak acids, and fats,

The material selected should be unaffected by any food material likely to be stored within it. This might include liquid foods, fruit juices which are weakly acid, and fats, which could all attack the material, ultimately causing its collapse. At the same time it should remain waterproof to external moisture, such as might be encountered during preparation for freezing.

Retention of properties at low temperatures

The properties for which packaging materials are selected must be maintained at low temperature; for example, films should remain flexible, adhesives effective, and laminates intact, at temperatures as low as $-20°F$ ($-29°C$).

The points above are not exhaustive, but are the main requirements for good packaging materials; consideration will now be given to the available materials which meet these requirements.

MATERIALS AVAILABLE FOR PACKAGING FROZEN FOODS

The materials available or suitable for packaging frozen foods can be classified into six groups as follows:

(a) Paper derivatives such as waxed cartons.

(b) Cellulose derivatives, such as cellophane.
(c) 'Plastics' such as polyethylene.
(d) Rubber derivatives, such as pliofilm.
(e) Aluminium, such as aluminium containers.
(f) Laminates, such as waxed paper and polythene.

Each group will be considered in turn, illustrating the various advantages of the many materials available in these groups.

Paper derivatives—waxed paper cartons and tubs

Untreated paper is completely permeable to moisture vapour and most gases. However, paper, preformed tubs, and cartons may be given a coating of wax in order to decrease the permeability. Two methods of waxing are employed; wet waxing when both sides of the paper are coated, and dry waxing when the paper is completely impregnated with wax; for the best results a good-quality sulphite board should be used. Paraffin wax is the type of wax used and the more densely it is applied the more effective is the barrier produced.

Waxed tubs, cartons, and boxes are available in a variety of shapes and sizes for home freezing. Some tubs have a screw-on lid which gives a perfectly satisfactory seal, but 'freezer tape' should be used with most cartons and boxes. Waxed containers may be stored flat, or stacked inside each other, thus taking up little space when not in use. Waxed containers should not be handled roughly in the freezer, as some of the protective wax will be removed and the effectiveness reduced. For this same reason the containers should not be filled with hot foods or washed in hot water, otherwise the wax will be melted.

Although not providing the ideal protection for frozen food, these waxed containers are convenient in use, and re-usable, particularly if used with a 'liner'—usually of polythene—which has the additional advantage of giving increased protection to the frozen product. The use of waxed paper on its own is not recommended for long storage periods, as it does not provide a very efficient barrier against the loss of moisture vapour, and oxidation of the product. Folding and creasing the waxed paper cracks the wax film and increases permeability.

Waxed containers have been used extensively by the large-scale frozen-food producers, although various disadvantages have led to them being superseded in many instances by laminates, frequently incorporating waxed board. Waxed cartons cannot be effectively heat-sealed without the incorporation of a special adhesive, which involves an additional handling process. 'Scuffing' is a term used to describe marks made on the outside of the containers by machinery during the freezing processes, sometimes resulting in the removal of a portion of the wax coating. In order to improve the effectiveness of the pack as well as the appearance presented to the consumer, an illustrated fully labelled overwrap is frequently added.

Cellulose derivatives: cellophane

Cellophane is derived from cellulose, which is found in the cell walls of most plants and which is largely responsible for the structure of the plant. The cellulose is dissolved out by a solvent and then re-precipitated in sheet form. Untreated, the material is highly permeable to both moisture vapour and oxygen, and is not heat-sealable.

However, a material which provides an effective barrier to moisture vapour and oxygen is produced when the film is coated with a moisture-resistant lacquer. The substances used are a combination of waxes and cellulose nitrate and may be applied either to one or both sides of the cellulose film. The application of lacquer may also be used to bestow other properties on the cellophane. It may render it heat-sealable, it may give additional wet strength so that the cellulose film may be used to package goods which are in a liquid state prior to freezing. Some cellophane manufacturers have instituted a code to indicate which properties are possessed by their product; for example, the letter 'M' indicates that the cellulose is moisture-vapour-proof.

Cellophane expressly prepared for use in freezers provides an effective protection for frozen food and can be conveniently heat-sealed. However, cellophane should not be used in a freezer without an overwrap as it may become punctured, particularly as it has a tendency to become brittle in a dry atmosphere.

A special application for cellophane for both home users and frozen-food manufacturers, is in the separation of individual portions within a container. A double thickness of cellophane between chops or fish fillets makes them easy to separate as and when necessary.

'Plastics'

Plastic materials are widely used in film form as well as in semi-rigid container form for frozen foods. The 'plastics' family is made up of several members, each member is in turn made up of a large number of simple carbon-based compounds joined together to form a chain. Such a chain is called a polymer, and the formation of such a chain is known as polymerization.

The polymer takes its name from the basic units of which it is composed, for example polyethylene (or polythene as it is more commonly called) is probably the best known polymer and is made up from some 2,000 ethylene molecules, as follows:

$$\text{n (CH}_2 = \text{CH}_2) \quad \cdots -\overset{\displaystyle \underset{|}{\text{H}}\ \underset{|}{\text{H}}\ \underset{|}{\text{H}}\ \underset{|}{\text{H}}\ \underset{|}{\text{H}}\ \underset{|}{\text{H}}\ \underset{|}{\text{H}}\ \underset{|}{\text{H}}}{}\text{C}-\text{C}-\text{C}-\text{C}-\text{C}-\text{C}-\text{C}-\text{C}-\cdots$$

H H H H H H H H
| | | | | | | |
—C—C—C—C—C—C—C—C— ············
| | | | | | | |
H H H H H H H H

(n being equal to about 2,000)

The properties of each member of the plastics family are specific to that member; thus the temperature at which it melts, the degree of permeability to various gases, the ease or difficulty with which it may be handled (often called the 'slip' characteristics), are specific to each plastic material.

As newer techniques for producing plastic materials have been developed it has been possible to modify some of these existing properties, and to bestow other properties which are quite new. For instance, various ingredients may be added to the raw chemical before it is extruded as a film, or moulded into semi-rigid containers to make it less permeable to certain gases, or more resistant to moisture vapour. Likewise the plastic film may be stretched and re-orientated as it is extruded, thus giving it added resilience.

It is also possible to combine two or more plastic materials together, or to combine a plastic film together with another substance such as paper, in the form of a laminate, and thus achieve an improved packaging material.

The characteristics of five of the more commonly used plastics will be considered in turn.

(a) Polyethylene

Polyethylene, or polythene, is formed from the polymerization of ethylene. It was first produced by what is normally termed the high-pressure reaction—the bulk polymerization of ethylene at pressures of 1,000–3,000 atmospheres and at temperatures of 80–300°C, using trace oxygen as a catalyst. Pellets of the polymer are subsequently extruded as a film.

The structure of the polythene molecule produced by this method is not completely regular or crystalline, and parts of the chain show considerable branching. Polythene produced by this method is called low-density or conventional polythene. The density describes the amount of crystallinity in the molecule and in a low-density polythene it is of the order of 65 per cent; the remaining 35 per cent being described as the amorphous phase.

From the beginning polythene film was found to be waterproof, an effective barrier to moisture vapour, and heat-sealable. At the same time it provided a poor barrier to oxygen and some volatile oils. It had poor 'slip' characteristics, making it difficult to handle, and it did not readily take prints.

However, some of these disadvantages can be overcome by the use of various additives to the polymer pellets before extrusion. A film which provides a more effective barrier to oxygen can be produced by the addition of antioxidants to the polymer pellets; polythenes produced for the domestic market are not usually so treated, possibly because it also results in a loss of clarity. Pinholing may still persist in very thin films, although a scheme for fusing together two films during production may prove a satisfactory solution to this problem.

Mention should also be made of high-density polythene produced by a low-pressure catalytic process, developed by a Professor Ziegler and which takes his name. This polythene is more rigid and of greater mechanical strength, due to the increase of the crystalline phase. The consequential reduction in the amorphous phase is responsible for a decreased permeability to oxygen and moisture vapour. High-density polythene has widespread applications, including the semi-rigid kitchen containers which are quite satisfactory for use in the freezer.

Polythene film is probably the most widely used material in the domestic freezer. It is available in sheet and bag form, easily heat-sealable with a domestic iron (provided it is used at the right temperature and over paper), or with wire 'ties'. It is not easy to write on directly, but other effective methods of labelling are possible. When used to pack irregularly shaped food materials, the shape assumed by the bag and contents make it difficult to pack neatly within the freezer.

Polythene may be used on its own or as a liner in a carton. It is being increasingly used by the commercial frozen-food producer in laminated forms with paperboard and aluminium. Because polythene does not provide the most effective barrier to oxygen its use for the storage of those foods of a high fat content for long periods cannot be recommended.

It is worth mentioning one further property of polythene. Like nylon and terylene, it may be orientated as it is produced. It is extruded, stretched and cooled, so that stresses are built into it. Application of heat to such a product releases these stresses and cause it to shrink or return to its unstretched size. This property may be made use of when a close pack is required.

(b) Polypropylene

Polypropylene is produced as a polymerization of propylene (a homologue of ethylene) by the Ziegler process, similar to that used to produce high-density polythene. The properties are very similar to those of polythene, although they may be modified by orientating the film in one or both directions as it is extruded. It is a little stiffer than polythene and there is less 'creep' than with the rest of the petroleum derivatives. It is much less permeable to moisture vapour, but has about the same resistance to oxygen. It also has a high melting-point, and this fact combined with the quality of being heat-sealable has led to its use by manufacturers for the 'boil-in-the-bag' type of product. This is the latest development in convenience foods. A complete course, usually a speciality such as a goulash or fricassée, is prepared and cooked by the manufacturer and sealed in the polypropylene bag. The bag, complete with its contents, is put into a pan of boiling water and reheated. Thus an interesting dish is available without even the necessity of washing a pan.

(c) Polyvinylidene chloride with polyvinyl chloride: Saran and Cryorap

Saran and Cryorap are the names given to the copolymer produced from polyvinylidene chloride and polyvinyl chloride (P.V.C.). (P.V.C. on its own has a limited application as a food-packaging material.) The properties of the film produced may be varied by modifying the amount of copolymers present, and the degree of polymerization. Also, in some cases small amounts of plasticizers are added. In many respects the properties of Saran are better than those of polythene. It is a very effective barrier, both to moisture vapour and oxygen, and like polythene it is heat-sealable.

Saran film is very convenient to use, as it has a tendency to cling to the product, and this makes packaging more manageable.

The particular property of the Cryorap film is that it shrinks when heated to about 160°F (70°C). Advantage is taken of this to ensure a snug fit around irregularly shaped foods, such as chicken or awkward joints of meat. The chicken or joint of meat is put inside the bag formed from Cryorap, then entrapped air is withdrawn by means of a vacuumizer, and the resultant film is caused to fit closely over the food. The bag is sealed by a metal clip, as heat-sealing is not possible to the bunched film at the neck of the bag. The bag and contents are then immersed in hot water and the film shrinks and snugly fits the food inside it. The great advantage of this method of packaging for irregularly shaped foods, particularly chicken and joints of meat, is that oxidation is prevented from taking place in localized air pockets. Chicken, ducks, and turkeys presented in this pack can be seen daily in the supermarkets.

(d) Polyamide—nylon, rilsan

Nylon is the generic name by which the polyamides are usually known. It is usual to follow this by a number which denotes the number of carbon atoms in the basic unit from which the nylon is formed. Nylon 66, common nylon, is formed as a condensation product of hexamethylene diame and dicarboxylic acid. Nylon 11, or Rilsan, which is the name for this film when produced in France, is obtained as the result of polymerization of 11 amino undecanoic acid. Nylon 11, or Rilsan, has been found more suitable than Nylon 66 for the storage of frozen products.

Normally nylon does not provide a very effective barrier to water vapour, but it has low permeability to oxygen and most gases and odours. It also has excellent clarity. It may be heat-sealed and it takes print reasonably well. Nylon films are not greatly used, either domestically or by frozen-food producers at present, as the cost is two to three times greater than equivalent films, although Rilsan has been used for some time for 'boil-in-the-bag' products; for example, frozen kippers have been vacuum packed in seamless tubes of this film.

(e) Polyester: Mylor, Melinex, Hostophan, Scotchpak

Polyester film is produced from a type of polymerization of dimethyl terephthalate and ethylene glycol.

The film is usually stretched as it is produced, so that the molecules are orientated, thus achieving a very strong film, but it is not easy to heat-seal, as it has a high softening point and at these high temperatures the film also shrinks and becomes brittle. This difficulty in heat-sealing may be overcome by laminating the polyester film with polythene, and thus utilizing the very good heat-sealing properties of polythene.

The polyester film is clearly transparent, and has low permeability to water vapour and most gases; but it has poor slip and static characteristics.

Polyester film is marketed on the domestic market in England and America; in America a special electrically heated sealing unit is also available. The film is used commercially, particularly for the 'boil-in-the-bag' type of product, when its toughness and temperature range are of great value. The film will accept printing by the application of special inks.

Rubber derivatives: 'Pliofilm'

This rubber derivative is obtained by treating a solution of rubber in an organic solvent with hydrochloric acid gas. Plasticizers may be added before the film is formed and the amounts added naturally govern the properties of the film. The addition of plasticizers makes the film more elastic, and enables a snug fit to be achieved over the food product. It may be heat-sealed.

Pliofilm has low permeability to water vapour and most gases, and is a very satisfactory film for use at low temperatures. However, it is not available at the present time in this country.

Aluminium

Pure aluminium is produced in sheet form in varying thicknesses. Sheets of aluminium (of the order of 0·03 mm. thickness and above) are completely impermeable to moisture vapour and oxygen, making them the ideal material in which to store frozen products. However, as the thickness of the sheet is reduced, so minute pores are found; these are unavoidably produced in the manufacturing process. The number of pores produced in the manufacturing process per square metre is related to the thickness of the sheet, and their presence will naturally reduce the protective value of the wrapping. But not only is the thinner gauge aluminium more porous, it has much less mechanical strength. Unless handled very carefully during preparation and in the freezer, it may easily become punctured. Most of the rolls of aluminium foil sold for general domestic use are of a thinner gauge—usually 0·018 mm. thickness—and it is not recommended for use on its own in the freezer. The threat to the quality of the product is much less due to the slight permeability of this gauge than to its poor mechanical strength.

Aluminium on its own cannot be heat-sealed, though it is frequently laminated with other materials which can be heat-sealed. An air-tight package may be made quite simply by folding together the edges of the aluminium sheet.

Aluminium containers are available in a wide variety of shapes and sizes, as patty tins, pie dishes, tart cases and pudding basins. They are mostly pressed out into their final shape by the pressure of the die into a sheet of aluminium usually of between 0·03 and 0·1 mm. thickness. The effect is to give a smooth base and closely pleated or 'wrinkled' sides which, in fact, add strength to the container. The raw edge around the rim is usually turned under to give strength and prevent cuts. Some rectangular containers are formed by folding the aluminium sheet at the corners. In this case, although strong square corners are achieved, the sides are less rigid.

Aluminium is very resistant to weak acids, but less resistant to alkaline products. Salts tend to corrode it, and therefore it is unwise to use aluminium to pack foods with a high salt content. Frequently aluminium containers are supplied coated with lacquers, and polyvinyl plastics are also used for lining many containers such as aluminium pudding basins.

Aluminium is used extensively by frozen-food producers, who find the containers an attractive and convenient way in which to present their products. The preformed containers are easily and satisfactorily sealed by an aluminium lid. They may be closed quickly by 'crimping' together the edge of the lid to the rim of the container on a special machine.

Laminates

From this brief review of some of the principal materials available for packing frozen foods, it is evident that although some materials have several good qualities, their performance in this field is reduced by other inadequacies. However, these inadequacies may be overcome to some extent by the use of various coating materials, such as wax coating on paper cartons, or the application of lacquers such as those applied to cellulose. Another method being used increasingly is the formation of laminates from two or more materials which can be satisfactorily combined together. A material may thus be produced whose combined properties or characteristics render it more suitable for the packaging of frozen foods than either or any of the constituent materials.

There are, in fact, a great many laminates available today produced by one or other of the following methods:

(a) Adhesive laminations

Perhaps the most obvious method of producing a laminate is to stick together by means of an adhesive the two or more materials selected. This method is beset by difficulties, principally the selection of an appropriate adhesive. It must be a substance which is acceptable to both surfaces, and not affected by the various substances with which it may come into contact. At the same time it must be non-toxic, odourless, and effective at low temperatures. Sometimes the selection of an appropriate adhesive may add certain properties of its own, and thereby increase the effectiveness of the laminate it helps to form.

(b) *Extrusion laminations*

By this method the two materials are laminated or bonded together by extrusion of one substance in molecular contact with the other. Obviously for this method to be successful the two materials must be compatible. Polythene is the most notable example of a substance which may be successfully extruded in contact with another. It may be used with all the basic materials such as paper, paperboard, aluminium, and cellulose.

The following example will serve to show how the properties of the substances laminated together combine to produce a much-improved material. When waxed paper and polythene are laminated together, the polythene provides a more effective barrier to moisture vapour than the waxed paper on its own. At the same time, waxed paper adds strength to the polythene and prevents it from becoming punctured if handled roughly in the freezer.

REFERENCES

Modern Food Packaging Film Technology. J. W. Selby. The British Food Manufacturing Industries Research Association. Scientific and Technical Surveys, No. 39. November 1961.

6

Choosing the Right Packaging Material

THE new owner of a freezer may well wonder which of all the materials available for packaging frozen foods reviewed in the previous chapter will give the best results. To assist in the matter of selection a brief summary of those materials commonly available on the domestic market is given in the table in Fig. 8

In later sections of this book dealing with the preparation and freezing of fresh produce, meat, fish, fruit and vegetables, pre-cooked and prepared foods, suitable packaging materials have been suggested where possible. However, the user of a domestic freezer must also take into account other factors, and the following practical considerations are given so that the correct choice of packaging materials can be made.

PRACTICAL CONSIDERATIONS

Economy of freezer space

There are usually many demands for the space within a freezer; thus it is important that the ratio of packaging material to food stored within is kept as small as possible. Space can be wasted in a freezer by selecting containers which are bulky in themselves, being made of thick materials, or manufactured with a false base. The amount of space between containers must also be kept to a minimum; thus round or irregularly shaped packages are more wasteful of freezer space than are square or rectangular shapes. It has been shown that each cubic foot of freezer space will hold about 40 pints of rectangular containers, but only about 30 pints of cylindrical cartons. Two or three chickens will be found to monopolize a lot of space in a small-capacity freezer which might have been more economically employed for storing vegetables more regularly packed.

94

Suitability for food stored

Although some packaging materials are more economical in storage space than others, there may be other overriding considerations. The type of food to be frozen is equally important. Thus those foods which need to be packed in a liquid, brine, or syrup, are best catered for in tall, upright containers; soft fruits, usually packed dry in sugar, are best accommodated in tubs, while most vegetables pack very compactly in flat boxes. Pre-cooked and prepared foods are suitably stored in aluminium containers so that they can be put directly from the freezer into the oven. Thus it is often necessary to use a range of containers when storing a variety of foods, even though some of them may not be the most economical in space.

There are some foods, i.e. poultry, joints of meat, etc., which cannot be accommodated in proprietary containers. These foods are best made into parcels wrapped and sealed in some form of plastic film; such packages must be as compact as possible. It is also essential to eliminate any pockets of air within the film, because where these occur a small local area of dehydration may exist inside the plastic film.

How easy to manage

Some types of packaging materials and the methods of making them air-tight are easier to manage than others. For some it is simply necessary to screw on a lid, while others must be secured with ties or sealed with tape or heat-sealed. Although one method of storage is more suitable for certain foods, there is sometimes an alternative method—which may be quicker or less complicated—and it is logical to exercise a preference. In fact, there is scope for experimenting with the different packaging materials in order to discover which ones are easier to handle and give the best results.

Unfortunately the full range of materials is not available in every town, and it may be necessary to write to the manufacturers for supplies. It is always worth while to keep a small stock of containers and wrappers ready to use in the freezer. There is not time to shop around when the peas are ready for picking and the raspberries are ripening on the canes.

Economic unit size

The size of the family and the number of portions required also govern the choice of containers for storage in a freezer. It is neither practical to attempt to divide up solidly frozen blocks of food, nor

Figure 8

BRIEF SUMMARY OF MATERIALS COMMONLY AVAILABLE ON THE DOMESTIC MARKET

	Permeability to moisture vapour and oxygen	How easy to manage	Some uses	How sealed	Whether re-usable and approximate costs	Some manufacturers
Waxed drums and cartons	Provides an adequate barrier; use of a plastic liner increases effectiveness	Easy to fill and close. Tall upright containers particularly suitable for purées and liquid foods	Vegetables. Soft fruits. Soups and sauces	Lids may be screwed on or sealed with freezer tape	With careful handling may be re-used. Costs vary with size and method of closure; from 4d. to 1s. 6d. each	Frigicold, 10 Manchester Square, London, W.1. (distributors)
Cellophane	Specially lacquered cellophane provides an adequate barrier	Only skill required is that to make a neat parcel. Overwrapping is advisable	For irregular-shaped foods such as chicken, also baked products. Double thickness of cellophane is useful for separating portions of meat, fish, etc.	May be heat-sealed or sealed with freezer tape	Re-use not particularly recommended, although may be used a second time as a 'liner' or to separate portions of food. Cost about 1d. per square foot.	British Cellophane
Polythene	A good barrier to moisture vapour, but not such a good barrier to oxygen	Necessary to take care when filling to prevent food sticking on the mouth of the bag, which prevents satisfactory sealing. Advisable to use a funnel with liquid foods	Vegetables. Fruits. Meats and cooked foods with low fat content	Easily heat-sealed. May also be fastened with freezer tape or wire ties	May be re-used if intact and clean. Gussetted bags and simple pouches from 1d. each. Available in roll form at 50 yards for 2s. 6d. Polytape 36 yards × 1 inch for 10s. 6d.	Bx. Plastics Ltd. I.C.I. Ltd.

FIGURE 8 (contd.)

	Permeability to moisture vapour and oxygen	How easy to manage	Some uses	How sealed	Whether re-usable and approximate Costs	Some manufacturers
Aluminium	Heavy-duty aluminium foil, used for shaped containers, is completely impermeable. But the rolls of aluminium foil available domestically are slightly permeable	In sheet form may be moulded around irregular-shaped foods, although care is necessary not to puncture it. Pre-formed basins and pie dishes very easy to fill and seal	All pre-cooked dishes and un-baked pies	Effectively sealed by folding the edges closely together. Polytape may also be used	In sheet form not usually suitable for re-use, but basins and pie plates may be re-used several times. In roll form 15 feet for 2s. 6d. Plates and basins from about 2d. to 1s. each	Impalco Foils Ltd.
Saran	An excellent barrier	The film has a tendency to cling to the product, thus making packaging a simple process	For irregular-shaped joints. Chicken. Baked products	May be heat-sealed or sealed with freezer tape	Re-use is possible when completely intact and clean. In roll form at 50 feet for 5s. 6d.	Dewey and Almy Ltd. (Cryorap) The Dow Chemical Company, Mid-land, Michigan (Saran)
Polyester	A good barrier to water vapour (equivalent to Polythene). Permeability to oxygen is low	Only skill required is that to make a neat parcel	For irregular-shaped joints. Chicken. Baked products	Cannot be heat-sealed without a special electric-ally heated unit, not available in England. May be sealed with tape	Is very strong and may be re-used if thoroughly cleaned	I.C.I., as Melinex

wise to thaw a block of food in order to divide it. Once food has been removed from the freezer and thawed, it should not be refrozen, but consumed as soon as possible.

There is an additional advantage to be gained from packaging food in small quantities, such portions can be frozen more quickly and result in a better quality product.

Certain types of foods, for example garden peas, and a variety of pre-cooked foods, may be frozen prior to packaging. These foods remain loose and can therefore be removed from the carton in small quantities or individual portions if required. Selecting a carton with a screw-on or clip-on lid solves the problem of resealing the pack each time a portion of food is removed.

Cost of containers

The cost of the different containers is important. The least expensive containers are probably polythene bags, but they must be carefully checked for holes before attempting to use them a second time. Waxed containers are more expensive, but with careful use they may be expected to serve for several years. However, it is frequently recommended that waxed containers are used with polythene liners, so that only the additional cost of liners arises each time.

Saran is probably the most expensive film, but in view of its special qualities of clinging both to the food being packaged and to itself, it provides a very snug fit and in many instances is the best choice.

Thus the decision whether to use a cheap or more expensive container will depend on what foods are being stored and for what length of time.

UTILIZATION OF EXISTING EQUIPMENT

Several items of everyday kitchen equipment may be used in the freezer and for many pre-cooked and prepared dishes one of them may be the most suitable choice.

Oven glassware

Most manufacturers of oven glassware claim that their product may be safely stored in the freezer; this applies to all casseroles, pie plates, and jugs. For dishes which need to be reheated before use, it is advisable to allow glass containers to come up to room temperature before returning them to the oven, otherwise the glass may crack. One manufacturer claims that this warming-up process is not necessary, as his product is able to withstand extremes of temperatures of some 500°F (260°C).

However, pre-cooked and all other foods should always be cooled before putting them into the freezer, whatever the nature of their container, in order to prevent a temperature rise within the freezer, and of the already frozen products.

For casseroles with lids it is advisable to seal around the rim with freezer tape—however tightly the lid appears to fit. Those dishes without lids must be satisfactorily wrapped, or sealed over with a plastic film.

'Casserole moulds'

It may not always be convenient to have some casseroles or ovenware dishes out of circulation while they and their contents are in the freezer. This difficulty may be overcome by lining the casserole or dish with aluminium foil. The cooked food, for example Courgettes Provençale, or Flemish Beef, can be cooked, cooled, and arranged in the dish, which is then frozen. When the food is solidly frozen it may be removed complete with the aluminium foil lining and packaged in the usual way. When required, the aluminium foil can be peeled off and the frozen block of food returned to the casserole in which it was moulded to be reheated, or to complete cooking prior to serving.

Baking tins

Cake, bread, sandwich, and patty tins, and their contents, may be stored in the freezer providing they are satisfactorily sealed in a moisture-vapour-proof wrapper. Tins likely to rust when containing a raw mixture should be lined first with greaseproof paper.

Some prepared foods, particularly uncooked pastry in the form of tarts or mince pies, may be frozen overnight in their tins without wrappings. They can then be removed when quite firm on the following day and packed more compactly; this then frees the baking tins for everyday use.

Semi-rigid plastic containers

A wide selection of semi-rigid plastic containers with tightly fitting lids is available for general storage purposes in the kitchen. They are frequently produced from injection mouldings of polythene, although other materials are also used. These containers are quite suitable for storage in a freezer and provide an effective protection for the frozen foods.

There is no limit to the number of times these containers can be used, although using this type exclusively would involve a considerable initial outlay. They are particularly useful, however, for storing foods for short periods, as they are obtainable in a wide variety of shapes and sizes. There are large containers which accommodate cooked joints or chickens; shallow ones suitable for tarts and pies; and there are jugs complete with lids suitable for the storage of soups. Plastic basins which may be boiled are also available and are therefore very suitable for the storage of puddings and similar dishes.

Glass jars

Glass jars may be used in the freezer providing they have a neck wide enough to enable the contents to be removed while still frozen. Also adequate headspace must be allowed, particularly with liquid foods, otherwise the jar will burst in the freezer as the food expands. Most preserving jars are suitable for use in a freezer, but they are not particularly economical in freezer space.

SCOPE FOR IMPROVISATION

Re-use of containers in which frozen foods were purchased

It is possible to re-use boxes and cartons in which frozen foods have been purchased, providing they are opened carefully. They should be washed and thoroughly dried. Afterwards they may be stored flat until they are needed. It will be advisable to re-use these cartons with a polythene liner when there is any uncertainty about their composition; some are waxed, some are laminated with polythene, while others which have relied on an inner wrapping to provide protection for the frozen product, have merely been coated with a varnish. In all cases it is necessary to seal carefully with freezer tape.

Re-use of cream and cheese tubs

Some of these tubs are made of cardboard which has received only a light waxing for protection. They do not give adequate protection in a freezer for more than two weeks, unless a liner is added. They do have a function for storing such items as small quantities of sauce or grated cheese which will be used up quickly.

Some cream and cheese tubs which are semi-rigid are made from polystyrene. While this material when produced as a film is not suitable for use in a freezer, in its moulded form although slightly brittle, it may be used satisfactorily for the storage of small quantities of foods such as baby food and herbs.

Re-use of tins

Lacquered tins in which foods have been purchased make good containers for frozen foods. Large coffee tins, particularly those supplied with plastic airtight lids, are excellent for this purpose, but only after the smell of coffee has disappeared! However, it is unwise to use unlacquered tins for frozen storage.

Frozen-food producers at the beginning of their history were reluctant to use tins to present their products, because they feared the consumer would fail to differentiate between the frozen foods and those which were heat-sterilized. Thus frozen foods might mistakenly be stored on the shelf of the larder or food cupboard, and not in the freezer where they belonged. Now that the consumer has become accustomed to using frozen foods, the producers are beginning to use tins for some of their products, frozen concentrated orange juice being one of the first products to be presented in this way.

Cautionary note

Plasticizers are sometimes used in the manufacture of various containers to improve their properties. It is possible that some of these plasticizers are toxic, and therefore containers should not be used in a freezer unless their composition is known, or unless they are specifically recommended for use at sub-zero temperatures. It could be that the effect of low temperatures on some plasticizers cause their breakdown and release of toxin.

METHODS OF SEALING

Films of cellophane, polythene and Saran may be satisfactorily heat-sealed with a domestic iron set for 'rayon' and used over thin paper, but this method is not suitable for polyester films. When a plastic bag or film is to be heat-sealed it is important that food is not spilt on the surfaces to be joined together, otherwise a satisfactory seal may be prevented. In order to open a heat-sealed pouch it is necessary to cut off the sealed portion of the bag; this, of course, reduces its size for further use.

Freezer tape, which is made of polythene plus adhesive, is very effective for sealing all types of packaging materials. Ties of plastic-covered wire or paper-covered wire of the type used for fastening up plants in the garden give a satisfactory seal when twisted round the top of a plastic bag. It is not advisable to use rubber bands for long-term storage, as they will perish.

Overwraps

Because of the danger of both aluminium and some plastic films becoming punctured in the freezer, they may be given an additional wrapping of strong paper or stockinette. Stockinette is particularly suitable for this purpose as it does not add much bulk. It is available in rolls as a knitted seamless tube from which pieces may be cut off as required; each end can be secured with a knot. When cutting up such pieces, allowance has to be made for stretching and the consequential loss of length.

General note

It has been shown that with careful handling many of the containers and packaging materials may be used a second time. The first requirement is that they are still completely functional. Most waxed containers, providing they are not washed in too hot water, are reusable. Aluminium foil, if it has been removed from the food without splits or tears, may be washed, dried, and used again. However, it is never advisable to use plastic bags and film again if there is any doubt about their still being intact.

An equally important requirement is that any container used a second time is scrupulously clean. This is not always easy to achieve with some of the paperboard derivatives. Aluminium foil presents no problems, but great care is necessary with plastic bags—they must be turned inside out and thoroughly washed and dried if a musty odour is to be avoided.

LABELLING

The value of clear and effective labelling cannot be overstressed. The freezer door or lid should be opened as little as possible; therefore it is important that the produce be quickly located and easily identifiable. After some years' experience of freezing foods it is likely that a certain pattern will be followed. For example, one type of container may be found most suitable for vegetable storage and another for fruits. Or the use of different coloured ties and crayons for fruit, meat, vegetables, pre-cooked dishes and so on are helpful for quick location in the freezer. However, all produce should be labelled clearly with the following information: food, weight, number of portions, date, and a note of preparation or special treatment. A specimen label would be as follows:

FRENCH BEANS   September 5th, 1966.

1 lb.   Blanched 2 minutes in
4 portions.   boiling water.

The method of labelling will obviously depend on the packaging material used. China-graph pencils, or a cheap lipstick, write effectively on waxed containers, glass jars, and aluminium foil; they write more effectively if the surface is slightly warmed first. For plastic film or bags, the label may be slipped under the surface just before sealing. Adhesive labels are available which adhere securely at low temperature. Another simple and effective method is to fix the label beneath a piece of transparent freezer-tape.

As well as labelling it is valuable to keep a record of food stored in the freezer in a notebook, or more conveniently on index cards. Each addition of food to the freezer is recorded, and it is helpful for the records to show how this quantity of food has been broken down. Thus if 5 lb. of garden peas are put into the freezer on May 1st, in 1 lb. and $\frac{1}{2}$ lb. packs, the entry would be as follows:

GARDEN PEAS

May 1st: 1 lb., 1 lb., 1 lb., $\frac{1}{2}$ lb., $\frac{1}{2}$ lb., $\frac{1}{2}$ lb., $\frac{1}{2}$ lb.
May 15th: 1 lb., 1 lb., $\frac{1}{2}$ lb., $\frac{1}{2}$ lb.
June 1st: 1 lb., 1 lb.

Entries for May 15th and June 1st record further additions of garden peas to the freezer.

When items are removed from the freezer for consumption it is simply necessary to strike through the items concerned.

The index cards perform the same function as a bank statement and provide a means of recording deposits, withdrawals, and remaining credit.

At the same time on the backs of the cards a note may be made of special treatments, and the quality of the frozen produce, to serve as a helpful reference in the future.

PACKAGING ON A LARGE SCALE

In concluding a section on packaging problems it is worth considering the large-scale producer of frozen foods. He experiences most of the difficulties which arise domestically, but on a much greater scale. The way he deals with some of his problems may well prove a useful guide to the user of a domestic freezer.

Costs

As far as the producer is concerned, there is no possibility of using food containers a second time, therefore the ratio of the cost of packaging to the cost of the food contained in it must be realistic. But it is not only the material cost of the container which is important. Maintaining a product at a sub-zero temperature is expensive and the producer must ensure that the minimum space is taken up by his packaging materials.

Economic unit size

Most of the products we see in retail cabinets are appropriate to the family needs; vegetable packs are labelled with the number of servings; the number of portions of meat is always indicated. The large selection of convenience foods available today which need only to be reheated, notably complete meals-for-one, indicates how closely the producer of frozen foods is in touch with the pattern of living, and is ready to accommodate his product to our requirements.

Ease of handling

The frozen-food producer choses packaging material which may be handled easily on automatic machinery. Some of the earlier containers were 'hand erected and closed'; this is time consuming and has a retarding effect on the rate of production. Heat-sealing of containers is no doubt a quicker method of closure. Some plastic films are not easily handled by machinery, because they are subject to the development of electrostatic charges. This may result in the plastic container being difficult to fill, particularly with food in a semi-liquid state.

The type of material selected for the product may depend upon whether the food is to be frozen before or after packaging, and upon the method of freezing adopted. For example, vegetables which are to be plate-frozen are put into their containers and sealed before freezing. Several minutes may elapse after filling and before being frozen. During this time the pack may remain in a damp environment and deterioration of the pack would certainly result were it not adequately waterproof. Many commercial containers are coated with varnish, which avoids this problem.

Presentation and labelling

An important factor to the frozen-food producer is that his product should be well displayed. The addition of an overwrap on waxed cartons used to be very prevalent; it served to give additional protection to the contents, to provide a means of labelling, and not least to cover any 'scuff' marks the container received during the freezing process.

Some of the newer materials do not take print readily and special techniques are necessary in order to label and illustrate the contents of the frozen pack. Where laminates are used it is sometimes necessary to print the labelling on to the surface of one of the laminae before the lamination process is complete. Other techniques involve the inclusion of a small viewing window.

It is hoped that the above comments will be helpful in explaining why commercial techniques can differ from domestic practice, yet both can be logical in their own context.

7

Commercial Aspects

A BOOK about deep freezing would not be complete without a brief description of the frozen-food industry as it exists today. It is difficult to realize that this industry which makes such a big contribution to the diets, and even the way of life of so many people, has been largely developed since the Second World War. It is true that the frozen-food industry was established in America in the 1930s, and no doubt had the war not intervened the attempts being made to develop the industry in England at the end of that decade would have been successful.

Establishing a new industry was not easy in the early postwar years. Licences were necessary for new building, materials were in short supply, and fuel was rationed. The physical problems of factory production was one main difficulty which had to be overcome; inducing the housewife to buy frozen products was a problem of equal importance. The poor quality of some of the frozen fish supplied during the wartime had resulted in a prejudice against frozen foods which had to be overcome. However, because some of the early frozen products represented supplements to food rationing which continued for some years after the war, some consumers may have been encouraged to give frozen products a second chance.

But before the consumer could be invited to sample frozen foods the local retailer had to be convinced that frozen foods were a good proposition. In order to stock them, he had either to buy or be loaned a frozen-food cabinet, and some of the early models took up a great deal of valuable floor space without displaying the produce very effectively. All this development was taking place without the help of television advertising which boosts new products launched in present times.

The situation is very different today, when few provision or delicatessen shops are without a frozen-food cabinet, and where there is a large variety of foods from which to choose. In the early days of the industry the range of products was very small, consisting only of certain fruit, vegetables, and fish. This meant that the processing plants were very busy for short spells of six to eight weeks when the crops were harvested, and at other periods of the year freezing demand was slight and employees were discharged. With the increase in the variety of foods produced, these slack periods have been eliminated. Each year new products are to be found in the food cabinets, including pre-cooked and prepared foods; for example, there are now chicken pies, sponge rolls, puff pastry ready for shaping, and fish fingers ready for frying. 'Convenience foods' is the term used to describe these products and this approach reaches perfection in the form of 'dinner-for-one', a complete meal on an aluminium tray requiring only to be reheated.

At the beginning of the growth of the industry, as with most new products, many producers attempted to get into the frozen-food business, not all of whom had the highest standards. Fortunately the exacting requirements for the production and distribution of frozen foods did not permit this type of producer to flourish. The pattern of the frozen-food industry today is of a few large producers who have control of the product from the beginning to the end; thus frozen-vegetable producers provide the seeds for the crops and exercise control of the product through growing, harvesting, preparation, to final distribution.

Obviously different techniques are developed for gathering and preparing the range of fruits and vegetables for freezing according to their specific problems. However, the emphasis is always on speed, so that the quality of the product can be maintained.

Some of the large frozen-food producers raise their own chickens in order to have a regular supply of poultry to keep their plant fully operative.

In the fish industry, the supply of fish is always less predictable, but instead of buying fish for their requirements when it is auctioned at the quayside, the tendency is for frozen-food producers to have a contract with trawler-owners. In some cases the frozen-food producers own trawlers of their own. Plant for freezing fish is naturally sited at the fishing ports.

The large producers have their own frozen-storage depots, transport, and methods of distribution. There are, however, many smaller producers who do not compete in the large-scale production of fruit and vegetables, and are not able to control large areas of farmland. These smaller producers confine their activities to a few specialist lines, frequently of pre-cooked and prepared foods. These producers do not undertake the storage and distribution of their products, but supply them to a wholesaler who provides frozen storage and undertakes the distribution to the retailers.

Whether the storage and distribution of the frozen foods are undertaken by the producer himself or by a middle-man, most of the attendant problems are unchanged.

PROVISION OF COLD STORAGES

Some of the first cold stores were, in fact, merely converted warehouses which had been lined with insulating materials. Others had been built originally as sharp freezers. This method of freezing had been in widespread use for many years, mainly for freezing meats. The food to be frozen was spread out on shelves, and frozen by cold air at temperatures of 5°F to −20°F (−15 to −29°C) passing over it.

Modern cold stores are built exclusively for the purpose of storing frozen foods. Economic evaluation is given to attaining the right balance between the size of the refrigeration plant and the thickness of the insulation walls. Obviously the thicker the insulation the lower will be the heat leakage into the cold store. But insulation is costly and a compromise is usually drawn between the choice of thin insulation with a large refrigeration plant, and thick insulation with a small plant. The former is cheaper to construct, but more liable to produce fluctuations of temperature; the latter is more costly to construct, but able to provide a relatively steady temperature.

The shape of the store is also important; the more it resembles a cube the lower will be the heat leakage per cubic foot of store, and thus the more efficient in operating costs. Cork, sawdust, fibre-glass were some of the insulating materials used, sometimes in sheet form and sometimes loose. These are being superseded at the present time by plastic materials such as expanded polystyrene and expanded polyurethane which when blown into a cavity expand as a foam and completely fill it.

This method of insulation is particularly suitable for a method of construction referred to as the 'curtain wall' construction. This resembles two boxes, one within the other, with the insulating material sandwiched in between. This method of construction stands up well to any stresses and strains produced by shrinkage of the building materials, without the production of cracks, as the inner and outer shells may move independently of each other.

Heat may be introduced into a cold store through the opening of the doors, by people entering it, and through the walls. One cooling system uses cooling pipes carried just below the ceiling and along the walls; this is very effective, as the refrigerating source is near to the source of heat. An alternative method which is less costly to install is achieved by circulating around the chamber, air which has just been passed over a cooling unit. Although the temperature of the cooling coils in such a cooling unit may be very low, this method does not provide the same 'wall of cold' produced by the first method and therefore permits a greater variation of temperature.

TRANSPORT

Cold stores which cover certain areas of the country receive frozen food from the factory or plant producing them, and then undertake the distribution to the retailer. Transport is obviously an important item in the distribution of frozen foods.

Whether it is the long-distance lorries which take the food from the factories to the cold stores, or the short-haul lorries calling on retailers, they must be capable of maintaining the food at 0°F (−18°C). However, each presents a different problem. The long-distance lorry once loaded is not opened until it reaches its final destination, which may be twenty-four hours later; whereas a van used for retail deliveries may be opened at least twenty-five times in the space of eight hours.

The designers of refrigerated transport have been called upon to provide a vehicle of good capacity which can maintain a suitable temperature in the face of all weather conditions, and which must also comply with road transport regulations. At the same time the construction must be strong enough to withstand the movement and hazards of road travel. These problems have been simplified by recent developments in materials and methods. Moulded reinforced plastics are used to provide the inner and outer shell of the food container, and expanded plastic foam used as the insulating material between.

Using power from the engine of the vehicle to run the refrigeration unit in the food container presents technical difficulties. This is because the speed at which the compressor functions would vary with the speed of the engine, and would cease to function altogether when the engine was stationary.

One method of cooling the food compartment which is independent of whether the vehicle is in motion or not is by the use of 'hold-over' plates. In the same way that a domestic hot-water bottle, filled with hot water, is a retained source of warmth, so a hold-over plate may be 'charged with cold' and is a retained source of cooling. These plates are slim flat tanks in which run evaporator coils (similar to those found in a domestic refrigerator). The tank itself is filled with an eutectic solution freezing at about $-12°F$ ($-24°C$) which surrounds the coils. When the hold-over plates are plugged into the vehicle's compressor the evaporator coils produce their cooling effect on the eutectic solution and it freezes. The compressor is switched off when the vehicle is in motion, and the hold-over plates absorb heat from the food compartment as long as the eutectic solution remains frozen. The compressor is operated from the electricity mains and it must be possible to link up the vehicle to a power supply when it is at rest in order to 'recharge the plates'. Frequently the plates are such that the time taken to recharge them is about the same as the time for which they are effective, so the vehicle must be at rest for twelve in every twenty-four hours.

Hold-over plates are produced which do not contain evaporator coils, and are not therefore dependent on an electrically powered compressor to freeze the eutectic solution. These 'portable' plates may be cooled by immersion (see reference to immersion freezing on page 112). If the vehicle is able to exchange spent hold-over plates for recharged ones at a depot, so the useful refrigerating hours of the vehicle can be increased.

Probably the most popular method of cooling found in refrigerated transport is that using solid carbon dioxide. Carbon dioxide absorbs heat at the rate of 273 BTU's/lb., and so it is possible to calculate the weight necessary for the duration of any journey undertaken.

Not all frozen foods are transported by road, some travel by rail, sea or air. Special containers are designed which are completely independent, and may be fitted on to a lorry or railway wagon as convenient. They are usually fitted with an attachment for lifting and they may be transferred from one means of transport to another.

Methods of Freezing

(a) Blast freezing

Food frozen by this method is simply brought into contact with a blast of cold air. The temperature of the air is in the region of −20°F (−29°C) and the velocity may vary from 300 to over 1,000 feet/sec. The freezing may be carried out in a small insulated room or on a larger scale in a tunnel through which the food may be passed. In both techniques fans circulate the air over the cooling coils and then over the food, which may be loose or already packed.

This method has several advantages; it can be used to handle foods of all shapes and sizes, and when a tunnel is used it is possible to achieve continuous throughput. If a free-flowing product is required, as with peas and some fruits, they may be placed on mesh trays or mesh belts and air blown through from below.

This method is probably the slowest of the commercial methods of freezing, and for maximum efficiency food must be carefully spaced out, so that air may circulate freely. Careful control is also necessary so that the system is not overloaded.

Probably the main disadvantage of this method is the loss of moisture from unpacked foods, with the twofold disadvantage of desiccation of the product, and a build-up of ice on the evaporator coils.

The underlying physical phenomenon is that as air is warmed so its moisture-holding capacity is increased. When it is afterwards cooled it can no longer hold the same amount of moisture, and this is deposited on a cold surface. The cold air in a blast freezer is warmed by contact with the food and takes up its moisture; this moisture is afterwards shed when the air is cooled by contact with the evaporator coils. This process of sublimation is the same as that which occurs on a smaller scale in a domestic freezer.

Various measures are adopted to deal with this problem of moisture loss. Staging is one approach, where, instead of circulating air of the same temperature throughout the chamber, warmer air is used when the product is first introduced. As the product passes through the system the air with which it comes into contact is progressively cooler. The air temperature is lower than that of the food at every stage, but by keeping the temperature differential small the moisture loss is low also. Another device to achieve this same objective is to blow cold air through the tunnel in the reverse direction to the one in which the food is travelling.

111

(b) Contact or plate freezing

A contact freezer resembles an insulated cupboard with a number of hollow shelves, not in a fixed position, but able to move up and down in a vertical slide. Each shelf is connected up by flexible tubes to a condenser, and refrigerant is circulated through the shelves so that each one functions as a cooling surface.

The food to be frozen, packed into regular flat boxes, is loaded on to the shelves. Then by means of a hydraulic press applied either from above or below, the plates are moved together so that the food is closely sandwiched between the shelf on to which it was loaded and the one above. Heat is removed from the food by the refrigerant circulating in the plates, and freezing by this means is quite rapid.

This method provides well-shaped packages of food with little or no dehydration. The plate freezers are independent units which do not occupy much floor space and which may be moved when required to suit a new plant layout.

A disadvantage of this method is that it may only be used for slim, flat packs of food, such as vegetables, fish, and sliced meat. With foods of an irregular shape, such as chicken joints, it is difficult to achieve the same close contact with the shelves. This system cannot be used to provide a free-flowing product, thus packs of vegetables or fruits are squeezed together into a solid 'cake' in the freezer.

The main criticism made of this method of freezing by the producer is that it is not possible to achieve a continuous throughput. Labour is required for loading and unloading. Some plate freezers have been introduced which have associated equipment for loading and unloading the freezers, but such equipment results in a larger, more expensive, and less mobile unit.

(c) Immersion freezing

Immersion freezing brings the product being frozen into direct contact with the refrigerant, which is itself in contact with the cooling coils.

The refrigerant is normally circulated by a pump in a large shallow bath through which the food can be passed. At the same time the refrigerant may also be sprayed on to the food from above; this method is particularly suitable for poultry.

The main advantages of this method of freezing are that it is very rapid and can be used to freeze foods of all shapes and sizes. Foods are normally passed through the bath before they are packed, and thus can be presented as a free-flowing product, instead of the solid block produced in the plate freezer.

The problems associated with this freezing technique lie in the selection of an appropriate refrigerant. Common salt is used fairly extensively for vegetables, fish, and poultry. Although should the salt adhere to the surface, or penetrate the tissue of fish and poultry as a result of osmosis, it will act as a catalyst for oxidation of the fat content and rancidity will be accelerated. Osmosis may work in the opposite direction and result in a loss of weight from the food as the natural salts diffuse out into the refrigerant. Obviously strict control must be kept on the concentration and temperature of the refrigerant.

Salt solution cannot be used for fruits, and sugar solution and glycerol may be substituted. The sugar solution which surrounds the individual fruits helps to preserve a good colour and flavour. When the temperatures of sugar solutions are lowered they become viscous and unmanageable, so that here, too, the concentration and temperature must be carefully controlled; the cleanliness of the bath must also be watched if it is to be kept free of contamination.

Immersion freezing can also be used for packed foods, but obviously only for those packed in water-tight containers. One of the most successful applications of immersion freezing is for concentrated fruit juices, which are packed in round tins and then rolled along through a bath of alcohol brine.

THE PEA HARVEST

Frozen peas continue to be the most popular frozen product on the market today, and it is therefore interesting to follow the course of this vegetable before it reaches the table of the consumer.

Formerly the frozen-food producers used to buy the crops of peas when harvested by the farmer. This method was not satisfactory for large-scale production, because the producer could regulate neither the flow nor the quality of peas into his plant.

A more satisfactory system is in operation today whereby the farmer leases his land to the producer, who in turn supplies him with seed for the crop. The seed supplied is that which has been developed by the frozen-food producer to give a high yield of peas of a regular size and good colour. Different varieties are selected and the planting staggered so that the harvest may be spread over as long a period as

113

possible. Fieldsmen employed by the producer are in continual attendance, giving advice on sprays and fertilizers where necessary. By keeping a record of soil temperature the fieldsmen are able to estimate when the crop will be ready for harvesting.

Frozen peas are sweet and succulent, because they are harvested before they are fully matured; as peas mature, more of the sugar content is changed into starch, with a resultant loss of sweetness. Thus the time chosen for harvesting is most critical. Radio-telephones in the fieldsmen's cars keep them in constant touch with the factory. As the peas approach maturity, samples are tested regularly in a 'tenderometer'—this is a machine able to assess the texture of peas fed into it. As soon as the appropriate reading is recorded harvesting begins without further delay. Mechanization has been applied to harvesting. The complete vines are cut and loaded into lorries from which they are tipped into the viners, which are stationed near to the fields. In the same way that a threshing machine releases grain from the chaff, so a viner releases peas from the pods while they are still attached to the vines. Perforations in a revolving drum allow the peas to pass into a container in which they are taken to the factory.

Mobile vining machines have been introduced which compare closely with a combine harvester; they follow the 'cutters', gather up the pea vines, shell out the peas, which can then be rapidly transported to the factory.

The plant or factory receiving the harvested peas is situated as near as possible to the fields, and it is the proud claim that peas can be fed into the freezer within ninety minutes of leaving the viner. In the factory the peas are graded, washed and carefully inspected, then blanched and cooled before being frozen. Formerly freezing was carried out in a plate freezer, but today the trend is to use a blast freezer resulting in a free-flowing product which is considered more acceptable by most consumers.

THE FISHING INDUSTRY

This chapter began with a short description of the development and operation of the frozen-food industry, and it is appropriate to give some indication of the changing trends in the fishing industry. As well as fillets and cutlets prepared ready for cooking, products containing fish in some form or other feature prominently in the lines offered for sale by the frozen-food producers. Thus such producers compete along with caterers, fish friers, and fishmongers for the landings of fish.

The fishing industry has been faced with demands for regular supplies of fish of good quality, but because some foreign countries have extended their territorial waters, and because of an increasing scarcity of fish, the trawlers are forced to travel further in order to fill their holds. However, the length of time the boats can remain at sea is limited by the length of time the fish may be kept on ice. This allowance must always include the time taken for the return journey to port. The first-caught fish go into the bottom of the hold, and it may be that when the return journey is longer than estimated the first-caught fish is not fit to be offered for human consumption and must be sold for fish meal. Only the last-caught fish which forms the top layer in the hold of the trawler is suitable to be frozen on the return to port, and if the homeward journey of the trawler is delayed by storms, then even the quality of this fish may not be of a sufficiently high standard for it to be frozen. Increasing the speed of the boats has not resulted in a significant improvement in the quality of the fish landed, and the increase in space taken up by the larger engine has led to a reduction in the size of the hold.

Freezing at sea

Freezing the catch on the trawlers would seem to be the obvious way of increasing the landings of high-quality fish, and several such schemes have been put into practice. One compromise solution was carried out experimentally some years ago in the trawler, the *Northern Wave*. This trawler was fitted with freezing equipment and cold storage enabling it to freeze the first quarter of the fish landed, and bring it back in frozen storage to port. By doing this the length of time the first-caught fish must be kept on ice is shortened, and the losses of fish going for fish meal is reduced, or alternatively the length of time spent by the trawler on the fishing grounds is increased.

Experiments have also been made in using a 'mother ship' to escort a fleet of trawlers in order to take off their catch of fish and to freeze it.

The trend for the future will probably be to build trawlers which are able to stay at sea for seven to eight weeks and which freeze the complete catch on board. On these ships the fish may also be filleted before being frozen. Obviously trawlers which carry equipment for

115

freezing and storing the complete cargo, and carry the additional crew members necessary to prepare and freeze the fish, will be larger and more costly to produce. The landing of a big catch of fish may overload the filleting and freezing facilities, but in such a case the fish may be kept on ice for up to three days before being frozen. As would be expected, there will be a noticeable quality difference in fish frozen one and three days after being landed. It is the general opinion that better results are achieved with fish which are frozen post-rigor; this will be some thirty hours after being caught (apparently fillets from fish frozen pre-rigor are not as satisfactory for curing or smoking). This practice is very difficult to achieve on board a trawler.

Vertical-plate freezers have been found the most satisfactory to operate on board ship. The fish to be frozen is first gutted and washed, and frequently the heads are also removed before being dropped into the cavities between the freezer plates. It has been found that where fish are frozen without gutting them first, the flesh becomes discoloured and is likely to be rejected by most consumers.

The size of the cavity between the vertical freezer plates is based on the average size of the fish to be frozen. Usually it is cod and the cavity is about 4–5 inches. Excessively long or thick fish are blast frozen or sharp frozen on shelves in the cold store.

In order to facilitate the release of the frozen blocks of fish from the vertical-plate freezer, the usual method employed is to circulate hot gas in place of the refrigerant through the cooling plates. The resulting blocks of fish may be in the region of 60–100 lb. each and will need careful handling both in the cold store on board ship and when ultimately transferred from the ship in port, if they are to be kept intact and undamaged.

Frozen fish may be kept for many months at −20°F (−29°C), but where the temperature is allowed to fluctuate above this temperature, the storage life and ultimate quality will be much reduced. There is a tendency for the outer layers to become desiccated, but this can be largely prevented by glazing. This is a method of giving the outside of a block of fish a thin coating of ice, by successively pouring over it quantities of ice-cold water which freezes and provides a continuous covering. In cases of prolonged storage it is necessary to repeat this glazing at intervals. Other treatments given before freezing include a brine dip for some lean fish to prevent the loss of fluid or drip when thawed. Fat fish may be treated with an antioxidant such as ascorbic acid to prevent the development of rancidity.

116

Thawing of fish

The thawing of large blocks of fish will obviously take some time. Separating out the blocks of fish on shelves and allowing them to thaw at normal air temperature is one method, but this is very slow; for quicker results water also is used. By flowing water over frozen fish the normal conduction of heat process applies and fish can be thawed more quickly then by using air conduction.

A very effective postwar development for thawing fish is to use radio-frequency energy in dielectric thawers. Frozen fish are passed on conveyors through a series of ovens containing valve oscillators which generate radio waves. The energy of these waves is dissipated in the frozen fish in the form of heat and thaws out the fish. To avoid any cooking of the surface fish, and to attain a more even thaw, the frozen slabs of fish pass along the conveyor in rigid polythene trays containing some water and covered with a polythene lid. This method uses electrical energy and is therefore more expensive than air or water methods, but its speed of thawing, compactness, and its aid to mechanization are obvious advantages, and it will be the normal thawing method in the future in sophisticated areas and countries.

Distribution of fish

Fish prices, like those of most commodities depend to some extent on whether supplies are plentiful or scarce. Thus when landings are large prices are low so that if part of the catch can be withheld, which is possible with frozen fish, it may be thawed and offered for sale many months later when fish is scarce. This obviously involevs an increase in cold-storage facilities at the ports.

It has been shown that firm fillets may be cut from thawed whole fish, and these fillets result in a satisfactory product when taken for smoking or curing. Considerable prejudice has always accompanied frozen fish, although the confidence of the caterers, the frozen-food producers, and fish friers has now been secured. The average fishmonger, however, has remained the most difficult to convince. This is no doubt because fish, when laying on a fishmonger's slab, is judged by its appearance, and fillets from frozen fish lack the characteristic 'bloom' associated with fresh fish. It is, in fact, possible to re-create this bloom by briefly dipping the fillets into a brine solution.

It can only remain a matter of time before the fishmonger and his customers accept frozen fish as being as good, and frequently a great deal fresher, than the unfrozen equivalent offered for sale in most inland towns.

ACCELERATED FREEZE DRYING AND DEHYDRO-FREEZING TECHNIQUES

Accelerated freeze drying and dehydro-freezing are two further techniques for treating foodstuffs, both involving the freezing of the product at one stage in its preparation.

Accelerated freeze drying

Freeze drying has been used as a method of preservation for many years, but it was during the war that its full potential was realized as a method of preserving blood plasma. Freeze drying is a method of producing under controlled conditions, a condition we seek to prevent in domestic freezers. Temperature fluctuations in a freezer result in a loss of moisture vapour from inadequately wrapped foods. This is caused by the sublimation of ice from the surface of the food.

The principle involved in freeze drying is the sublimation of ice from the frozen food held under vacuum by the application of heat, and the removal of water vapour from the drying layer of the food and vapour space surrounding it.

The heat necessary for sublimation may be supplied by direct contact with heater plates or by radiant heaters, the product being considered dry when the surface and the internal temperatures are the same. Obviously the heat input must not be such as to cause the surface of the food to become denatured.

The unit used for drying is like a very large box or cabinet in which a vacuum can be produced; there is a condensing unit either inside or outside the cabinet, which is able to take up the full quantity of moisture vapour produced from the subliming ice crystals.

The speed at which the food is dried depends not only on the rate at which heat is applied, but also on the speed with which moisture vapour can escape from the centre of the product through the dry layer, and also upon the ability of the condenser to keep pace with the removal of moisture vapour produced from the food.

An increase in the drying rate, or *accelerated freeze drying*, is achieved by inserting a sheet of expanded metal between the heater and the food, thus enabling the moisture vapour to escape more rapidly.

It has been shown that the size of crystals in the frozen food influences the drying rate. Quickly frozen foods containing small crystals dry more slowly than slowly frozen foods containing larger crystals. At the same time, freeze-dried foods produced from slowly frozen foods seem to reabsorb moisture more quickly, although it

118

does appear that the reabsorption is not always complete, and it is possible to 'squeeze out' the moisture from such foods.

A factor not affecting the drying rate, but which affects the speed at which the system can be operated is the necessity for drying to take place in a vacuum. Because of the problems involved in maintaining a vacuum, the system must be batch-loaded, and thus it is not possible to achieve a continuous throughput.

Most foods are suitable for freeze drying, although the product size is a limiting factor; this method is thus not suitable for large items such as joints of meat; but thinner cuts of meat and fish are suitable, both raw and cooked. Fatty foods are not suitable, as when the heat is applied to the surface of the food the fat melts and forms a waterproof layer, inhibiting the diffusion of moisture vapour from the surface. For the best results when freeze drying, vegetables should be blanched before being frozen. On the whole, foods with a high sugar content are unsuitable for this method, as they become viscous and difficult to handle when the heat for drying is applied.

One example of a food for which freeze drying has proved outstandingly suitable is the smaller shellfish, prawns and shrimps. Because of their convoluted appearance, unless given special protection in the form of a vacuum pack when frozen they rapidly show signs of desiccation. The freeze-dried product, after a few minutes rehydration in water, is almost indistinguishable from the fresh equivalent.

Accelerated freeze drying produces a very satisfactory product which may be quickly reconstituted, and which shows little shrinkage. There is little migration of the soluble constituents.

Obviously preservation of food by this method is more costly than by freezing alone. But because the products do not have to be kept in a frozen condition, and because there is a reduction in weight and volume, considerable savings are possible in storage and transport. There should therefore be little difference in the retail price at which A.F.D. products and frozen products are offered for sale; greater advantages accrue in hot countries and where longer haulage is involved.

The shelf life of A.F.D. products is good, providing the foods are properly packed. Packaging materials which provide an effective barrier to moisture vapour and oxygen are obviously essential.

Dehydro-freezing

Dehydro-freezing is a technique which has been developed, but

119

which has never featured on the retail market and therefore only a brief explanation of it is given here.

It was observed that in the normal process of preserving foods by drying the bulk of the water was removed very rapidly, but the last few per cent much more slowly. It is thought that the loss of quality observed in some dried foods is caused by the prolongation of the drying cycle necessary to render the food 100 per cent dry.

Dehydro-frozen foods are those which are frozen after the bulk of the water has been removed by the normal drying process. Naturally they weigh less, occupy a smaller volume than their frozen equivalent, but they still need to be stored at zero Fahrenheit, as they still contain a small percentage of water.

REFERENCES

'Quick Frozen Foods.' John L. Rogers. *Food Trade Press*. 1958.
'Frozen Foods—the Growth of an Industry.' R. W. D. MacIntosh. *Refrigeration Press*.
The Freezing Preservation of Foods, Vol. I. Tressler and Evers. A.V.I. Publishing Co. Inc. 1957.
Report on an Experiment into the Freezing of Fish at Sea. White Fish Authority. 1965.
'The Preservation of Fish at Sea.' G. C. Eddie. *Recent Advances in Food Science*, Vol. II. Butterworth. 1962.
'The Accelerated Freeze Drying Process.' J. C. Forrest. *Recent Advances in Food Science*, Vol. II. Butterworth. 1962.
'Accelerated Freeze Drying of Food Stuffs.' J. D. Mellor. *Food Preservation Quarterly*, Vol. 22, No. 2. 1962.

Part Two

8

Freezing Fresh Fruit and Vegetables

MERITS OF FREEZING FRUIT AND VEGETABLES

THERE can be no doubt when assessing the various methods of preserving fruit and vegetables, that frozen storage is the only one yielding a food completely indistinguishable from the fresh equivalent. However, it must be stressed that there is nothing magical about freezing, and the quality of the food removed from the freezer is no better than the food put into it. Thus overmature stringy beans will not emerge young and succulent some six months later.

Two advantages must surely commend freezing as a method of preservation in the home. The first is that it is less time consuming and probably less complicated than either bottling or canning. Also while bottling and canning are not usually worth undertaking unless a sizeable quantity of fruit or vegetables is to be processed, the odd pound or two of garden produce may be frozen with little interruption of the normal timetable. In fact, freezing in small quantities is to be recommended, partly because the produce can then be handled more quickly and there is no possibility of overloading the freezer with unfrozen food.

An attempt is made in this section to give directions for dealing with most fruit and vegetables which may be frozen. However, because freezer space is precious, it is more valuably employed for those fruits and vegetables which have a short season or a widely fluctuating price. Frozen raspberries and asparagus spears represent a better investment than plums or sliced carrots.

To give some guidance in the planning of freezer space the calendars in Figs. 9 and 10 set out when fruit and vegetables are in season and likely to be at their best for freezing.

123

FIGURE 9

FRUIT CALENDAR

This calendar shows when home-grown fruits are likely to be in season and when imported varieties are at their best.

NOTE: Imported varieties are shown in *italics*

January – February	Rhubarb (forced)
February – March	*Peaches*, *Grapefruit*
March – June	Rhubarb (outdoor)
May	Gooseberries (cooking)
June – July	Strawberries
June – August	Gooseberries, *Apricots*
June – October	Peaches
July – August	Peaches, Raspberries, Blackcurrants, Redcurrants
July – August	Cherries, Plums, Nectarines, Blackberries
August – October	Apples, Pears, Bilberries
September	Greengages
September – October	Damsons
October – December	*Cranberries*

The following imported fruits are available all through the year:

Apples (and some home-grown varieties) *Avocados* *Bananas* *Grapes* *Grapefruit* *Melons* *Oranges* *Pineapples*

VEGETABLE CALENDAR

This calendar shows when the various varieties of vegetables are in season and when imported varieties are likely to be at their best.

NOTE: Imported varieties are shown in italics.

January
February
March
April
May
June
July
August
September
October
November
December

Celery

Purple Sprouting Broccoli

Broad Beans

Brussels Sprouts Celery

Brussels Sprouts

Asparagus

Peas Courgettes Tomatoes

Parsnips

New Potatoes

Beetroot French Beans

Carrots Corn Runner Beans Aubergines Turnips

The following home-grown and imported varieties are available all through the year:

Beetroot Broccoli/Cauliflower Cabbage Spinach Mushrooms Potatoes *Aubergines Courgettes Peppers Tomatoes*

VARIETIES TO BE FROZEN

As far as possible varieties of fruit and vegetables are selected for their retention of colour, flavour, and texture throughout frozen storage. Obviously for the frozen-food producer the selection of the right variety is of utmost importance. The nurseryman will select varieties for their potential yield and resistance to disease.

Two organizations whose work includes the growing and assessment of varieties suitable for freezing are:

The Department of Agriculture and Horticulture, Research Station, Long Ashton, Bristol, and
The Fruit and Vegetable Canning and Quick Freezing Research Association, Chipping Campden, Gloucestershire.

Here varieties are bred, grown, and tested over a number of years before they are recommended. Occasionally a variety which has been successful over a long period of time fails, maybe due to an inability to resist a particular disease, or due to genetic breakdown, and then a suitable substitute is found.

An account of the work at these research stations may be found in their annual reports. Obviously anyone aiming to start a large fruit and vegetable garden with a view to freezing much of the produce would be advised to contact them. A list of some recommended varieties of fruit and vegetables published by the Research Station, Long Ashton, is included in Fig 11.

The owner of a small, successful, and established fruit garden would not be advised to replace his fruit bushes and trees with new stock; on the other hand, if replacements or additions are needed, the freezing potential of the new variety might be considered before purchasing. Because the vegetable garden is sown annually, there is scope for experiment with different seeds. It might be worth while to consult the various seed manufacturers for their advice when selecting seed.

MATURITY STATE FOR FREEZING

It is important that both fruit and vegetables are frozen when at the correct stage of maturity. For fruits this is when they are just ripe for eating. Underripe fruit has neither the full flavour nor the sweetness of fully ripened fruit, also preparation such as the removal of the skin, stone, or calyx is more difficult. Fruits which are just overripe or

FIGURE 11

SOME VARIETIES OF FRUITS AND VEGETABLES RECOMMENDED FOR HOME FREEZING

RECOMMENDED FRUITS

Blackberries	Cultivated or wild, if large, juicy and ripe.
Cherries	Black varieties.
Loganberries	Red ripe
Raspberries	Early Red, Lloyd George, Malling Promise, Norfolk Giant.
Strawberries	Cambridge, Prizewinner, Cambridge Vigour, Royal Sovereign, if sliced.
Whortleberries	Wild or cultivated Blueberries.

RECOMMENDED VEGETABLES

Asparagus	
Beans, Broad	All varieties when young, especially Aquadulce Claudia, Beck's Dwarf Gem, Bunyard's Exhibition, Eclipse, Harlington White, Leviathan, Six-seeded King of Beans, Sussex Wonder, Taylor's Windsor White.
Dwarf	Divil Fin Precoce, Granada, Perpetual, Tenier Green, The Prince, The Victory.
Runner	Cookham Dene Improved, Kelvedon Wonder, Scarlet Emperor, White Monarch.
Broccoli	Green Sprouting (Calabrese), Purple Sprouting, White Sprouting.
Brussels Sprouts	Small firm sprouts preferred. Cambridge varieties, Noisette, Sanda.
Carrots	Shorthorn varieties with good orange colour, young. Amsterdam Forcing, Early Nantes, Perfect Gem.
Cauliflower	Divide into sprigs. Majestic.
Peas	Early June, Kelvedon Wonder, Lincoln, Newburgh Gem, Onward, Perfected Freeze, Peter Pan, Phenomenon, Thomas Laxton.
Spinach	Giant Savoy Leaf, Goliath, New Zealand, Perpetual, Prickly New Giant, Zenith XXX.
Sweet Corn	Canada Cross, Earliking, Golden Bantom, John Innes.

The above is reproduced by the courtesy of the Research Station, Long Ashton, Bristol, and is an extract from the Annual Report, 1965.

slightly damaged may be used for purées. Vegetables are also just right for freezing when at their best for eating; with most vegetables, particularly of the leguminous variety, this is before they are fully matured. The carbohydrate food store is largely in the form of sugar in young peas and beans, but it changes into starch as the pods ripen.

Because this change-over takes only a few days, the time for picking is quite critical. The texture of most vegetables deteriorates as they pass their prime; frequently they become tough and stringy.

PICKING AND PURCHASING

It is not sufficient merely to pick or purchase fruit and vegetables when the condition is just right for freezing if there is to be a delay in getting them into the freezer. This delay may be avoided if the produce is supplied from the garden. As far as possible, fruit and vegetables are best gathered early in the morning and never in the heat of the day. If, however, the supplier is the local nurseryman or tradesman, it would be worth while to check with him when he expects to be picking or receiving fresh supplies. Some advance preparation is thus possible. Packaging materials should be checked, and where sugar syrups are to be used these may be made up and refrigerated several days ahead. Whether picking or purchasing, it is important to bear in mind the capacity of the freezer, and how much unfrozen food may be added per twenty-four hours.

All produce awaiting preparation for freezing should be kept in a cool place, and not allowed to lie about in a warm kitchen until it can be prepared and packed for the freezer. If possible, the produce should be tipped out from the containers and spread out on trays. This is because fruit and vegetables continue to respire and produce heat even after they are gathered. Left together in a large container, the amount of heat produced, particularly in the centre, accelerates the loss of Vitamin C, and generally results in a deterioration of quality.

PACKAGING FOR FRUIT AND VEGETABLES

If it is intended to preserve a large quantity of fruit and vegetables annually, the choice of suitable containers is of importance. There is a wide choice of shapes and sizes in both the semi-rigid plastic containers and those made of waxed cardboard (for more details see the chapters on packaging).

It is, however, worth drawing attention again to the following points:

(a) As far as possible choose a rectangular shape which will pack easily and neatly in the freezer.

5a. Some examples of aluminium packing materials and
containers

5b. Some examples of packaging materials and
containers

6a. Blanching in boiling water

PREPARATION OF VEGETABLES FOR FREEZING

6b. Cooling under running water

7a. Drying in kitchen paper

PREPARATION OF VEGETABLES FOR FREEZING

7b. Packing into freezer container

8a. Holdover plates installed in refrigerated trucks absorb heat from the food compartment and maintain the food at sub-zero temperatures

8b. Dielectric fish thawer made by Radyne Ltd. for Stirk Bros. of Hull

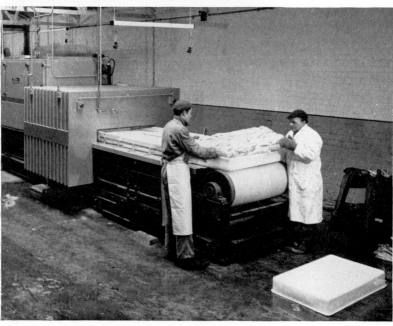

(b) Select the size which will hold the requisite number of portions for the family; for example, an adult portion of prepared vegetables is 3–4 oz.

(c) Where there is a possibility of the container becoming stained by the fruit or vegetables contained in it, a polythene liner should be used.

(d) Use a funnel when filling a container to avoid spilling food over that section of the container which is to be sealed. This is particularly necessary when heat-sealing plastic bags or using freezer tape.

Headspace allowance for fruit and vegetables

For produce packed without added liquid and which does not pack very compactly, such as florets of cauliflower, no headspace is necessary. But for most other dry packs a small allowance should be made. Where the produce is packed in syrup, sugar or brine, or as a purée of either fruit or vegetables, an adequate allowance must be made. Those containers having a narrow top require a bigger headspace than those with a wide one.

Approximate allowance:

Dry packs of fruit and vegetables (packed loosely)	— no headspace necessary
Dry packs of fruit and vegetables (packed compactly)	— allow $\frac{1}{2}$ inch for all containers
Wet packs of fruit and vegetables (in sugar, syrup, water or brine, and purées of fruit and vegetables)	— allow $\frac{3}{4}$ inch per pint for narrow-topped containers, but allow $\frac{1}{2}$ inch per pint for wide-topped containers

Block freezing

Plastic bags which provide a cheap method of packaging fruit and vegetables are frequently discounted because the resulting irregular shapes cannot be packed neatly in the freezer. This disadvantage may be overcome by freezing the produce in a regular container, such as an ice-cube tray, before packaging. The resulting block of fruit or vegetables can be packaged in a plastic bag or film on the following day. This method is particularly suitable for freezing small quantities of fruit and vegetables.

Loose Packs

The usual method of freezing in a domestic freezer produces fruit and vegetables in solid blocks similar to those produced commercially in a plate freezer. It is possible, however, to produce a free-flowing product similar to that produced commercially by immersion freezing, in a domestic freezer. The small-berried fruits, peas and diced vegetables are most satisfactory for this treatment. After preparation, blackcurrants or peas—to take two suitable examples—are spread out on baking trays or large dishes and put into the freezer over-night. The next day the individually frozen berries or peas may be packed into containers. Because expansion of the water content of the produce has taken place during freezing, no headspace allowance is necessary when packing. The short time the produce remains unprotected in the freezer does not result in a significant loss of moisture. In most domestic freezers there is insufficient space to freeze a large proportion of the produce by this method; however, to have some packs filled with loosely frozen fruit and vegetables will be found very convenient when only a small quantity is required for a garnish or a single portion. The required portion can be tipped from the container, which is afterwards resealed.

REFERENCES

Domestic Preservation of Fruit and Vegetables. Ministry of Agriculture, Fisheries and Food Bulletin No. 21. *HMSO*. 1962.

Home Freezing of Fruit and Vegetables. Ministry of Agriculture, Fisheries and Food Advisory Leaflet 434. *HMSO*. 1962.

Home Freezing of Fruits and Vegetables. Home and Garden Bulletin No. 10. United States Department of Agriculture. 1965.

Horticultural Research for the Quick Freezing Industry. V. D. Arthey. The Fruit and Vegetable Canning and Quick Freezing Research Association Technical Memorandum No. 55. 1964.

9

Fruit

Two considerations must be taken into account when selecting the method by which fruit is to be packed; the first one is the type of fruit being preserved. Fruits which make their own juice, such as raspberries and strawberries, are most successfully packed in dry sugar. Mild-flavoured, firm-textured fruits, from which juice does not flow liberally, such as peaches and apricots, give good results when frozen in syrup. Finally, those fruits possessing a fairly tough unbroken skin can be frozen without any additions.

The second consideration is the use for which the fruit is intended, whether it is to be used directly for dessert, in a pie, subsequently for jam making, or for someone following a special diet.

According to these considerations, one of the following methods of preparing the fruit will be selected.

PACKING WITH DRY SUGAR

After washing and preparing the fruit in the usual way, it is combined with castor sugar in a shallow bowl. Some fruits, such as raspberries, are normally left whole, but others, such as strawberries, may be sliced. The juice which flows from the fruit combines with the sugar to make a natural syrup. When adding the sugar to the fruit, work in small quantities so that the fruit is not crushed under its own weight. Sprinkle the sugar over the fruit in a shallow dish and leave to stand for a few minutes until the juices begin to flow from the fruit. The fruit and sugar may be gently stirred together, or the dish lightly shaken. The fruit liberally coated by its own syrup may then be packed into containers allowing $\frac{1}{2}$ inch headspace.

Amount of sugar required. Allow approximately 1 lb. sugar to 4 lb. of

131

fruit. This quantity may be varied slightly according to the sweetness of the fruit.

Freezing fruit by this method is very quick and simple, and provides fruit suitable for desserts, or for use in pies.

Packing in Sugar Syrup

A syrup may be referred to in general terms as heavy, medium, or light, or more directly according to the percentage of sugar present in the solution; thus a 50 per cent, 40 per cent, or 30 per cent syrup may be specified. The choice of an appropriate syrup is made according to the sweetness or otherwise of the fruit to be frozen, and according to the personal preference of the consumers. Light syrup, however, is recommended for those fruits with a particularly delicate or subtle flavour.

Because the water must be heated to dissolve the sugar, it is good practice to prepare the syrup a day ahead and refrigerate it until required. The syrup is thoroughly cold when required and this prevents delays when preparing the fruit.

The proportions for preparing the syrup are given below in both English weights and measures and in American cup measures:

Approximate percentage of sugar in syrup	Weight of sugar (lb.)	Volume of water (pts.)	Approximate yield of syrup (pts.)
20	$\frac{1}{2}$	2	$2\frac{1}{4}$
30	1	2	$2\frac{1}{2}$
40	$1\frac{1}{2}$	2	$2\frac{3}{4}$
50	$2\frac{1}{4}$	2	$3\frac{1}{8}$

Approximate percentage of sugar syrup	Amount of sugar (cups)	Amount of water (cups)	Approximate yield of syrup (cups)
20	1	4	$4\frac{1}{2}$
30	2	4	5
40	3	4	$5\frac{1}{4}$
50	$4\frac{3}{4}$	4	$6\frac{1}{2}$

Amount of syrup required. When making up the syrup allow $\frac{1}{2}$ pint syrup (or 1–$1\frac{1}{2}$ cups) per lb. of fruit. Obviously the amount actually required depends on the type of fruit and the shape and size of the container used.

Keeping the fruit submerged. Because fruit floats in syrup, the upper layer may not remain completely covered by the syrup, and consequently may discolour. A piece of crumpled cellophane or greaseproof paper can be used to keep the fruit submerged.

Corn syrup and honey. The colour of some fruits preserved in syrup in which up to 25 per cent of the sugar has been replaced by corn syrup is thought to be retained better than when sugar alone is used. It is also possible to substitute honey for part of the sugar content, but the resultant honey flavour may not be suitable for all fruits.

PACKING WITHOUT SUGAR OR SYRUP

This method is suitable for those fruits with a tough skin such as cranberries, currants, and gooseberries. After washing and drying, the fruit is packed into containers leaving about $\frac{1}{2}$ inch headspace.

This is a very useful method of dealing with a quantity of fruit intended for jam making. When the fruit is ripening on the bushes or available in the shops there is not always sufficient time to spare for jam making. Preparing the fruit for freezing by this method takes very little time, and the actual jam making can be completed when it is more convenient.

The syrup and sugar normally added to fruit when it is frozen function as a preservative and help the fruit to retain colour and flavour. Fruit frozen with no additions may not therefore be kept as long in frozen storage, but this is no handicap with fruit frozen for jam making when only a short frozen storage life is intended.

PACKING IN WATER

For persons who may not have added sugar in their diet, fruit can be satisfactorily frozen in water, although for those fruits which have a tendency to discolour an antioxidant should be added.

The prevention of discoloration is discussed in a later section.

PURÉES AND JUICES

The preparation of a purée for freezing is a useful way of utilizing slightly damaged or well-ripened fruit. After washing and drying, the fruit is passed through a sieve. Cooking is not required except for apples and tomatoes. Sugar may be added according to taste, about 2–4 oz. per lb. It is important to avoid incorporating air into the purée during the preparation, and when adding it to the container; also sufficient headspace should be allowed.

Purées pack compactly, and make economical use of the freezer space; they are excellent for sauces, for use with desserts—particularly ice cream—and for baby foods.

The juices from citrus fruits freeze well. Squeeze the juice from the half fruits, taking care not to press overhard, otherwise the pithy flavour and the natural oils from the skin may be incorporated and result in the development of an off flavour and a shortened storage life. Peel the citrus fruits if a 'Simplex' press or electrical juice extractor is used. Sugar may be added according to taste, and 300 milligrams of ascorbic acid added to each pint of juice.

Fruit juices are excellent as breakfast drinks and make a useful addition to a wine cup.

PREVENTION OF BROWNING IN FROZEN FRUIT

Certain enzymes present in some fruits, particularly peaches, apricots, apples, and pears, cause them to darken when in contact with oxygen.

This darkening or browning can be prevented to a certain extent by speedy methods of preparation, and the immediate transference of the fruit to a container of cold syrup. Some discoloration may still take place during the freezing and thawing periods as a result of oxygen trapped in the container, and ultimately on exposure to the air.

The addition of an antioxidant, such as ascorbic acid, is usually a more effective method of preserving the colour of frozen fruits.

Ascorbic acid

Ascorbic acid is usually available in crystalline form. Difficulty may be experienced in purchasing it in this form, but it is readily available in tablet form at all chemists under its more familiar name of Vitamin C. When ascorbic acid is required in a recipe, the quantity called for is usually given in teaspoons. It is calculated that 1 teaspoon of ascorbic acid is equivalent to 3,000 milligrams. Proprietary bottles of ascorbic acid tablets disclose the number of milligrams present in each tablet (usually it is 50 or 100), thus it is possible to calculate the number of tablets required to make up the appropriate quantity of ascorbic acid.

One difficulty of using tablets is that they are more difficult to dissolve than crystalline ascorbic acid, although crushing them first partly overcomes this difficulty. The filler present in the tablets usually causes the solution to be cloudy.

Equivalent quantities of ascorbic acid in crystalline and tablet form

	Crystalline	Tablets	
375 milligrams	⅛ teaspoon	4	Assuming ascorbic acid content of each tablet to be 100 milligrams
750 milligrams	¼ teaspoon	7½	
1,500 milligrams	½ teaspoon	15	
2,250 milligrams	¾ teaspoon	22½	
3,000 milligrams	1 teaspoon	30	

Whether in crystalline or tablet form, ascorbic acid should be dissolved first in a little cold water.

The ascorbic acid solution may be used either for dipping, or sprinkling directly over fruits to be packed in sugar, or fruits packed with no additional sugar. The solution should be stirred into the syrup used for syrup packs, and added directly into fruit purées. Ascorbic acid crystals may be added directly to fruit juices, but if the tablet form is used, the tablets should be finely crushed and dissolved in a little water before being added to the juice.

Lemon Juice

Lemon juice consists of a mixture of ascorbic and citric acids; it is effective in preventing darkening. However, to be effective it must be added in such large quantities that its flavour dominates that of the fruit to which it is added. Even so, a satisfactory method of retaining the colour of some fruits is to dip them for a short period— say up to 15 seconds—into a solution of lemon juice. To prepare the solution dilute the juice of one lemon with 2 pints of cold water.

Blanching in Steam

This method is particularly suitable for fruits to be packed with no additional sweetening. After preparation the fruit should be steamed for 2–3 minutes. In order to achieve thorough penetration, the portions must be adequately spread out, and not densely piled in the top of the steamer. After cooling and chilling, the fruit may be packed by any of the suggested methods.

Blanching in Syrup

The prepared fruit should be immersed in gently boiling 30–40 per cent syrup for 1–3 minutes, then removed, cooled, and chilled. Fruit prepared in this way is suitable for sugar or syrup packs.

Recommended maximum storage life at 0°F (−18°C)

Most fruits packed in syrup or sugar	9–12 months
(Possible exception pineapple 3–4 months)	
Most fruits packed alone or in water	6–8 months
Fruit purées	6–8 months
Fruit juices	4–6 months

UNFREEZABLES

The following fruits do not respond well to freezing:

Bananas—avocados—pears—melons.

It will be noticed that these fruits have a bland flavour and texture which is not completely restored on thawing. However, suggestions are included in the section dealing with individual fruits for the best methods of preparing them.

THAWING FRUIT

Fruit to be served for dessert should be thawed to that stage when a few ice crystals still remain. Some fruits when fully thawed have a tendency to collapse, and the natural juices leak out; also the texture is flabby and there is a loss of colour. Thus when fruit is to be served as a last course in a meal the moment when it is removed from the freezer must be very carefully judged. To overcome this problem, thawing is best carried out slowly in a domestic refrigerator. In the event of fruit being needed quickly, the container (providing, of course, it is waterproof) may be placed in a bowl of cold water.

Approximate thawing times for 1 lb. fruit packed in syrup or sugar

In a refrigerator	6–8 hours
At room temperature	2–4 hours
In cold water	$\frac{1}{2}$–1 hour

For dense packs such as fruit purées the thawing times should be increased about 50 per cent.

FRUIT

All fruits must be thawed in their wrappings and left in them until required. This is particularly important for those fruits with a tendency to discolour in contact with oxygen. For the same reason no more than is required for immediate use should be thawed at one time.

Fruit which is to be used in fruit pies need only be thawed sufficiently to separate it for spreading onto the pastry or in the base of the dish. Fruit to be stewed or used for jam can be put directly into the pan, providing a gentle heat is applied until there is sufficient liquid to prevent the fruit sticking to the base of the pan.

APPLES

Some apples give better results than others; therefore before freezing in quantity try a few experimental packs.

Select apples which are free of blemishes; peel, core and slice them into twelfths. To prevent discoloration during preparation of large quantities, slice the apples into a salt solution, 1 tablespoon salt to 2 quarts water. Do not leave them in this solution longer than 10 minutes, and rinse well afterwards. It is advisable to treat the apples to prevent browning.

Syrup. Use 30–40 per cent syrup.

Add ascorbic acid solution, 750 milligrams in $\frac{1}{8}$ pint cold water to each pint of syrup.

Apples packed in syrup are suitable for dessert.

Sugar. $\frac{1}{4}$ lb. sugar to 1 lb. fruit.

> *Either:* Steam blanch the slices $1\frac{1}{2}$–2 minutes, cool and pack,
> *or:* Pour ascorbic acid solution over the prepared slices. For each lb. of fruit use 500 milligrams of ascorbic acid dissolved in $\frac{1}{4}$ pint of water.

Apples packed in sugar are suitable for pies and flans.

Unsweetened. Steam blanch, or pour ascorbic acid over the slices, as per sugar pack, and pack dry.

Unsweetened packs of apples are suitable for pies and preserves.

Purée. Peel and cook apples in sufficient water to prevent sticking. Cool, mash or sieve, and pack into containers. Sugar may be added to taste. No antioxidant is necessary.

137

APRICOTS

Select firm uniformly coloured fruits. Wash, but do not peel. Apricots packed in syrup give the best results; the addition of an antioxidant is recommended to prevent browning. Less satisfactory results are obtained with apricots packed in sugar even when treated with an antioxidant.

Syrup. Use 30–40 per cent solution.

Add ascorbic acid solution 1,000 milligrams in $\frac{1}{8}$ pint water, to each pint of syrup.

Apricots packed in syrup are suitable for dessert.

Sugar. $\frac{1}{4}$ lb. sugar to 1 lb. fruit.

>   *Either:* Scald in gently boiling syrup for 2 minutes, cool—the skins may easily be removed at this stage if preferred—then pack.
>
>   *or:* Pour ascorbic acid solution over the prepared fruits. For each lb. of fruit use 500 milligrams of ascorbic acid dissolved in $\frac{1}{4}$ pint of water.

Apricots packed in sugar are suitable for pies.

AVOCADOS

The whole fruit does not freeze well, but the pulp may be removed and made into a purée.

Cut into halves, remove and mash the pulp. Add seasoning or sweetenings to taste.

For *Savoury* dips or spreads, mix with lemon juice—1 tablespoon lemon juice to each fruit. Add salt and pepper to taste.

For use as a *Sweet*, add lemon juice as above and mix with sugar to taste.

Pack in cartons; to prevent discoloration at the surface, cover with a piece of cellophane before sealing.

BANANAS

Freezing of bananas is not recommended, as they are available all the year round and they do not respond well to frozen storage. However, in the event of having a surplus just before a holiday, unpeeled fruits may be wrapped individually in moisture-vapour-proof film,

FRUIT

when, although the skin may darken, the fruit remains a good colour. Alternatively, the peeled fruit may be sliced or mashed, and packed in containers with lemon juice and sugar. In each case the texture and flavour is reminiscent of the overripe fruit and is only suitable for mashing and mixing with other ingredients to form a spread or dessert.

BILBERRIES

Remove the stalks, wash thoroughly, and drain.

Syrup. Use a 40 per cent syrup. Suitable for dessert.

Unsweetened. Pack dry. Suitable for use in pies, sauces and jellies.

BLACKBERRIES

Choose fully ripe berries. Remove the stalks, wash, and drain.

Syrup. Use 40–50 per cent syrup. Syrup packs are suitable for dessert fruits.

Sugar. ¼ lb. sugar to 1 lb. fruit. Suitable for pies and tarts.

Unsweetened. Pack dry. Suitable for jams and most cooking purposes.

CHERRIES

Fully ripened red and black varieties freeze well. Remove the stalks and wash well. Pit the cherries if desired. If the stones are allowed to remain, the fruit develops an almond-like flavour, and the storage life may be shortened.

Syrup. Use 30–40 per cent syrup. Suitable for dessert fruits.

Unsweetened. Pack dry. Suitable for jams and cooking.

CRANBERRIES

Select firm, well-ripened berries. Remove stalks, wash and drain. Cranberries are very useful frozen as a sauce or purée.

Syrup. Use a 50 per cent syrup for dessert fruits.

139

Unsweetened. Pack dry. In this form they are suitable for all cooking purposes.

Purée. Prepare a purée in the usual way, and sweeten to taste.

CURRANTS—RED AND BLACK

Select currants which are fully ripe, but not soft. Remove from the stalks, wash and drain.

Sugar. ¼ lb. sugar to 1 lb. fruit. Suitable for use in pies and tarts.

Unsweetened. Pack dry for use in jam and jelly making, and all cooking purposes.

DAMSONS

See note for Plums.

GOOSEBERRIES

Select firm, ripe berries, whether of the large dessert varieties or the smaller cooking ones. Top and tail, then wash and drain the fruits.

Syrup. Use a 50 per cent syrup for the large dessert berries.

Unsweetened. Pack the smaller cooking varieties dry. Unsweetened packs are suitable for jam and jelly making and most cooking purposes.

GRAPES

Although grapes are usually available throughout the year, the price fluctuates considerably. The preparation of fresh grapes is both time consuming and tedious, and therefore packed in small quantities they are a useful asset in the freezer. Chose grapes which are firm but ripe. Wash and remove the stalks, skins and pips.

Syrup. Use a 30 per cent syrup. Pack in small quantities for use in fruit salads and special dishes.

GRAPEFRUIT

Grapefruit are always available, but subject to wide price fluctuations. They are time consuming to prepare and therefore justify freezing.

Choose fruits which are fully ripe, and if possible with few pips. Peel thickly, removing the outer membrane with the peel. Remove the segments whole, separating them from the membranes on either side with a sharp knife.

Syrup. Use a 40–50 per cent syrup, according to taste, for fruit to be used at breakfast and in fruit salads.

GREENGAGES

See note for Plums.

MELONS

The texture and flavour may not be so fine after freezing, but in small portions melon is worth freezing for use in fruit salads. Choose melons which are fully ripe for eating. The remainder of a melon which proves too large for immediate requirements may conveniently be frozen. Scoop out melon balls with a Parisienne spoon or cut into cubes.

Syrup. Use a 40 per cent syrup for melon to be used as a dessert.

NECTARINES

See note for Peaches.

ORANGES

See note for Grapefruit.

PEACHES

Freezing of peaches is most rewarding, as the fruit retains the flavour and texture lost in the canned equivalent. Peaches are liable to darken unless specially treated.

If the skin can be removed without first treating the fruit in boiling water, the flesh will remain firm. If this cannot be done, submerge the peaches in boiling water for 30 seconds, and immediately transfer them to cold water in order to remove the skins. Remove the stone, and either pack the peaches in halves or slices, but whether preserving the fruit in syrup or sugar, it is essential to work quickly and cover the fruit as soon as possible.

Syrup. Use a 40 per cent syrup.

Add ascorbic acid solution, 1,000 milligrams dissolved in ⅛ pint cold water to each pint of syrup.

Fruit packed in syrup is suitable for dessert.

Sugar. 3–4 oz. sugar to 1 lb. fruit.

Pour ascorbic acid solution over the prepared slices before adding the sugar. For each lb. of fruit use 500 milligrams of ascorbic acid dissolved in ¼ pint of water.

Peaches packed in sugar are suitable for pies, flans and dessert.

Unsweetened. For special diets, peaches may be packed in water to which ascorbic acid solution, 1,500 milligrams to 1 pint of water, is added.

PEARS

Pears do not freeze particularly well; the texture after thawing is sometimes disappointing. Some varieties give better results than others. Freeze as a purée those varieties which do not give good results when frozen whole—prepare as for apple purée, page 137.

Select firm but ripe pears. Peel, core, and cut into quarters or slices, depending on the size of the pears. Simmer gently in 40 per cent syrup for 1–2 minutes. Drain, cool and pack.

Syrup. Use a 40 per cent syrup. It is important that cold syrup is used for packing, not the syrup in which the fruit has been blanched.

Add ascorbic acid solution, 1,000 milligrams in ⅛ pint water to each pint of syrup as an additional measure to improve quality.

Pears packed in syrup may be served for dessert.

PINEAPPLE

There is considerable fluctuation in the price of fresh pineapples and they are not always available. They give good results, particularly when packed in a light syrup. However, they develop 'off flavours' after prolonged storage, and therefore the storage life should be limited to three months.

Select fully ripened fruits, peel and remove all the eyes. Take out the core, and either slice or cut into cubes.

142

Syrup. Use a 30 per cent syrup.

Frozen pineapple provdes a satisfactory substitute for fresh pineapple.

PLUMS

It may be felt that freezer space can be put to better use than for the storage of the more commonplace fruit such as plums and greengages. However, these fruits may be held temporarily in the freezer until they can be used for jam making. If the plums are to be stored for any length of time, it is better to remove the stones first, or the fruit will develop an almond-like flavour.

Select plums which are fully ripe but firm. Wash, and preferably remove the stones.

Syrup. Use a 50 per cent syrup for fruit to be used for dessert.

Ascorbic acid solution, 750 milligrams in ⅛ pint water, may be added to each pint of syrup for improved quality.

Unsweetened. Pack dry, if to be used for jam making.

RASPBERRIES

Raspberries are probably the most rewarding of the frozen fruit, and when packed in sugar one of the simplest to prepare.

Select well-ripened berries, and discard any which are imperfect. If the raspberries are home grown, and no insecticide sprays have been used on the canes, the fruit may be frozen without washing.

Syrup. Use a 40 per cent syrup.

This method is suitable for dessert fruit, but provides no advantage over packing in sugar.

Sugar. Use ¼ lb. sugar to 1 lb. raspberries. Mix gently without crushing the fruit until the juice begins to flow.

Suitable for dessert.

Unsweetened. Pack dry for use in jam and jelly making.

RHUBARB

Only young, tender stems are worth freezing.

Wash, trim the stems, and cut into suitable lengths of 1–2 inches. Blanching 1 minute in gently boiling water or syrup, and then cooling, helps to retain a good colour.

Syrup. Use a 50 per cent syrup for fruit to be used for dessert.

Unsweetened. Pack dry after washing and draining. Suitable for use in pies and preserves.

Purée. Prepare the purée by cooking the fruit in a minimum of water until tender. Cool, and press through a sieve, add sugar to taste.

STRAWBERRIES

Choose firm, small to medium size strawberries of a good colour. Large strawberries are more liable to collapse when thawed, but they can be successfully frozen provided they are cut into ¼ inch slices first.
  Remove hulls, wash and drain the fruit.

Syrup. Use a 40 per cent syrup for fruit to be served for dessert.

Sugar. ¼ lb. sugar to 1 lb. fruit. This is the simplest method of freezing strawberries for dessert.

Unsweetened. Pack dry for fruit to be made into jam.

Purée. Slightly damaged or soft fruit can be made into purée. Push the fruit through a sieve and sweeten to taste.

TOMATOES

Although strictly classified as a fruit, it seems more convenient to include tomatoes in the vegetable section.

10

Vegetables

THE preparation of vegetables for freezing is no more complicated than for normal meal preparation, and the same amount of clearing up is necessary whether a large or small quantity of produce is prepared.

Each vegetable is prepared according to its type. The leafy varieties must be thoroughly washed in cold running water; the root vegetables thinly peeled; peas and broad beans need not be washed provided that their pods are clean and in good condition. It is essential, however, that all preparation be carried out as quickly as possible.

Where there is a variation in the size of the vegetables, as occurs with carrots or asparagus, they should be sorted into like batches. This makes packing easier, and is essential when blanching, since the time required for blanching is related to the size of the vegetable.

BLANCHING

Nearly all vegetables must be blanched if they are to retain their quality for any length of time in the freezer. This is because frozen storage alone is not sufficient to inactivate the enzymes contained in the vegetables. Blanching has the additional advantage of reducing the natural turgidity of the vegetables particularly of the leafy varieties, thus making them easier to pack into the containers.

Blanching in boiling water

Equipment required. Very little is required in the way of special equipment apart from a blanching kettle of at least 1 gallon capacity, although if this is not available use either a deep-fat bath or a large lidded saucepan in which a mesh basket can be fitted. A colander and vegetable spoon are also useful. It is also important to have a clock with a second hand available in order to make an accurate time check.

Method. Wash and prepare the vegetables according to their specific needs. Sort into batches of uniform portions.

Have the water in the blanching kettle boiling rapidly. 6–8 pints of water are required to blanch 1 lb. vegetables, although the same water can be used for successive batches. In the case of those vegetables which impart a characteristic and strong flavour to the blanching water—such as cauliflower and spinach—it may be necessary to change the water more frequently. In all cases a kettle of boiling water should always be kept to hand for topping-up purposes.

Add the vegetables contained in the basket to the blanching kettle. Cover with a lid; it may also be necessary to use a mesh lid or enamel plate to keep the vegetables submerged.

The blanching time is measured from the moment that the water in the kettle returns to a full boil. This must not take more than 1 minute—check this carefully when adding the first batch of vegetables and if the recovery time is slower than 1 minute, reduce the weight of subsequent batches accordingly.

Shake the basket and contents frequently during blanching. When blanching is finished, drain the basket, cool and chill the vegetables as quickly as possible, either under running cold water or in a large bowl of cold water to which ice has been added. Prompt and thorough chilling is essential if a high-quality product is to be achieved. A quantity of ice may be prepared in advance for this purpose and kept in the freezer until required. When the vegetables feel cold to the touch—this should not take longer than 5 minutes— drain and pack into containers. Place in the freezer immediately.

If for some reason the vegetables cannot be put directly into the freezer they should be held temporarily in the domestic refrigerator.

Steam blanching

It has not been shown that there is a significant improvement in the nutritive value of steam-blanched vegetables over those blanched in boiling water. Thus the method of blanching selected is largely one of personal convenience. Steaming may be preferred for those vegetables which need to be handled carefully, such as cauliflower and asparagus. But steam blanching is not recommended for the leafy vegetables such as spinach, as the manner in which the leaves pack down prevents them from being thoroughly penetrated by the steam.

146

Method. Prepare the vegetables as for water blanching.

See that the water in the bottom of the steamer is boiling rapidly, to ensure that the top section is kept well filled with steam. Carefully add the vegetables to the top section of the steamer. It is important not to overload the steamer if a good circulation of steam is to be maintained and all the vegetables thoroughly blanched.

Steam blanching takes about half as long again as water blanching. Begin timing as soon as the lid has been replaced.

When blanching is finished, cool and pack the vegetables according to the directions given for water blanching.

METHOD OF PACKING

Most vegetables may be packed satisfactorily both dry and in brine. Vegetables packed in brine are usually more tender. However, the improvement in quality is not sufficiently marked to justify the more time-consuming method of preparation. In addition, the liquid contents of the pack require rigid containers which are more costly.

In later sections dealing with the preparation and freezing of individual vegetables, the method recommended is for dry packs. The method of packing in brine is as follows:

Packing in brine

Prepare a 2 per cent solution of brine—2 tablespoon of salt in 2 pints of water, and chill thoroughly before use.

Prepare, blanch and cool vegetables in the usual way.

Pack vegetables into the containers and cover with brine. Use crumpled paper to keep the top vegetables submerged—allow sufficient headspace, seal and freeze.

VEGETABLE PURÉES

A purée is a useful form in which to store many vegetables. The vegetable should be prepared and cooked in the usual way and either pressed through a sieve or blended in a liquidizer. Care must be taken when preparing the purée not to incorporate air into the mixture. Seasoning may be added to taste. Vegetable purée may be used as a basis for soups and sauces, or frozen in small quantities it is useful for

VEGETABLES

baby foods. Freezing the purée in an ice-cube tray with divisions is one way of preparing sufficiently small portions for a baby's appetite.

Recommended maximum storage life at 0°F (−18°C)

 Most vegetables   10–12 months
 Vegetable Purées   6–8 months

UNFREEZABLES

Salad vegetables to be eaten raw, for example lettuce, watercress, celery, and chicory, cannot be frozen successfully; the leaves wilt and the characteristic crispness is not restored on thawing. However, some of these vegetables may be frozen in order to be served at a later date as a cooked vegetable. Thus frozen celery hearts may be braised, or boiled, and served with a white sauce.

COOKING FROZEN VEGETABLES

It must be remembered that the blanching process given to vegetables before freezing not only has the effect of inactivating the enzymes, but is a preliminary cooking period; account must be taken of this when calculating future cooking times. The cooking time of most frozen vegetables is about half that of the fresh equivalent.

Conventional methods of cookery may be used for frozen foods and in most instances they should be cooked directly from the frozen state.

Boiling

With the exception of corn-on-the-cob, vegetables to be boiled need no thawing before cooking. As with all fresh vegetables, only a minimum of water should be used so that the loss of soluble nutrients is reduced. For each pound of vegetables allow between ¼ and ½ pint of water. A certain amount of moisture results from the ice and frost of the frozen vegetables. Bring the water to a rapid boil before adding the vegetables, and where they are in the form of a solid block, choose a wide-based pan so that the block may lie flat on the bottom. After 2–3 minutes separate out the portions of vegetables with a fork to ensure each portion cooks evenly. Continue cooking over a gentle heat until the vegetables are just tender. Corn-on-the-cob

148

should be thawed before cooking, otherwise the core will feel cold and unpleasant even though the kernel is cooked.

A pressure cooker may be used to cook frozen vegetables, although the time must be closely controlled to prevent overcooking. One minute or even half a minute over time at 15 lb. pressure represents serious overcooking. Separate the frozen block into smaller portions if possible before placing in the pressure cooker. About $\frac{1}{8}$ pint of boiling water is usually sufficient.

The following timetable gives approximate times for cooking the vegetables in a covered saucepan and in a domestic pressure cooker:

	Minutes measured after the water in the pan returns to the boil	Minutes in pressure cooker at 15 lb. pressure
Asparagus	5–8	$1–1\frac{1}{2}$
Beans Broad	8–10	$1–1\frac{1}{2}$
French	7–8	1
Runner	6–7	$\frac{1}{2}$
Broccoli	5–7	1
Brussels sprouts	5–9	$1–1\frac{1}{2}$
Cabbage	5–8	1
Carrots	5–10	$1\frac{1}{2}$
Cauliflower	5–8	1
Corn-on-the-cob	4–6	1–2
Peas	2–3	$\frac{1}{2}$
Potatoes	7–10	$1\frac{1}{2}–2$
Spinach	5–7	1

A fairly wide margin of time has been given for the saucepan method to allow for variations in the size and thickness of different packs.

Steaming

With the exception of the leafy varieties, steaming is a satisfactory method of cooking most frozen vegetables, and may be preferred for those vegetables where careful handling is necessary. It is advisable to thaw the vegetables sufficiently to separate the pieces before putting them into the steamer.

Oven cooking

Frozen vegetables can be cooked in a covered casserole with little or no added water. Thaw sufficiently to separate the pieces and put them into a well-greased casserole with a large knob of butter and seasoning. Cook at 350°F (175°C), Regulo 4, for 30–40 minutes, frequently shaking or stirring the contents of the casserole.

Excellent results are obtained with corn-on-the-cob cooked in the oven. Thaw the cob for 3–4 hours, then wrap individually in buttered foil and cook at 375°F (190°C), Regulo 5, for 30–40 minutes.

Gentle frying

Gentle frying of frozen vegetables in butter gives good results. Melt a little butter in a heavy pan with a lid and add the partially thawed vegetables, together with the necessary seasoning. Toss the contents over a gentle heat until tender. It may be necessary to remove the lid towards the end of cooking to evaporate off excess moisture.

ASPARAGUS

Select stalks which are young and tender, and not woody. Divide into batches of equal thickness and cut into 6-inch lengths. Wash thoroughly.

Blanch in boiling water:

Small stalks	2 minutes
Medium stalks	3 minutes
Thick stalks	4 minutes

Cool, drain, and pack into semi-rigid containers, or waxed boxes, to give some protection.

AUBERGINES

Choose firm aubergines with dark glossy skins but with tender seeds within.

Wash thoroughly, and with a stainless steel knife cut into slices, $\frac{1}{2}$–$\frac{3}{4}$ inch thick.

Blanch in boiling water for 4 minutes.

Cool, drain, and pack; separate the slices with a double thickness of cellophane.

BROAD BEANS

Pick the pods before they are fully mature, so that the beans are sweet and tender and not starchy. Shell the beans.

Blanch in boiling water for 3 minutes.

Cool, drain, and pack.

FRENCH AND RUNNER BEANS

Choose beans which snap cleanly and are not stringy. Wash and remove the ends.

Cut French beans obliquely into 2–3-inch pieces.

Slice runner beans longitudinally.

Blanch in boiling water:

French beans   2–3 minutes
Runner beans   1 minute
Cool, drain, and pack.

BEETROOT

Beetroot which has been frozen, although a little soft, is acceptable for use in salads. Choose young beetroot not more than 2–3 inches in diameter. Remove the leaves, but not so close that the beetroot bleeds during cooking. Thoroughly cook the beetroot for 10–13 minutes according to the size. Cool, rub off the skin. Cut into slices or cubes, although small beetroot may be left whole, and pack.

BROCCOLI

Choose tender young heads, discarding any which are near to blossoming. Wash thoroughly in salt water if necessary. Divide into sprigs 1–2 inches wide and about 3 inches long.

Blanch in boiling water for 3–4 minutes.

Cool, drain, and pack.

BRUSSELS SPROUTS

Choose firm, tight sprouts not more than 1 inch in diameter. Trim off any diseased or coarse outer leaves. Make a cut in the stems and wash thoroughly.

Blanch in boiling water for 3–4 minutes.

Cool, drain, and pack.

CABBAGE

Since cabbage is never out of season it is unlikely to be frozen in quantity. But where a cabbage proves too large for immediate consumption it is useful to be able to freeze the surplus.

151

Choose only the inner tender leaves. Wash thoroughly and cut into coarse shreds or small wedges.

Blanch in boiling water:

| Shredded cabbage | 1½ minutes |
| Cabbage cut into wedges | 2 minutes |

Cool, drain, and pack.

CARROTS

Although carrots are available all through the year, the quality is variable. It is most rewarding to be able to serve tender young carrots when the only ones available in the greengrocers' shops are coarse and woody.

Choose young carrots of finger thickness. Cut off the tops and wash.

Blanch in boiling water for 3–4 minutes.

Cool in running water when the skins may be rubbed off at the same time, drain and pack.

CAULIFLOWER

Select cauliflowers with firm white heads and surrounded by fresh green leaves. Separate into sprigs of no more than 1½ inches across. Wash thoroughly in salted water.

Blanch in boiling water for 3 minutes.

Cool, drain, and pack.

CELERY

Celery preserved by freezing is not suitable for use as a salad vegetable, but celery hearts may be frozen for service as a cooked vegetable. The addition of a stick of celery adds flavour to certain savoury dishes, and frozen celery is quite suitable for this purpose; blanching first is not essential for celery sticks used for flavouring.

Choose roots of celery which are crisp and tender, and not stringy. Remove the coarse outside stalks, and trim the heart to 3–4 inches, or cut the tender stalks into 1-inch pieces. Wash thoroughly.

Blanch in boiling water:

| Celery hearts | 6–8 minutes |
| Cut stalks | 3 minutes |

Cool, drain, and pack. Celery sticks to be used for flavouring may be frozen loose and subsequently packed into a plastic bag, from which they may be removed in suitable quantities when required.

CORN

Whether the corn is frozen on the cob or as whole kernel, it must be tender and not overripe, otherwise the kernel will be starchy.

On the cob. Remove the husk and silk and cut off the immature section at the top. Wash.

Blanch in boiling water for 4–8 minutes, depending upon the size, but not more than four cobs should be placed in the blanching kettle at one time.

Cool thoroughly making certain that the centre is completely cold before packing.

Whole kernel. Cut the kernels from the cob after blanching and cooling, and then pack.

COURGETTES

Choose tender young courgettes. Wash thoroughly and cut into 1-inch slices.

Blanch in boiling water for 3 minutes.

Cool, drain, and pack.

MUSHROOMS—FIELD AND CULTIVATED

Although cultivated mushrooms are available throughout the year, because they may be incorporated in so many savoury dishes and used for garnishing purposes, it is always worthwhile to keep two or three 4 or 8 oz. packs in the freezer. Freshly picked field mushrooms may also be frozen if available in any quantity.

Both field and cultivated mushrooms should be washed in cold water and drained. Only the stems of cultivated mushrooms need be trimmed, although field mushrooms must also be peeled. Where they are larger than 1 inch in diameter the mushrooms should be sliced or quartered. Blanching is not recommended for mushrooms, as little improvement in colour or texture is achieved.

Cooking the mushrooms first in a little butter before freezing gives good results; these mushrooms are then readily available for use in omelettes or for garnishing. It is important to pour off any excess fat and thoroughly cool the mushrooms before packing.

PARSNIPS

Parsnips are not a very popular vegetable and therefore unlikely to be frozen in quantity; however, because they are not always available it is worth freezing a few packs so that they can be included on the menu if desired.

Choose parsnips which are young and tender. Peel and quarter, or cut into slices.

Blanch in boiling water for 2 minutes.

Cool, drain, and pack.

PEAS

Peas are probably the most popular frozen vegetable. Only young peas are suitable for freezing, because as they approach maturity they become starchy and lose the initial sweetness and succulence associated with young peas.

Shell and check over the peas for any which are tough or discoloured.

Blanch in boiling water for 1–2 minutes.

Cool, drain, and pack.

PEPPERS—GREEN OR RED

Peppers may be packed singly, or wrapped individually and several packed together in larger containers for use in a variety of dishes.

Choose peppers which are firm, of good colour, and not rubbery. Halve and remove the seeds. Cut into slices if preferred. Peppers can be frozen whole if required for stuffing at a later date; cut only a lid from the top of each one in order to remove the seeds and pithy section. Peppers retain their attractive colour without the necessity of blanching and may be used directly in uncooked dishes.

New Potatoes

It cannot be claimed that after frozen storage new potatoes are indistinguishable from their fresh equivalent, but it may still be considered worth while to freeze a few packs for special occasions. Although the appearance of frozen potatoes after cooking may be satisfactory, the texture is often described as mealy and lacks the distinctive flavour characteristic of fresh new potatoes.

Choose potatoes which are about walnut size and free of blemishes. Wash thoroughly and remove the skins with a brush or by lightly scraping.

Blanch in boiling water for 3–5 minutes.

Cool, drain, and pack.

Spinach

Select only tender young leaves, and wash thoroughly in running water. Blanch in small quantities allowing about 8 pints of water for each ½ lb. of raw spinach. Shake the blanching basket frequently to ensure a thorough penetration.

Blanch in boiling water for 2 minutes.

Steam blanching is not recommended for spinach.

Cool, drain, and pack.

Tomatoes

Tomatoes frozen whole are not suitable for serving raw in salads, but may be used as a cooked vegetable. They should be cut in half on removal from the freezer and put directly into hot fat in a frying pan, or brushed with oil and grilled, when they are indistinguishable from the cooked fresh equivalent.

Choose small tomatoes, wipe, do not blanch them, and pack individually in moisture-vapour-proof film in a semi-rigid container.

Purée. Tomatoes are most satisfactorily frozen in the form of a purée.

Wash and cut into quarters, simmer in a minimum of water until tender. Cool and pass through a sieve. Pack in upright containers, allowing sufficient headspace.

Turnips

Although turnips are not usually frozen in any quantity, it is worth freezing them 'loose' or in small packs suitable for use in casserole dishes.

Choose turnips which are young and tender. Peel and cut into ½-inch cubes.

Blanch in boiling water for 2 minutes.

Cool, drain, and pack.

11

Meat

SELECTION

To buy a whole carcass or side of an animal for freezing may on first thoughts appear a sound and economical proposition. But probably only those living on farms and used to killing and butchering their own meat are sufficiently experienced to deal with meat in such quantities.

Probable yields of carcass Meat

From a side of beef	—	about 200 lb.
From a whole pig	—	about 150 lb.
From a whole lamb	—	about 40 lb.

It is obvious that a large freezer is necessary before meat can be bought in bulk, and even then space must be left for other commodities. There are two further considerations. Unless one is experienced in cutting meat, and possesses the right tools for the job, it is wise to call in the family butcher to deal with the carcass, which involves an additional expense. Also the complete carcass may include several cuts of meat not popular with the family and not purchased for them in the normal course of events. Therefore it cannot be considered economical to buy, much less to use freezer space, for meat not normally enjoyed.

However, retail prices vary considerably with supply and it is possible to make considerable savings by buying for the freezer when the prices are low. In many areas it is also possible to buy meat in catering packs, ready to use, at very economical prices.

Whether buying half a pig or a family-sized joint, only meat of the best quality should be taken for freezing. It is thought that freezing has the effect of tenderizing meat, but this does not mean that a piece of braising steak is suitable for grilling on removal from the freezer.

157

Preparation of the Carcass

Chilling and ageing

After slaughtering, the animal must be properly prepared. Pork and veal need only be chilled before freezing, but beef and lamb should be hung for several more days.

Chilling takes place at a temperature of little over freezing point and carcasses of veal and pork should reach a temperature of 40°F (5°C) in one to two days. Prompt and speedy chilling is important if spoilage from the activities of micro-organisms and enzymes is to be prevented. Veal and pork should immediately be jointed and frozen after chilling.

Beef and mutton are hung for additional periods of eight to ten and five to seven days respectively. During this period, usually referred to as 'ageing', the flavour of the meat is thought to improve; there is also considered to be a tenderizing effect.

The conditions in which meat is chilled or aged influence its keeping quality. These conditions include temperature, relative humidity, and accurate timing, and they are most satisfactorily supervised by the family butcher. If he is given prior notice of requirements for certain joints, he will no doubt be pleased to co-operate in supplying the meat at just the right condition for freezing. For example, pork frozen promptly after chilling is thought to have a longer storage life, and if consulted the butcher will be happy to supply this meat immediately chilling is completed.

Preparation of joints

Time spent in preparation not only simplifies packaging but also is time-saving when the meat is required. Diagrams of the cutting methods for carcasses of beef, lamb and pork are given in Figs. 12, 13 and 14 respectively, together with suggested cooking methods for the various joints.

Joints for roasting. Trim off any excess fat, and remove bone if possible, as this will save space in the freezer. If removing the bone is not desirable, it should be well protected with several thicknesses of paper to avoid puncturing the packaging material. String the joint to make it as compact a shape as possible.

Steak and chops for grilling and frying. Trim off any excess fat. Pack in quantities appropriate to the size of the family. Place a double thickness of cellophane between portions so that they may easily be separated when thawing.

158

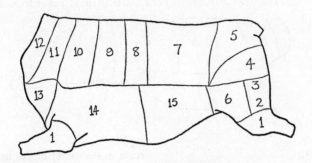

1. SHIN	stew—needs several hours.
2. SILVERSIDE	salt and boil.
3. TOPSIDE	roast or braise.
4. AITCHBONE	roast or braise.
5. RUMP	grill or fry.
6. TOP RUMP	roast or braise.
7. SIRLOIN	roast—most popular joint; or cut into steaks and grill or fry; 'fillet' usually separated and served alone grilled or fried.

8. BACK		
9. TOP	RIBS	bone, roll and roast; wing ribs (next
10. WING		to sirloin) are most tender.
11. CHUCK		braise.
12. STICKING PIECE		stew—needs several hours.
13. CLOD		
14. BRISKET		stew or cook slowly in casserole; or
15. FLANK		salt and boil.

FIGURE 12

CUTS OF BEEF

Meat for stews and casseroles. Prepare this meat as for cooking. Remove excess fat and skin and cut into cubes or bâtons. Weigh out and pack in quantities required for favourite dishes.

Offal—heart, liver, kidney, tongue and sweetbreads. Trim off any attached skin and blood vessels, and wash thoroughly in cold water. Hearts and liver may be frozen whole, or in slices separated by a double thickness of cellophane. Special care must be taken when packaging offal, as it is more prone to develop off flavours, and for the same reason a shortened storage life is advocated.

Mince and sausages. Prepare mince from lean meat, never from scraps.

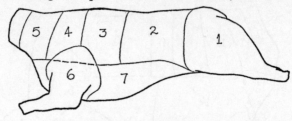

1. Leg	roast.
2. Loin	roast in the piece; or cut as chops and grill or fry (chump chops are cut from the part next to the leg).
3. Best End of Neck	cut as chops—grill, fry or braise.
4. Middle Neck }	
5. Scrag End }	stew or cook in casserole.
6. Shoulder	roast.
7. Breast	bone, stuff and roast.

FIGURE 12

CUTS OF LAMB

It may be frozen as it is, or formed into steaklets or meat balls. The addition of salt and other seasoning accelerates rancidity and reduces the storage life; they should be omitted if the mince is likely to remain in the freezer longer than a month. Seasoning should also be omitted from sausages if prepared at home.

Smoked or cured meats. Smoked or cured meats keep well in a cool larder or domestic refrigerator. Freezer storage is not recommended for them. However, a small quantity of bacon or ham may be put into the freezer before going away or as an emergency store.

Stock. Bones and trimmings resulting from the preparation of joints for freezing can be used to make stock which can also be stored in the freezer. The stock may be concentrated and frozen in ice-cube trays in order to provide a more convenient shape for packaging (refer to Preparation of Stocks and Soups, page 204).

PACKAGING

Select suitable packaging material

Most criticism made of the poor quality of frozen meat is a result of poor packaging. Where the material selected fails to provide an

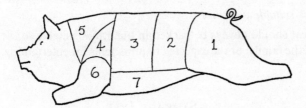

1. Leg   roast.
2. Loin with Kidney }   roast in the piece, or cut as chops
3. Fore-Loin   }   and grill or fry.
4. Blade Bone   bone, stuff and roast.
5. Spare Rib   roast or grill.
6. Hand   bone, roll and roast—the cheapest roasting joint.
7. Belly   salt and boil in the piece; or cut into rashers and grill or fry.

FIGURE 14

CUTS OF PORK

adequate barrier against oxygen, rancidity develops as a result of the fat content becoming oxidized. Also, where the material is not sufficiently moisture-vapour-proof, any fluctuation in the temperature of the air in the freezer may result in a movement of moisture away from the meat, and a dry product results, sometimes with evidence of freezer burn.

Form into a regular shape

Even when the film provides an adequate barrier against the movement of moisture vapour, if it does not fit snugly there may still be a loss of moisture from the meat. This moisture is deposited within the packaging in the form of cavity ice. This, however, can be largely prevented by forming the meat into a regular shape, wrapping closely in a suitable material and squeezing out as much air as possible before sealing it.

Provide adequate protection

When a joint still contains bones it is advisable to pack it with paper so that the bones do not puncture the packaging material. Also the packaging material may become punctured if subjected to rough handling within the freezer and it is advisable to overwrap with strong paper, or stockinette.

Pack in suitable quantities

Meat should always be packed in quantities suitable to the appetite of the family, or the expected requirements for entertaining.

STORAGE LIFE

As indicated in an earlier section, storage life is affected, among other things, by the condition of the meat at the time of freezing, and in the case of minced meat and sausages by whether or not seasonings are included.

The following approximate figures are given in order to assist the planning of freezer space, but it is worth repeating here that food put into a freezer is not expected to complete an endurance test. A frequent turnover of produce will provide food of a consistently high quality.

Recommended maximum storage life at 0°F (−18°C)

Beef Lamb	9–12 months
Veal Pork	4–6 months
Minced meat (unseasoned) Sausage (unseasoned)	3 months
Minced meat (seasoned) Sausage (seasoned)	1 month
Offal	2–3 months
Cured and smoked meats	1 month

COOKING FROZEN MEAT

When to thaw

All meats may be cooked directly from the frozen state, and in an emergency this will be the quickest way of getting a hot meal. However, for large joints, cooking from the frozen state requires more skill if the meat is not to be served cooked on the outside while still quite raw and cold in the centre. For this reason many people prefer to thaw out large joints completely prior to cooking, presumably being of the opinion that the results of their cooking are by this means more predictable.

Thawing before cooking is necessary when the meat requires further preparation prior to cooking, such as dusting with flour or coating with egg and breadcrumbs. Meat to be cooked in deep fat should be thawed first to ensure that the centre is thoroughly heat-penetrated by this short method of cookery.

How to thaw

Meat thawed slowly is considered to be juicier than meat thawed more rapidly. This is because there is a longer period for the reabsorption of the juices which were withdrawn from the tissue during freezing. Thawing in a refrigerator gives the most satisfactory result, but this method is slow. Thawing at room temperature is faster. When thawed meat is needed quickly it may be placed in front of a fan or in frequent changes of cold water. The latter two methods are not as satisfactory and should only be used in an emergency.

Whichever method is employed the wrappings must be kept on.

Once meat has been thawed, it must be prepared and cooked without further delay.

Approximate thawing times for joints of meat

In a refrigerator	—	6 hours per lb.
At room temperature	—	3 hours per lb.
Before a fan or in cold water	—	$\frac{1}{2}$–$\frac{3}{4}$ hour per lb.

Cooking straight from the freezer. The slow method of roasting gives the best results for the average-sized frozen joint. Set the oven at 300–350°F (150–175°C), Regulo 2–4, and increase the cooking time $1\frac{1}{2}$–2 times. If a meat thermometer is available, use this to tell when the centre is cooked, but to avoid damaging the thermometer, it should not be inserted until half-way through the estimated cooking period.

Chops, steaks, slices of liver, may be safely cooked from the frozen state whether by grilling or frying, but only a gentle heat should be applied until the meat is thawed through. When the meat is to be coated with egg and crumbs, or dusted with flour preparatory to cooking, partial thawing is necessary to enable the coating to adhere to the meat.

Minced meat and stewing meat cut up prior to freezing may be put directly into hot stock and the pieces separated over gentle heat.

12

Poultry and Game

FREEZING has greatly increased the popularity of all kinds of poultry, and varieties of game previously only available when 'in season', are now available all the year round. The gourmet may say that game which has been frozen is not equal to the fresh equivalent, but many people would have difficulty in discerning the difference.

Poultry is taken to include all the usual domestic or farmyard birds, as well as tame guinea-fowl, pigeons, and rabbits. Game is the term which embraces the wild birds; partridge, pheasant, grouse, wild pigeon, etc., as well as large game such as venison, and small game such as wild rabbit and hares.

To give guidance in the planning of freezer space, a Game Calendar is given in Fig. 15 which illustrates when suitable game is likely to be in season.

SELECTION OF DOMESTIC AND GAME BIRDS

It is obviously not possible to be selective of game birds from the sights of a gun, but when choosing domestic birds the following are suitable for freezing:

Poussins, 4–8 weeks old, weighing $1\frac{1}{2}$–2 lb.
Young chickens, weighing $2\frac{1}{2}$–$3\frac{1}{2}$ lb.
Capons, caponized young male birds, weighing 5–7 lb.
Ducklings, 10–12 weeks, weighing 2–3 lb.
Ducks, weighing 3–7 lb.
Geese, tender birds, 8–10 lb.
Turkeys, plump hen birds, weighing 8–15 lb.

When at the poulterers selecting domestic and game birds, remember that the best results are obtained with birds which are young and plump. Avoid game birds which have been badly shot.

164

FIGURE 15

GAME CALENDAR

For most game there is a close season, thus the period when it is likely to be available for freezing is very restricted.

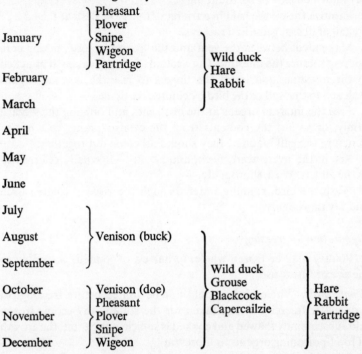

DOMESTIC BIRDS

Killing and Drawing

Domestic birds should be starved for twenty-four hours before killing, but given plenty of water to drink. They may be killed by dislocation of the neck or by 'sticking'. They should be hung by the feet with the head down.

Pluck the bird as soon as possible after killing. This enables the feathers to be removed easily. Continue to hang the bird for twenty-four hours. Remove any long hairs by holding the bird over a little lighted methylated spirits; this method is effective and does not produce any discoloration. Remove the feet, taking care not to cut through the sinews—these can be pulled out as the feet are removed. Trim the remaining pin feathers from the wings.

165

Cut off the head, leaving as much neck as possible on the bird. Cut off the neck close to the body and leave a big flap of skin. From the neck end loosen and remove the crop and windpipe and any fat which may be surrounding it.

Remove the oil sac held in a triangular piece of flesh at the base of the tail of ducks, geese and turkeys.

Make a cut between the vent and the 'parson's nose', taking care not to puncture the intestine. Cut around the vent, leaving it attached to the intestine, and insert the fingers to free the internal organs attached to the wall of the bird by connective tissue.

Free the internal organs at the neck end, and holding the gizzard firmly, draw out the contents from the cavity, (taking care not to puncture the gall bladder) they should all come out together.

Retain the liver, heart, neck, and gizzard—normally referred to as the giblets. Wash thoroughly.

Wash the bird, running water through the visceral cavity, drain and dry thoroughly.

Preparation for freezing

Poultry may be frozen whole, in halves, or jointed, according to the size or future use.

Whole birds. Whole birds should not be stuffed before freezing. In a domestic freezer the time taken for the stuffing to become frozen and subsequently thawed and cooked is sufficient to permit the growth of food-poisoning organisms if present.

Whole birds should be made as compact as possible for freezing. Tie the legs close together, covering any sharp ends of bone with several thicknesses of greaseproof paper. Lay the wings close to the body or twist them under the back and use them to secure the skin from the neck.

Half birds. Chicken and turkeys may simply be cut in half. Using poulterers' shears, cut along either side, and close to the backbone, for the length of the bird so that its neck and back strip can be removed. Then open out the bird and cut along through the breastbone.

Jointed birds. Most birds may be jointed into the pieces in which they are normally served. Fig. 16 shows the method of jointing a chicken. First cut through the skin between the leg and the carcass, then break the thigh joint by pulling the leg down and away from the body of the bird, the oyster piece should be removed at the same time.

166

FIGURE 16
JOINTING A CHICKEN

1. LEG
2. WING
3. WISHBONE
4. BREAST

Cut through the skin between
the leg and carcass

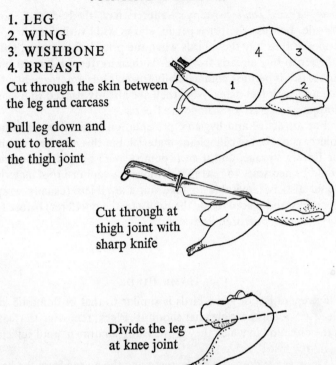

Pull leg down and
out to break
the thigh joint

Cut through at
thigh joint with
sharp knife

Divide the leg
at knee joint

Separate again at the knee joint if desired. Cut off the wings with a slice of breast attached. Cut off the wishbone and attached meat by cutting at right-angles to the end of the breastbone and directing the knife towards the neck end. Cut off the breast meat from each side close to the rib case.

It is possible to joint a bird in this manner without first drawing it if care is taken. After the joints have been removed, washed, dried, and packed, the viscera may be removed from the remaining skeleton, and the latter used for the preparation of the stock. To remove the viscera, lay the bird on its back and cut along each side at the ends of the ribs, just below the breastbone. Break open the carcass, but take care to leave the viscera intact. The giblets may be removed, washed, and frozen separately.

If several birds are prepared at the same time, the livers may be

167

collected, frozen together, and used at a later date for the preparation of liver pâté.

Ready-dressed poultry. Many poulterers have ready-dressed poultry for sale. Because the price per lb. shows wide variations it is often economical to buy these birds when the price is favourable. Often it is better to buy already frozen birds than to freeze a bought bird at home; this is because commercially frozen birds are frozen at a faster rate and with less delay. Also they are already packed in a suitable wrapping material for storage in a freezer.

For attractive and hygienic presentation, some dressed unfrozen poultry is packed in cellophane material, but this may not be suitable for freezer storage, and if such poultry were bought for freezing it would be necessary to rewrap it in moisture-vapour-proof material. It would also be advisable to take out the giblets (usually wrapped separately and placed inside the cavity by the poulterer) before freezing; they may be frozen separately.

GAME BIRDS

The preparation of game birds is similar to that of domestic kinds. Bleed as soon as possible after shooting. Pluck, removing the feathers in the direction in which they grow. Hang, undrawn, until sufficiently 'high' for taste.

Draw as for domestic birds, removing the oil sac from the base of the tail. Wash, drain, and dry.

Game birds may be frozen unhung, but the flavour of the flesh may be considered 'lacking' to some tastes. Game birds cannot be hung after freezing; they should be consumed immediately.

Water birds should be plucked, drawn, and frozen as soon as possible; otherwise the flesh may become contaminated by the fishy contents of the intestine.

LARGE AND SMALL GAME

Venison. Unless proficient in the art of butchering large animals, it would be advisable to enlist the services of the family butcher to deal with a carcass of venison. The preparation is similar to that for beef and it is usual to hang the carcass for up to six days after chilling, before jointing and freezing.

168

Rabbit and hare. Rabbit and hare should be beheaded and bled as soon as possible after killing, the skin and viscera removed and the carcass washed in cold water and dried. The carcass may be frozen whole, but because of the rather awkward shape it is more usual to joint it first. The carcass is divided into the two rear legs, back section, rib case, and two forelegs. Frequently only the meaty rear legs and back section are considered worth freezing. The less meaty upper portions may be used immediately for stews or for preparation of other dishes which may be cooked and frozen. If it is desired to hang hare, this should be done before freezing.

PACKAGING OF POULTRY AND GAME

The same points listed for the packaging of meat apply equally to the packaging of poultry and game. Young chickens and rabbits are not protected by fat as are most meats, and unless packed in packaging material which is highly impervious to moisture vapour, they are more prone to freezer burn.

Because of the risk of freezer burn, jointed poultry and rabbit should be wrapped individually before packaging, and the portions not merely separated by a double thickness of cellophane.

Because the storage life of the giblets from poultry and game birds is not more than three months, they should be packaged and stored apart from the carcass and not inside it.

Recommended maximum storage life at $0°F(-18°C)$

Chicken	12 months
Duck	
Goose	4–6 months
Turkey	
Giblets	3 months
Venison	12 months
Rabbit and hare	6 months
Offal	2–3 months

COOKING FROZEN POULTRY AND GAME

For venison follow the recommendations as for frozen meat given in the section entitled 'Cooking Frozen Meat' in the previous chapter.

169

For the best results with poultry, game birds, hare and rabbit, thawing in their original wrappings until pliable is recommended. Thawing is most satisfactorily carried out in a domestic refrigerator— a small bird will thaw out overnight. On the whole, poultry thaws at a slightly faster rate than that suggested for meat. The following times are given as a guide when thawing poultry in a domestic refrigerator:

Birds up to 4 lb.	up to 12 hours
Birds 4–12 lb.	up to 24 hours
Birds over 12 lb.	48–72 hours
(usually turkey)	
Joints of chicken,	
rabbit and hare	6 hours

At room temperature the time required for thawing is approximately half that in a domestic refrigerator, and for very large birds thawing in a cool room will prove the most practical solution. Birds and joints needed urgently may be thawed more quickly by placing them, still contained in their wrappings, in frequent changes of cold water.

After thawing the method of cookery is as for the fresh equivalent.

13

Fish

ONLY really fresh fish is worth freezing—ideally it should be frozen on the same day as it is caught. This limits the freezing of fish to those who live near enough to the sea to buy direct from the fishermen, and those fortunate enough to catch their own sea or freshwater fish. It is possible to return home from a summer holiday with fish for the freezer bought at the quayside that same morning, but it must be well packed in ice for the journey, and not piled unprotected into the hot boot of a car, if it is to retain anything of its freshness and quality on arrival home.

If really fresh fish is not available, it is better to buy packets of frozen fish produced by the frozen-food producers than attempt to freeze an inferior product. In fact, after making allowances for wastage and time spent in preparation, it may well prove a more economical product.

To give guidance in the planning of freezer space, a Fish Calendar is given in Fig. 17, which illustrates when suitable fish is likely to be in season.

PREPARATION OF WET FISH

Obviously the method of preparation will depend on the type of fish to be frozen and the form in which it is intended to be served. Fish may be frozen as caught, undrawn and simply washed before freezing, but this should only be done in an emergency, as there is a tendency for the flesh to become discoloured if the fish is not eviscerated first.

Unless presentation is important, as in the case of the larger fish such as salmon and salmon trout, and some of the smaller ones such as trout or whiting, it is more economical of freezer space and also time saving at a later date, to fillet or cut the fish into steaks before freezing.

171

FIGURE 17
FISH CALENDAR

This calendar shows when various fish are available and likely to be of a high quality.

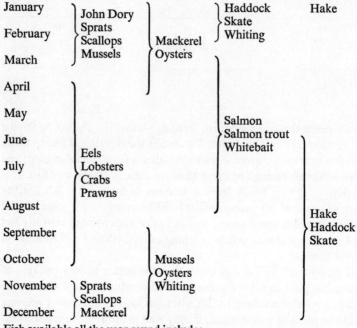

January			Haddock	Hake
February	John Dory, Sprats, Scallops, Mussels	Mackerel, Oysters	Skate, Whiting	
March				
April				
May				
June		Salmon, Salmon trout, Whitebait		
July	Eels, Lobsters, Crabs, Prawns			
August				Hake, Haddock, Skate
September				
October		Mussels, Oysters, Whiting		
November	Sprats, Scallops, Mackerel			
December				

Fish available all the year round include:

Cod
Halibut
Herring
Plaice
Shrimps
Sole
Trout
Turbot

The quality is likely to decline during spawning.

FISH FROZEN WHOLE

Round fish. Wash and remove scales by scraping the fish from tail to head with the blunt side of a knife; do this over newspaper.

Gut the fish by cutting the entire length of the belly and removing the viscera.

Remove the fins by cutting into the flesh around them; do not

172

remove with scissors or the base of the fin will remain behind—the dorsal fin and tail may sometimes be allowed to remain—they may be trimmed with scissors if their appearance is ragged. Remove the eyes. Wash thoroughly under running water, drain and dry on a clean cloth.

Flat fish. Small flat fish are sometimes frozen whole. Cut off the fins from around the edges with kitchen scissors.

To remove the viscera make a cut below the gills on the dark side and remove the intestine. Wash and dry as before.

Note on skinning whole fish

Round fish. Make a small cut through the skin only, behind the head and along the backbone. Gradually loosen the skin behind the gills, on one side; dip the fingers in a little salt so that the skin may be held more firmly, and taking care not to tear the flesh beneath, remove the skin from the head to the tail. Remove the skin from the other side of the fish in the same way.

Flat fish. Usually only the black skin is removed. Make a small cut in the skin just above the tail, and using the thumbs loosen the skin along each side close to the fins. Grip the skin firmly and tear it off from the tail to the head; press down with the other hand at the same time to prevent any of the flesh being removed with the skin.

To glaze large fish

Large fish are frequently glazed to give them protection in the freezer. Place the whole fish unwrapped into the freezer until it is solidly frozen. Take it from the freezer and dip it into a bowl of ice-cold water, when a thin film of ice will form over the entire fish. Immediately return the fish to the freezer. Repeat this process three or four times until the layer of ice or glaze formed is about $\frac{1}{8}$ inch thick. The fish must be handled carefully throughout this procedure to avoid cracking the glaze.

Finally, wrap the fish in suitable moisture-vapour-proof material, and if possible support large fish on a thin board of about the same size in the freezer.

FISH FROZEN AS FILLETS

Round fish. Cut down the back of the fish, (for example cod or haddock), from the head to tail. Then cut through to the backbone behind the gills on one side, and run the knife along in contact with the backbone for the length of the fish; lift off the fillet in one piece. Cut another fillet from the other side in the same way.

Flat fish. Lay the fish, for example plaice, on a board with the tail towards you. With a sharp knife cut along the line of the backbone; make a cut immediately behind the head and keeping the knife close to the backbone, and working from head to tail, remove the left-hand fillet. Turn the fish around so that the head is towards you and the remaining fillet on the left-hand side. Remove this fillet, working from tail to head. Turn the fish over and remove two more fillets from the other side of the fish, making a total of four in all.

Notes on skinning fish fillets (from round and flat fish)

Lay the fish on a board, skin side down, and tail towards you. With a sharp knife make a cut through the flesh, but not through the skin, about 1 inch from the end of the tail. Hold this tail section firmly between the finger and thumb and, keeping the knife almost flat, slide it between the skin and the flesh with a gentle side-to-side movement to separate the flesh from the skin.

FISH FROZEN AS STEAKS

Round fish. Cut across the backbone to give steaks 1½–2 inches thick.
Flat fish. Only large flat fish such as halibut and turbot are cut into steaks. Cut across the backbone at intervals of 1–1½ inches.

PACKAGING

Pack in suitable wrappings or containers in quantities appropriate to the family requirements. Separate fillets and steaks with a double thickness of cellophane.

SHELL FISH

Oysters and Scallops

Oysters and scallops should be frozen the same day they are taken from the water.

(a) Wash the outside of the shell thoroughly to remove traces of sand and dirt before opening.

FISH

(*b*) Open over a nylon strainer so that the liquor or 'juice' collects in a bowl beneath. Allow to drain for several minutes.

(*c*) Wash the oysters in a brine solution (1–2 tablespoons salt to 1 quart cold water).

(*d*) Drain and pack in semi-rigid plastic containers together with the retained liquor.

Crab and Lobster

Boil before freezing, and freeze only freshly killed and boiled crabs and lobsters.

Crab. To kill a crab drive a sharp knife through the underside of the head, or plunge the live crab into a pan of fast-boiling salted water or a court bouillon.

(*a*) Simmer in a court bouillon 10–15 minutes, drain and cool. Wipe on a cloth.

(*b*) Remove the big claws and put aside.

(*c*) Lay the crab on its back with the tail towards you. With the thumbs push up the tail until the body comes away from the shell.

(*d*) Turn the crab around until the head is facing, then remove and discard the small sac, the green matter, and the lungs (dead men's fingers) from the top of the shell.
(On request the fishmonger will always supply the crab with this part of the preparation completed.)

(*e*) Remove all the creamy brown meat from the shell, and pack in a container.

(*f*) Split open the body of the crab and the big claws and remove the white meat and pack in a separate container.

(*g*) Gently tap the inside edge of the shell, break it along the natural line to make a wider opening. The shell may be washed and oiled, and retained for serving the crab at a later date.

Lobster. To kill a lobster drive a sharp knife through the natural cross marked on the head, or plunge it into a pan of rapidly boiling water or a court bouillon.

(*a*) Simmer the lobster in a court bouillon 10–15 minutes per lb. Drain and cool.

(*b*) To split the lobster into two equal portions, insert the blade of a large knife into the cross on the head, bring the knife down through the tail, then, after turning the lobster, bring the knife down through the head.

(*c*) Remove the sac from the head, and the intestine which is a thin cord running along the tail.

(*d*) Crack open the claws and remove the meat, together with the meat from the tail, and pack into a suitable container. The meat may be used in dishes prepared from cooked lobster meat such as Lobster Newburg.

(*e*) Clean and retain the shell if it is intended to serve lobster mayonnaise.

The coral, bright red when cooked, may be frozen and used in a sauce or soup.

Prawns and Shrimps

Prawns and shrimps may be frozen raw or cooked, although it is thought that cooked prawns and shrimps toughen during storage.

Raw. Remove the heads, but not the shells. Wash well in a weak brine solution,—1 tablespoon salt to 1 quart of cold water—drain pack, and freeze.

Cooked. Boil in salted water 3–4 minutes. Cool in the liquor, drain, and 'pick'. Pack closely in suitable containers, protecting the top layer with moisture-vapour-proof film. Seal, leaving a little head-space, and freeze.

Recommended maximum storage life at 0°F (−18°C)

White fish—cod, whiting, plaice, sole	6 months
Fat fish—halibut, mackerel, herring, trout, salmon	3–4 months
Crab and lobster	3 months
Oysters and scallops	1–3 months
Raw prawns and shrimps	3 months
Cooked prawns and shrimps	1 month

COOKING FROZEN FISH

When and how to thaw. Large pieces of fish should be at least partially thawed before cooking. Thin fillets and steaks may be cooked from the frozen state, although if it is intended to give any type of coating, such as egg and breadcrumbs or a batter, they will also need to be partially thawed first in order that the coating can adhere to the fish. Cooked shellfish should be thawed, but served while still chilled. Uncooked shrimps, prawns, and scallops may be put into a saucepan of water or court bouillon whilst still frozen, and thawed and cooked at the same time.

For the best results wet fish should be thawed out in a refrigerator while still retained in their wrappings. The thawing rate is approximately doubled at room temperature. Wet fish required quickly may be thawed in cold water.

If excessive 'drip' or leakage is to be prevented, wet fish must be used promptly after thawing.

Approximate thawing times in a domestic refrigerator

Whole fish over 4 lb.	24–36 hours
Whole fish 2–4 lb.	12–18 hours
Steaks and fillets	4–6 hours
Packages of crab and lobster meat	8–12 hours
½ lb. packs of oysters, scallops, prawns and shrimps	12 hours

The usual methods of cookery and presentation may be used for fish previously stored in a freezer.

REFERENCES

Freezing Meat and Fish in the Home. Home and Garden Bulletin No. 93. United States Department of Agriculture. 1963.
Home Freezing of Poultry. Home and Garden Bulletin, No. 70. United States Department of Agriculture. 1960.

14

Dairy Produce

THE benefits of freezing dairy produce do not immediately come to mind, because all the commodities included under this general title are available all the year round, with little fluctuation in price. Eggs are probably the only exception, as their price rises and falls considerably as the conditions of supply vary. It is, however, extremely useful to have a small reserve of milk, butter, cheese, and eggs in the freezer which can be used in the event of omitting such items from the shopping list, or on returning home from holiday to an empty refrigerator.

In fact, when going away for two or more weeks it is a good policy to transfer the contents of the refrigerator, properly wrapped, to the freezer. Quite frequently bacon and fats become rancid if left too long in a domestic refrigerator.

MILK

Homogenized milk is the most satisfactory for freezing, as it does not separate out on thawing.

It may be frozen in the waxed carton in which it can be purchased, as there is sufficient headspace to accommodate the expansion which takes place as the milk freezes. However, in case of accidents and in order to give added protection, the carton itself should be placed within a polythene bag.

Thawing at room temperature takes six to eight hours, so that if a shortage of milk for breakfast is anticipated, the carton should be removed from the freezer overnight. When time does not permit thawing at room temperature the solid milk may be heated gently in a saucepan.

CREAM

Only really fresh pasteurized cream which has 40 per cent or more butter fat should be frozen—this is sometimes referred to as double or heavy cream. Single or pouring cream, having a butter fat content below 40 per cent, separates when thawed.

Cream is sometimes supplied in lightly waxed cartons and these are not suitable for freezing. The cream must be tipped into a small drum or carton which provides adequate protection and where sufficient headspace can be allowed. Where the cream is supplied in cartons suitable for freezing it is advisable to take out a little cream to make sufficient headspace, then reseal.

ICE CREAM

If there is the possibility of a bought pack of ice cream remaining in the freezer any length of time, it is advisable to give it added protection by overwrapping with a moisture-vapour-proof material.

BUTTER

Commercially produced butter may be satisfactorily frozen, but it is advisable to buy from fresh stock. If buying home-made butter direct from a small dairy, confirm that it has been made from pasteurized cream. Unsalted butter has a longer freezer life than salted varieties. Pack only in ½ lb. quantities and overwrap with moisture-vapour-proof film.

Lard and margarine made from vegetable oils may be kept in reserve in the freezer providing they are given adequate protection.

CHEESE

After a wine and cheese party, a quantity of uneaten cheese usually provides a more serious problem than a quantity of undrunk wine. This is a good example of how the freezer can come to the rescue. Hard cheese like Cheddar, blue cheeses such as Stilton and Gorgonzola, soft cheeses such as Camembert and Brie, may all be satisfactorily stored in the freezer. Cheese, particularly the blue and soft varieties, should be stored when fully mature. But, to prevent drying

179

out and loss of quality, the selection of a packaging material which is highly impervious to moisture vapour is very important.

Thawing is most satisfactorily carried out in the refrigerator. Blue cheeses may be a little crumbly when cut, but there is no other loss of quality, and the soft cheeses can be taken from the freezer at just the same degree of ripeness or maturity as when they were stored. Cottage cheese may also be stored in the freezer in small quantities; small foil cartons are suitable for this purpose. There is a tendency for cottage cheese to become slightly grainy, but this may be corrected by mixing with a little cream after thawing.

Eggs

Freezing eggs in quantity can be an economical proposition; the summer prices when eggs are plentiful are often substantially lower than those of the winter months. The bakery and confectionery trades make use of large quantities of frozen eggs and this illustrates the usefulness of this frozen product. (Because eggs, and in particular duck eggs, are a potential source of salmonellae organisms, bulk frozen eggs must be pasteurized prior to freezing.)

Individual eggs. Eggs cannot be frozen in their shells, because the expansion of the contents during freezing would crack the shells. Clean the shells thoroughly to prevent risk of contamination of the contents as they are broken, and then break the eggs into small waxed paper cases of the type used when making cakes and buns, one egg in each case.

Place on a wire tray, and cover each egg with a circle of greaseproof or other suitable material, leave in the freezer overnight. Remove from the freezer and stack the waxed cups one on top of the other. Pack in suitable quantities in a moisture-vapour-proof bag.

Eggs frozen in this manner can be used for poaching or frying directly from the frozen state, although they must be cooked more slowly if the white is not to become leathery before the yolk is cooked. For a more satisfactory result, thaw first.

Whole eggs in quantity. Several whole eggs may be frozen together in the same container for use in cake making or in egg dishes such as scrambled eggs or omelette. Break the eggs separately into a cup and check for freshness. Combine in suitable quantities and mix lightly with a fork just sufficiently to break up the yolk and combine together with the white—not so much as to aerate the mixture.

180

During frozen storage the egg yolk becomes thick and gummy as a result of coagulation of the solids, and this condition persists when thawed. The addition of a small quantity of salt or sugar prevents this from taking place.

To every 5–6 eggs —Add 1 teaspoon salt for
(depending on the size of the eggs) use in savoury dishes or 1 tablespoon sugar for use in sweet dishes.

Pack in waterproof containers and label clearly, giving the number of eggs and whether salt or sugar has been added.

Separated whites and yolks. Quite frequently a recipe calls for whites or yolks only, and the other part of the egg is put into a small container in the refrigerator, eventually to be discovered many days later unfit for use. Freezing unused whites and yolks is therefore to be recommended, when they can be utilized in another dish at a later date.

Whites

Freeze in useful quantities; no additions are necessary. Uses for whites include meringue cases, soft meringues, macaroons, and icings.

Yolk

Add salt or sugar as for whole egg. To every 5–6 yolks add 1 teaspoon salt or 1 tablespoon sugar.

Uses for yolks include mayonnaise, hollandaise and Sabayon sauces, Zabaglione.

It is worth keeping to hand small straight-sided pots for the purpose of storing egg whites and yolks.

Label clearly with the contents and quantity, and in the case of egg yolks, state whether salt or sugar has been added.

Approximate measures

1 tablespoon egg yolk is equivalent to 1 egg yolk.
2 tablespoons egg white is equivalent to 1 egg white.

Recommended maximum storage life at $0°F(-18°C)$

Milk	3 months
Cream	3 months
Unsalted butter	9–12 months

Salted butter	6 months
Lard	9–12 months
Margarine	9–12 months
Cheese	6 months
Cottage cheese	3 months
Eggs	9 months

Part Three

Part Three

Cooking for the Freezer

IT is not intended that this recipe section should be a complete coverage of all foods which may be prepared, or pre-cooked, before freezing, but it is hoped that by giving examples of foods which have been found to freeze and thaw satisfactorily the reader will be encouraged to try variations of her own.

Some of the considerations which have influenced the selection of recipes are given below:

Dishes worthy of freezer space

Dishes which require a lengthy and sometimes tedious preparation.
Dishes which take a long time to cook.
Dishes where preparation requires a special skill and which may only be attempted when the timetable is not too crowded.
Dishes which are made from seasonal foods.
Convenience foods which may be found invaluable at a future date, in the event of illness or unexpected guests.
Dishes for which freezing is particularly successful.
Dishes which may be prepared in larger quantities involving little more work.
Dishes for a special occasion, such as a party, a dinner party, Christmas.

Dishes and foods not for the freezer

As well as the few fruits and vegetables listed in the appropriate sections, the following foods do not give satisfactory results when frozen:

Hardboiled eggs.
Boiled icings.

185

Sour creams
Single cream
(with less than 40 per cent butter fat)
} Separate when frozen on their own, although may be combined with other ingredients.

Custard pies.
Soft meringues.
Mayonnaise   Separates on its own, although may be used in small quantities in the preparation of sandwich fillings.

Flavourings

Synthetic vanilla may become strong during storage—where this flavour is required use vanilla pods or vanilla sugar.

Cloves also become strong during storage and where possible this flavour should be introduced prior to serving.

The use of garlic is sometimes thought to result in the development of an 'off flavour' in foods. Where the flavour of garlic is desired, squeeze in a little juice before serving, or rub the dish used for serving or reheating with a cut clove of garlic.

Number of portions

Where possible, and with the exception of some baked foods and savouries, the quantities given in recipes are sufficient for eight portions. The quantities in which these foods are packaged and frozen will obviously depend upon the size of the family. (Many people find it convenient to double or treble the quantities normally used in their favourite recipes when preparing certain dishes. One part—one-half or one-third—is for immediate consumption, and the remainder may be frozen for use at a later date.)

Equipment

The limiting factor to the quantity of certain dishes which may be prepared at one time may be the size of the utensils available, for example saucepans, casseroles, or steamers. It may well prove a sound investment to purchase utensils of larger capacities.

Cooking temperatures and times

When preparing and cooking foods for the freezer many dishes are completed before freezing, while others are frozen after the preparation, but prior to cooking. It will be the practice in the following recipe section to give the *cooking temperature* and *cooking time* for the *completed dish* at the beginning of each recipe. Where freezing in the raw state is recommended, the time and temperature required to cook the food from a frozen condition is given in a note at the end of the recipe. It will be appreciated that to cook a dish direct from the freezer may require a slightly higher cooking temperature and usually a longer cooking period than normal.

SOME HINTS AND REMINDERS

On preparation and freezing

(a) The need for a high standard of hygiene in the kitchen at all times is extremely important, and never more so than when food is being prepared for freezing.

(b) Use ingredients which are fresh and of good quality. The choice of fats used is particularly important; on the whole the use of dripping is not recommended unless recently clarified.

(c) Cool foods promptly and quickly prior to freezing; this is probably best achieved by standing the food in its container in a bowl of cold water and ice-cubes. If it is decided to freeze a dish left over from a meal, for example some soup, or a meat casserole, it should be re-boiled and promptly cooled before freezing.

(d) Remove any surplus fat from foods before freezing. This may best be achieved by thoroughly chilling the food, when any excess fat solidifies on the surface in the form of fat globules, which can easily be removed.

(e) Fried foods need to be thoroughly drained on absorbent paper before freezing; they must be quite cold before packaging if a soggy result is to be avoided.

(f) With the exception of baked foods, slightly undercook foods which are to be frozen by up to 30 minutes, as for example, when preparing meat stews and casseroles; this is particularly important when the ingredients include a pasta which is liable to become very soft if cooked too long. On the whole, peas and potatoes are best added when the dish is reheated before serving.

On packaging

(a) Keep in reserve a supply of several different types of packaging materials so that all types of pre-cooked and prepared dishes may be packaged satisfactorily.

187

(*b*)   Freezing before packaging is recommended in some cases where there is risk of the surface or decoration of the food becoming damaged by the packaging material, for example unbaked pastry-covered pies, piped duchesse potatoes, uncooked fish or meat croquettes, decorated cakes and sweets. For some items freezing may take two or three hours, but others may need to be left unwrapped overnight before they become hard enough to pack.

(*c*)   When packaging foods in layers it is advisable to separate each layer with a double thickness of cellophane or other suitable freezer wrapping. When packing a large quantity of a food which packs solidly it is advisable to insert a double thickness of cellophane at intervals in the pack, so that the block may be separated into smaller portions for faster thawing.

(*d*)   Allow headspace, see note on page 129.

(*e*)   Whatever packaging material is used, exclude as much air as possible before sealing.

(*f*)   Use the packaging materials suggested below:

(*Refer to Chapters 5 and 6 for more details of packaging materials.*)

Hors-d'oeuvres	Small waxed cartons, tubs, semi-rigid plastic containers, foil dishes and trays.
Soups and stock	Waterproof containers with straight sides and wide necks, preserving jars, heavily waxed cartons preferably used with a polythene liner. Concentrated stocks and bouillon may be frozen in ice-cubes.
Sauces	As for soups and stocks, although sauces are frequently frozen in smaller quantities of ¼ and ½ pints when small cartons and tubs are suitable.
Fish *Meat* *Poultry* *Game*	Shaped mixtures such as croquettes and fish cakes may be frozen without wrappings and then packed in aluminium foil trays; separate layers if necessary. Meat stews and casseroles in the original casseroles if they can be spared, or in casserole moulds. Pies in ovenware or foil dishes.
Vegetables	Most potato mixtures may be frozen without wrappings and afterwards packed in foil trays. Vegetables with sauces in casseroles, casserole moulds, wide-necked jars, or cartons.

Cakes	Freeze without wrappings raw pastries, decor-
Yeast mixtures	ated cakes and sweets, then wrap neatly,
Pastries	protect in a box where possible. Use foil trays,
Hot and cold	plates and patty cases for flans and tarts.
Sweets	Freeze soufflés and creams in dishes in which
	they will be served.
Savouries	Freeze without wrappings for 1–2 hours, then
	pack in layers in waxed boxes.

On thawing and serving

As far as possible suggestions have been included for thawing and serving foods with each recipe. Because of the variations in the size, shape and texture of foods, as well as the variations in the ambient temperature (even the temperature within a refrigerator will fluctuate with the number of times the door is opened) it is difficult to give an exact time.

(*a*) Thaw soups, meat, fish and poultry dishes to be served cold, in the refrigerator. Because the rate of thawing for these items is very slow, remove them from the freezer the day before they are required. Cold sweets, savouries and hors-d'oeuvres are also better thawed in the refrigerator when time permits. In any case, they should be kept chilled until served.

(*b*) All foods thawed at room temperature and in the refrigerator should be retained in their freezer wrappings. When necessary these may be loosened as with decorated cakes and sweets. Aluminium foil is found to retard the thawing rate and may be replaced with another material where suitable.

(*c*) Transfer unbaked pies directly from the freezer to a preheated oven. Cut the vent for the escape of steam in the pastry after it has begun to thaw. Casserole dishes containing meat, fish, poultry, or vegetables may be placed in a cold oven which is then set at the temperature required. Although this may take a little longer, the mixture is less likely to become scorched around the edges of the dish. Stir the mixture from time to time to accelerate thawing, but care must be taken to avoid breaking up those ingredients intended to be kept whole, for example button onions and mushrooms. Add any topping of breadcrumbs or cheese after the mixture has thawed.

(*d*) The question of whether or not to thaw certain dishes prior to reheating is largely one of convenience and the amount of time available. Thawing before reheating is always recommended for those dishes in which eggs and cream have been used as thickening agents.

189

To prevent the mixture curdling in these cases it is advisable to use a water bath when reheating. Partial thawing may be necessary to facilitate the removal of some foods from their containers. This may be achieved by standing the container (with the exception of the waxed varieties) in hot water for a few minutes.

(e) Sauces and soups served hot may be thawed and reheated in a saucepan beginning with a low heat and stirring all the time. Where the sauce or soup is milk based, the use of a double boiler is recommended. In the event of a sauce or soup appearing to separate, a thorough whisking during reheating is usually sufficient to restore the consistency.

(f) The seasoning and consistency of all dishes should be checked and where necessary adjusted prior to serving.

On storage life

The final quality of the food on thawing is not determined entirely by the length of time spent in the freezer, but by a combination of factors, such as:

(a) the initial quality of the food prepared and frozen;
(b) the rate of freezing;
(c) the efficiency of the packaging material and method of sealing;
(d) fluctuations in temperature within the freezer;
(e) the attention given during thawing.

Recommended maximum storage life at 0°F (−18°C) for pre-cooked and prepared dishes

Hors-d'oeuvres } Savouries	1 month
Sandwiches	1–3 months
Soups } Sauces	3–4 months
Fish dishes	3 months
Meats:	
Roast meat and poultry	4–6 months
Leftover meat (sliced)	1 month
Casserole dishes with meat and poultry	3–4 months
Meat pies	3–4 months
Meat loaves } Fried meats	2–3 months
Vegetables	3 months

Hot and cold sweets:

Baked pies	6 months
Unbaked pies	4 months
Completed chiffon pies ⎫ *Open flans* ⎭	1–2 months
Steamed and baked puddings	3 months

Cakes and biscuits:

Raw cake mixture	3 months
Undecorated cakes	6 months
Decorated cakes	3 months
Baked biscuits ⎫ *Unbaked biscuit mixture* ⎭	6 months

Yeast mixtures:

Baked loaves ⎫ *Risen dough* ⎬ *Tea breads, savarins, etc.* ⎭	3 months
Unbaked yeast pastry	3 months
Baked yeast pastries	1 month

Reminder

Before embarking on an extensive programme of preparation and freezing, the reader is recommended to spend a little time planning and listing those dishes likely to be most in demand and the quantities in which they are required.

As far as possible a regular turnover of the freezer contents should be maintained, and this may best be achieved by working to a two- or three-month cycle.

The management of a freezer is, after all, very much an individual affair, and the suggestions included in this section are intended only as a general guide.

SOME MISCELLANEOUS USES OF THE FREEZER

Babyfoods

Prepare fruit, vegetables, meat, and fish suitable for a young baby in larger quantities than required for one meal. Blend in a liquidizer, or pass through a sterilized sieve. Freeze in an ice-cube tray until hard and pack the cubes in labelled containers. One or two cubes may be removed as required.

Special diets

Prepare meals for someone following a special diet (for example, diabetic, salt-free, fat-free) in larger quantities, and freeze in individual portions. By this means the need for duplicate cooking at each meal-time is avoided, and it may be possible to achieve a little more variety for the member of the household required to follow the diet.

Plate meals

These need not be prepared in accordance with a special diet, but may be just a form of convenience food. They provide a means for the housewife to cater for her husband or other members of the family during an enforced absence from home. Plate meals may be prepared with no extra work involved other than serving and freezing one extra portion at meal-times.

The following precautions should be taken:

(a) Choose ingredients making up the meal which thaw and re-heat at the same rate.

(b) Protect the protein portion—meat or fish—with a sauce or gravy to prevent drying out.

(c) Whether using a special 'meals-for-one' foil tray, or an ordinary plate or dish, exclude as much air as possible when packaging.

Packed meals

It is not easy to be original and resourceful every day when preparing a packed meal to be taken to eat at work or school. Sandwiches, small pies, and cakes keep in the freezer for several weeks and they may be prepared in quantity and packed in individual portions with very little extra work involved. Removed from the freezer at breakfast-time, these items will be thawed (within their wrapping) and ready to eat at lunch-time.

For the best results when preparing sandwiches:

(a) Choose one-day-old bread.

(b) Spread the butter to the very edge of each slice.

(c) Avoid those fillings which do not freeze satisfactorily—i.e. hardboiled egg, lettuce, watercress—these may be prepared

192

and packed separately each day as required. (Suggestions for suitable fillings will be found on page 410.)

(*d*) Package carefully in moisture-vapour-proof film to prevent drying out.

Breadcrumbs

Prepare fresh breadcrumbs in quantity and pack in suitable cartons. They will remain very fresh, separate, and ready to use in puddings, stuffings, or for coating certain fried foods. This also provides an excellent way of using up a surplus of bread.

Herbs, bouquets garnis

Fresh herbs, for example thyme, sage, rosemary, parsley, may all be frozen and keep their fragrance and ability to add flavour for many months. The appearance and colour may deteriorate, but this is true of dried herbs anyway, which remain the only other alternative.

Wash and dry the herbs. Wrap in small bundles in moisture-vapour-proof film and place in a screw-top jar. Plan to keep a variety of fresh herbs readily available in labelled jars at the top of the freezer in the same way that a variety of dried herbs is kept readily available in a well-stocked store cupboard. On removal from the freezer the frozen herbs are very brittle and may easily be chopped if necessary.

Make up a number of bouquets garnis when fresh herbs are available, and time permits, and keep to hand in a suitable jar or container at the top of the freezer.

Prepare chopped parsley in quantity (it does not require blanching), and place in a container in an accessible position in the freezer, where it may be quickly available for garnishes and sauce preparation.

Commercially frozen foods

There are some items produced by the frozen-food producer with which the housewife is not able to compete, as in the following examples. The town housewife may not be able to purchase and freeze her own fruit and vegetables more economically than she can purchase them ready frozen. Unless living near to the sea, the freezing of fresh marine fish is virtually impossible. The preparation of ice cream may seem a waste of time when the family prefers the commercial alternative.

It is therefore a wise policy to allocate some space in the freezer for certain commercially frozen products. There is still scope for the housewife to put on her own brand of originality in the way in which she chooses to serve some of the products; for example, the addition of a suitable sauce when serving frozen plaice fillets, or the addition of fruit purée, nuts and cream to a plain vanilla ice. Where possible, an attempt should be made to buy commercially produced foods in bulk. It is usually possible to buy in bulk such items as fish fingers, hamburgers, as well as fruit, vegetables, and ice cream.

Emergency store

The freezer may function very efficiently as an emergency store, for example when an item has been omitted from the shopping list, or on returning from holiday to an empty refrigerator.

It is worth making a list of some everyday items for which no substitute exists—many of these are items of dairy produce:

Milk (in cartons)
Eggs
Butter
Cheese
Bacon
Sausages
Bread (freshly baked)

(Details of freezing dairy produce are found in Chapter 14.)

These items and any others which may be considered indispensable should be used and replenished at regular intervals of, say, one to two months; particularly such items as bacon and sausages, which have a short freezer-life.

Some Terms Used in the Recipe Section

beurre manié	butter and flour mixed together as a paste—added to sauces and soups as a thickening agent.
bouquet garni	a mixture of herbs usually including parsley, thyme, and bay leaf, tied together in muslin, added to sauces, soups, and casseroles for flavouring.
to blanch	to put into boiling water for a minute or two to reduce pungent or strong flavours, to remove excess salt, or to par-cook (and in the case of vegetables to inactivate the enzymes).

194

bread raspings	usually made from crusts lightly browned in a slow oven crumbled beneath a rolling pin and sieved—used for coating fried foods.
court bouillon	liquid for cooking fish, containing wine or vinegar, seasoning, some vegetable and a bouquet garni.
croûte	a small shape of fried or toasted bread served with an entrée, or as a base for a canapé.
croûtons	small dice $\frac{1}{4}-\frac{1}{2}$-inch cubes of toast or fried bread served with soups.
dried breadcrumbs	soft breadcrumbs which have been allowed to dry thoroughly without colouring—used for coating fried foods.
egg wash	mixture of egg with milk or water brushed on to pastry prior to baking.
farce	stuffing.
to infuse	to extract flavour into a liquid from certain added ingredients, usually over a gentle heat; for example, the addition of an onion stuck with cloves in warmed milk.
macédoine	diced vegetables or fruits.
a purée	a fruit, vegetable, or mixture of foods which has been sieved (or blended in a liquidizer).
to reduce	to effect a reduction of volume by continued boiling.
ricer (potato)	a hand-operated mashing machine.
roux	a mixture of butter (or other fat) and flour, heated together, used as the thickening element in sauces.
sauter	to cook in hot oil or butter—gentle frying.
vanilla sugar	sugar flavoured with vanilla pods (one or two vanilla pods are kept in a jar of castor sugar, which gradually takes up their flavour).

EXPLANATORY NOTES

Measurements

In all cases where an ingredient is measured with a spoon (i.e. teaspoon, dessertspoon, or tablespoon), a *level* spoonful is intended.

Type of flour

Except for recipes calling for *bread* flour, *plain* flour should be used in all recipes. Should the reader wish to substitute *self-raising* flour, the manufacturers' instructions should be followed and an adjustment made in the quantity of baking powder used.

Oven settings

Oven settings have been given in degrees Fahrenheit,
degrees Centigrade,
gas Regulo setting.

Fahrenheit	Centigrade	Regulo	Description of the oven
250	120	½	Very cool
275	135	1	Very cool
300	150	2	Very cool
325	165	3	Cool
350	175	4	Moderate
375	190	5	Fairly hot
400	205	6	Fairly hot
425	220	7	Hot
450	230	8	Very hot
475	245	9	Very hot

(The Centigrade temperature has been approximated to the nearest 5°.)

Hors-d'oeuvres

MIXED HORS-D'OEUVRES

MIXED hors-d'oeuvres, or hors-d'oeuvres variés, may consist of a choice of six or more dishes containing individual vegetables, meat or fish in a dressing. Some of these are quickly and easily prepared and therefore not worth freezing, for example gherkins, anchovy fillets, mushrooms in lemon-flavoured vinaigrette; however, freezing is recommended for others which are more tedious and time consuming to prepare.

CAULIFLOWER PORTUGAISE

8 oz. cauliflower sprigs
2 tablespoons tomato paste
1 tablespoon olive oil
1 tablespoon vinegar
salt and pepper
4 tablespoons tomato juice or 2 frozen cubes tomato purée (page 230) with 2 tablespoons water.

Place all the ingredients in a small covered pan.
Bring to the boil, then simmer gently until the cauliflower is just tender.

To freeze: Cool rapidly and place in a semi-rigid container.

To serve: Thaw for 6 hours in the refrigerator. Pile on to a serving dish, garnish with chopped parsley and strips of tomato.

SELECTED VEGETABLES À LA GREQUE

young artichokes—in quarters
celery—small hearts

197

leeks—2½–3-inch pieces
cucumber—in chunks
mushrooms—caps only
1 pint water
⅛ pint olive oil
juice of ½ lemon
pinch salt
5 coriander seeds
5 peppercorns
large bouquet garni including   parsley
                                      thyme
                                      bay leaf

Boil together all the ingredients, except the selected vegetables, in a pan for 5 minutes.

Prepare and add the selected vegetables.

Cook artichokes, celery, leeks, for 18–20 minutes. Cook mushrooms and cucumber 10 minutes.

To freeze: Strain, cool, and pack in cartons.

To serve: Thaw in the refrigerator for about 6 hours and serve.

CHICKEN À LA REINE

diced cooked chicken breast or white meat   } equal
diced raw mushrooms   } quantities

vinaigrette

3 parts oil
1 part mixed vinegars *and/or* lemon juice (a higher proportion of vinegars and lemon juice may be used if preferred)
salt and pepper
sugar
mixed mustard
garlic—add on serving

Prepare the vinaigrette, adding seasonings to taste.

Use to moisten the diced chicken and mushrooms

To freeze: Pack in a carton and freeze.

To serve: Thaw in a refrigerator for 6–8 hours. Rub the serving dish with a cut clove of garlic if desired before adding the chicken mixture. Serve chilled.

198

DRESSED CRAB MEAT
(page 260)

Moisten the crab meat with a little mayonnaise when serving.

SOUSED HERRINGS
(page 257)

These can be prepared in small rolls each of half a fillet, or the fillets may be cut into flat pieces and soused for half the cooking time. Large rolls can be cut in half or sliced for serving.

SAVOURY HAM KEBABS
(page 431)

Serve the ham mixture shaped into balls.
Moisten with a little sauce, for example Sauce Diable (page 227).

CHICKEN LIVER KEBABS
(page 432)

Serve the chicken liver mixture shaped into balls or cork shapes.
Moisten with a little sauce, for example Sauce Piquante (page 228).

FRUIT HORS-D'OEUVRES

GRAPEFRUIT

THIS is easy, though time consuming to prepare, but it is worth doing in quantity when grapefruits are cheap and of good quality.
Peel the fruit thickly, removing the outside membrane with the peel. With a sharp knife separate the segments from the radiating membranes and remove them whole.

To freeze: Pack in 40 per cent or 50 per cent syrup according to taste, as on page 140.

To serve: Thaw overnight, or at least 8 hours in the refrigerator, and serve chilled.

199

MELON AND GRAPE-MINT COCKTAIL

8 oz. skinned and pipped green grapes   *8 portions*
1 lb. melon balls (2–2½ lb. melon)
2 tablespoons corn oil or olive oil
3 teaspoons lemon juice
1 teaspoon white vinegar
½ teaspoon sugar
salt and pepper
4 teaspoons chopped mint

Whisk together the oil, lemon juice, vinegar, sugar, and seasoning, until thick.
Add the chopped mint.
Pour over the prepared melon and grapes.

To freeze: Pack in suitable amounts and freeze.
Freeze quarter melon skins if required for serving.

To serve: Thaw overnight, or for at least 8 hours in a refrigerator.
Strain slightly if too much juice.
Serve chilled, piled upon melon skins or in glass dishes.

COMPLETE HORS-D'OEUVRES

THESE are dishes served on their own, and include such items as smoked salmon and caviare, for which little preparation is involved. Also in this category are pâtés and terrines, which take a long time to prepare, but which freeze satisfactorily.

CHICKEN LIVER PÂTÉ

Pâtés tend to change texture slightly during freezing, but if not smooth enough, they can be beaten before serving. This also makes an excellent sandwich filling.

4 oz. fat bacon   *16 portions as hors-d'oeuvre*
1 lb. chicken livers   *Oven: 300°F (150°C),*
4 oz. chopped onion   *Regulo 2 Time: 1 hour*
bouquet garni including   2 sprigs parsley
                       ½ bay leaf
                       ¼ teaspoon dried thyme

¼ pint chicken stock
¼ teaspoon ground nutmeg
¼ teaspoon ground cloves
4 eggs
12 oz. butter

Remove the rinds from the bacon and chop it coarsely.

Place the chicken livers, bacon, chopped onion and bouquet garni, in a pan. Add sufficient stock to moisten, and simmer covered for 1 hour. Add more stock during cooking if necessary.

Remove bouquet garni.

Blend in a liquidizer, adding a minimum of stock. Or mince and then pass through a sieve.

Season with the ground spices.

Soften the butter and gradually beat into the liver mixture. Cool.

Beat in the beaten eggs a little at a time.

Place in a greased dish or foil container, cover with a lid or sheet of foil. Place in a cool oven and bake for 1 hour.

To freeze: Cool rapidly. Pack in suitable quantities. Seal and freeze.

To serve: Thaw overnight, or for at least 8 hours in the refrigerator. Serve in slices garnished with salad vegetables.

Serve hot brown toast, and butter separately.

TERRINE

This has a coarser texture than pâté and freezes well.

10 rashers streaky bacon or pork fat   *12–16 portions*
2 oz. onion   *Oven: 350°F (175°C),*
½ oz. butter   *Regulo 4 Time: 1½ hours*
¾ lb. liver—lamb's or pig's
¾ lb. sausage meat
1 teaspoon salt
⅛ teaspoon pepper
pinch allspice
¼ teaspoon dried thyme
⅛ pint Madeira, brandy, or port
1 lightly beaten egg
¾ lb. lean veal or pork
1 bay leaf

Sometimes blanching the bacon is advised to give a better flavour.

201

(It should be noted that blanched bacon is more difficult to use when lining a mould than raw rashers.)

Remove the bacon rinds. Line the mould with rashers of bacon or thin slices of pork fat.

Chop the onion finely, and cook in the butter until tender.

Mince the liver and combine with the sausage meat. Seaon well.

Add the beaten egg, thyme, allspice, wine, and the softened onion.

Place a layer of this liver farce at the bottom of the lined mould.

Cut the lean veal or pork into thin strips across the grain of the muscle.

Place a layer of strips over the farce.

Alternate layers of farce and meat strips, finishing with a layer of farce.

Place a bay leaf in the centre of the top.

To ensure a perfect seal, mix a paste from flour and water and spread this thickly on to the rim of the dish before putting on the lid.

Cook in a water bath in a preheated oven for $1\frac{1}{2}$ hours.

Remove the lid, and the bay leaf. Place foil on top of the terrine and press lightly with, say, a 2 lb. weight.

Cool rapidly.

To freeze: Pack in suitable quantities, seal, and freeze.

To serve: Thaw for at least 8 hours in the refrigerator, or overnight. Serve chilled with a garnish of salad vegetables.

SAVOURY PANCAKES

Small filled pancakes, in a sauce, may be served on their own.

5-inch pancakes (page 401)   *Allow 2 or 3 per portion*
for suitable fillings see page 421 (vol-au-vents)
                    page 426 (choux pastry)
Sauce Italienne (page 227)—add when serving
Parmesan—add when serving

Spread the pancakes with the filling selected.
Fold in half twice, to form a quarter of a circle.

To freeze: Separate with moisture-proof film, and pack in suitable quantities.

To serve: Reheat in the sauce and sprinkle with grated cheese.

The following dishes are also suitable to be served on their own as a complete hors-d'oeuvre.

Beignets Soufflés (page 427)
Cheese and Vegetable Tartlets (page 402). Make in 2½–3-inch patty tins.
D'Artois (page 424)
Kromeskies (page 429)
Miniature Quiche Lorraine (page 416)
Puff Pastry Patties (page 425)
Stuffed Tomatoes (page 245)

Stocks and Soups

MOST soups freeze well and are very convenient to use, requiring only to be reheated prior to serving. The preparation and cooking of soup is normally tedious and long; short cuts are not possible if a good flavour is to be achieved.

Soup is a very good example of a food which may be prepared in quantity, involving very little extra work, providing saucepans and utensils of sufficient capacity are available. Soup may be made economically when vegetables or other main ingredients are in season or plentiful supply.

However, soup is a bulky food to store and this may present a problem in the smaller freezer. Concentrating some soups prior to freezing is one way of reducing the bulk, although this is not suitable in every case without a loss in quality. Sometimes it is possible to withhold a proportion of the liquid before freezing which can be made up when the soup is thawed and reheated.

An alternative method is to prepare and freeze independently some of the basic ingredients. These include stock, bouillon, sauces such as béchamel or velouté, and vegetable purées. Thus it is possible to prepare a variety of soups quickly by reheating and combining together certain of these ingredients. These ingredients are also available for other dishes if required.

Neither stock nor bouillon is likely to suffer any loss of quality as a result of being concentrated prior to freezing. Recipes for these will be included in the following section, together with recipes for soups which can be frozen successfully; at least one example for each category of soup has been given. Many of the accompaniments served with soups may be frozen, and suggestions for these are also included.

STOCKS AND BOUILLONS

To Concentrate Stock and Bouillon

To save freezer space the stock or bouillon may be concentrated before freezing.

After straining, boil the completed stock or bouillon in an open pan until the volume is reduced by half.

Pressure Cooker Method. Place the ingredients for the preparation of stock or bouillon in the pressure cooker with a little over half the quantity of water given in the recipe. Cook under 15 lb. pressure for at least 30 minutes. Strain in the usual way.

Concentrated stock and bouillon may be conveniently frozen in ice-cube trays, and the separated cubes packed in a sealed polythene bag from which they may be removed as required.

CHICKEN STOCK

bones and carcasses of 2 cooked chickens   *Yield: 3 pints*
2 oz. carrot
6 sticks celery
2 onions
bouquet garni including   2 sprigs parsley
                            ½ bay leaf
                            2 cloves
                            good pinch thyme
2 thin strips lemon rind
2 teaspoons salt
12 peppercorns
4 pints water

Put the bones, vegetables, bouquet garni, lemon rind and seasoning in the pan with the water and bring to the boil.
Simmer for 2 hours in an uncovered pan. Strain.

To freeze: Reduce to double strength if desired, cool, remove any fat from the surface of the stock, pack in suitable quantities, seal and freeze.

VEGETABLE STOCK

4 oz. carrot   *Yield: 4 pints*
3 oz. onion

2 oz. turnip
1 stick celery
2 oz. butter
2 tomatoes
1 blade mace
12 peppercorns
2 cloves
1 teaspoon salt
4 pints water
bouquet garni including   1 sprig parsley
                                  ¼ teaspoon thyme
                                  ¼ bay leaf

Slice the vegetables and celery thinly.
Melt the butter in a stewpan and fry the vegetables gently without browning for 30 minutes in the covered pan.
Add the tomatoes, bouquet garni, herbs, seasoning, and water, and bring to the boil.
Simmer in the covered pan for 1½ hours. Strain.

To freeze: Reduce to double strength if desired, cool, remove any fat from the surface of the stock, and pack in suitable quantities, seal and freeze.

BOUILLON

Bouillon differs from a household stock in that meat (beef) is also included in the ingredients from which it is prepared. It may also be called a *Beef Broth*. It is the necessary first step in the preparation of a consommé, although it may be used as an expensive substitute in recipes calling for a more simple household stock.

BOUILLON FOR CONSOMMÉ

1½ lb. shin of beef   *Yield: 3 pints*
1½ chicken carcasses
3 oz. turnip
6 oz. onion
6 oz. carrot
little oil as necessary

bouquet garni including   ½ bay leaf
                                       blade mace
                                       3 cloves
                                       9 peppercorns

3 teaspoons salt
6 pints water

Cut the meat into cubes. Break the chicken carcasses into pieces.
Peel and slice the vegetables.
Place the meat, carcasses, and onion into a stewpan and toss over
heat until brown; a little oil may be necessary if the meat is very lean.
Add the turnip and carrot, bouquet garni, salt and water.
Bring to the boil, then simmer uncovered for 3–4 hours. Strain.

To freeze: Reduce to double strength if desired. Cool, remove any
fat from the surface, pack in suitable quantities, seal and freeze.

THIN SOUPS

Consommé

3 pints bouillon   *Yield: 2 pints*
12 oz. minced shin of beef   *6–8 portions*
⅛ pint sherry
2 egg whites

Place the bouillon, beef and sherry into a saucepan.
Whisk the egg whites to a froth, add to the bouillon, and continue
to whisk over heat until the liquid boils.
Allow to boil without whisking until the froth rises in the pan, then
remove from the heat.
Return the pan to the heat and allow the froth to rise once more;
then simmer for 1 hour.
Clarify the consommé by passing the contents of the saucepan
through a jelly bag or a double thickness of muslin.
The consommé should be passed through the bag twice, or until a
clear liquid is obtained.

To freeze: Cool rapidly and skim off any globules of fat. Pour into
1- or 2-pint containers as required.

To serve: Thaw overnight in a refrigerator or over hot water sufficiently to free the contents from the container. Transfer to a saucepan
and complete thawing over gentle heat. Bring back to boiling-point.
Adjust seasoning if necessary, and garnish as required.

GARNISHES

For Consommé Julienne	For Consommé Brunoise
1 tablespoon shredded carrot	1 tablespoon diced carrot
1 tablespoon shredded turnip	1 tablespoon diced turnip
1 tablespoon shredded leek	1 tablespoon diced celery
$\frac{1}{2}$ tablespoon shredded onion	1 tablespoon diced leek
$\frac{1}{2}$ tablespoon shredded celery	
1 tablespoon shredded cabbage.	

Prepare the vegetables, and cut into very fine shreds, or neat dice, as required.

Boil in salted water until tender.

Drain and place in the tureen in which the consommé is served.

The garnish may be prepared and frozen in a small carton when preparing and freezing the soup.

CONSOMMÉ MADRILÈNE

This consommé requires a good chicken stock to give a set to the consommé when it is served cold (en gelée). Otherwise gelatine must be added.

1 lb. tomatoes

$\frac{1}{2}$ lb. lean minced beef

$\frac{1}{8}$ pint sherry

1 strip lemon peel

2 egg whites

2 pints chicken stock

(1 teaspoon gelatine dissolved in $\frac{1}{8}$ pint water)

Yield: 2 pints
8 portions

Chop the tomatoes coarsely and place in a saucepan with the beef, sherry, lemon peel, and stock.

Whisk the egg whites to a froth and add to the consommé.

Whisk over heat until boiling-point is reached, then allow the froth to rise in the pan.

Lower the heat and simmer for $\frac{3}{4}$ hour.

Add the gelatine if necessary. Clarify through a jelly bag as for the consommé.

To freeze: Cool rapidly and remove any globules of fat from the surface. If the consommé is to be served 'en gelée', it can be frozen overnight in small dishes, marmites or waxed cases, which may then be packed and sealed altogether in a polythene bag. Otherwise the consommé may be frozen in 1- or 2-pint containers.

To serve 'en gelée': Thaw in the refrigerator about 18 hours, and garnish with tomato and pimento pieces cooked in stock.

To serve hot: Transfer to a saucepan, thaw over gentle heat, and return to boiling-point. Adjust seasoning if necessary, and serve. Garnish as for 'en gelée'.

Consommé Mulligatawny

As this consommé is so heavily spiced, the starting bouillon may be made from commercial bouillon cubes if this is more convenient.

½ lb. minced beef
2 tablespoons Madeira
½ oz. coriander
½ oz. cardamom
¼ oz. cumin
¼ oz. fenugreek
12 peppercorns
1 bay leaf
thinly peeled lemon rind
2 egg whites
3 pints beef bouillon
juice of a clove of garlic—add after thawing

Yield: 2½ pints
8 portions

garnish: cooked rice flavoured and coloured with turmeric

Bruise the spices in a mortar and tie in a piece of muslin.
Place the spices, herbs, and lemon rind in the bouillon, bring to the boil, then simmer for 40 minutes.
Add the beef and Madeira to the bouillon, and the egg whites whisked to a froth.
Continue to whisk the bouillon over heat until boiling-point is reached.
Allow the froth to rise to the top of the saucepan, then lower heat and simmer for 30 minutes.
Clarify the consommé by passing through a jelly bag as for basic consommé recipe.

To freeze: Cool rapidly and remove any globules of fat from the surface. Pack in 1- or 2-pint containers, seal and freeze.

To serve: Transfer to a saucepan and thaw over gentle heat. Add the juice from a clove of garlic. Bring back to the boiling-point, adjust seasoning, and serve. Garnish with cooked rice flavoured and coloured with a pinch of turmeric.

FRENCH ONION SOUP

1½ lb. onions
2 oz. butter
2 tablespoons flour
3 pints stock (preferably beef stock)
black pepper
salt
slices of French bread or toasted croûtes
grated cheese

Yield: 2½ pints
8 portions

Finely slice the onions and toss them in butter over gentle heat until soft and golden.
Sprinkle in the flour.
Gradually stir in the stock.
Season with salt and pepper.
Bring to the boil and simmer for 20 minutes.

To freeze: Cool rapidly, and remove any fat solidifying on the surface. Pour into 1- or 2-pint containers. Seal and freeze.

To serve: Transfer to a saucepan, thaw over gentle heat, and gradually bring back to the boiling-point. Adjust seasoning if necessary. Pour into a heatproof tureen.
Spread slices of French bread or toasted croûtes with butter, and cover with grated cheese.
Place in a hot oven at 450°F (230°C), Regulo 8, for 5 minutes, or beneath a hot grill, to melt the cheese.

THICK SOUPS

PURÉE SOUPS

CELERY SOUP

THIS is a recipe for a simple purée soup which may be adapted according to the vegetable available.

210

2 lb. celery (other suitable vegetables: *Yield: 2 pints*
artichokes, carrots, cauliflower)   *8 portions*
2 oz. butter
2 medium-sized onions
2 pints stock
bouquet garni
seasoning
1 oz. cornflour
½ pint stock (or for cream soups, use ½ pint milk)
For richer cream soup: Mix together ¼ pint single
                cream and 2–3 lightly   add when
                beaten egg yolks if   serving
                available

Wash and cut up the celery into 1–2-inch pieces, roughly slice the onions.
Melt the butter in a pan and add the celery and onions.
Fry lightly for 5–10 minutes, without allowing to colour.
Add 2 pints of stock, bouquet garni, and seasoning, bring to the boil and simmer until cooked—about 30 minutes.
Rub through a nylon or hair sieve (blend first in a liquidizer if available), and return to a clean pan.
Stir in the cornflour blended with the ½ pint stock or milk.
Bring to the boil, stirring all the time, and cook for 5 minutes.

To freeze: Cool rapidly, remove any fat from the surface, pour into 1- or 2-pint containers, seal and freeze.

To serve: Transfer to a saucepan, thaw and gradually bring back to boiling-point. Adjust seasoning, add cream and egg yolks if available.

TOMATO SOUP

1 lb. tomatoes   *Yield: 2 pints*
4 oz. onion   *8 portions*
2 oz. carrot
¾ oz. butter
1 dessertspoon concentrated tomato purée
½ teaspoon salt
½ pint tomato juice
½ bay leaf
1 blade mace

1½ pints beef stock
1 teaspoon arrowroot
1 teaspoon sugar
¼ teaspoon paprika
¼ clove garlic—add when serving

Peel and slice the onion and carrot.
Cut the tomatoes into halves and squeeze out the pips. Strain juice from pips.
Melt the butter in a stewpan, add the tomatoes, onion, carrot, and salt and cook in the covered pan over a low heat for 40 minutes.
Remove the carrot and rub the remaining ingredients through a sieve.
Mix the sieved purée with the tomato juice, juice from the pips, and concentrated tomato purée.
Add the spices and stock, bring to the boil, then simmer for 20 minutes.
Adjust the seasoning, remove the bay leaf and mace.
Mix the arrowroot with a little tomato juice and add to the soup with the sugar and paprika. Boil for a few minutes.

To freeze: Cool rapidly, remove any fat from the surface of the soup. Pack in 1- or 2-pint containers. Seal and freeze.

To serve: Rub the saucepan with a clove of garlic if desired before adding the frozen soup. Thaw over gentle heat and gradually bring the soup up to boiling-point before serving.

CREAM (PURÉE) SOUPS

CREAM OF SPINACH

A simple cream soup using a white sauce as a thickening agent, suitable for soups made from green vegetables.

2 lb. fresh spinach
2 small onions
2 pints vegetable stock
2 oz. butter or margarine
2 oz. flour
1 pint milk
salt and pepper
⅛ pint cream } add when serving if available
egg yolks

Yield: 2½ pints
8 portions

212

Thoroughly wash and drain the spinach, and chop coarsely.

Finely slice the onions.

Bring the stock to boiling-point and add the prepared spinach and onions. Continue to boil with the lid on the saucepan until the spinach is tender, about 30 minutes.

Blend in a liquidizer, if available, before passing through a nylon sieve.

Melt the butter or margarine in a saucepan, add the flour and fry together without colouring, remove from the heat and gradually stir in the milk.

Return to the heat and bring to the boil to thicken the sauce.

Stir in the sieved spinach, season with salt and pepper. Heat to boiling-point.

To freeze: Cool rapidly, remove any fat from the surface. Pack in 1- or 2-pint containers, seal and freeze.

To serve: Partially thaw the soup in the refrigerator, or stand the container in hot water before transferring it to a saucepan. Reheat to boiling-point, remove the pan from the heat, and stir in the cream and egg yolks.

Hot Cucumber Soup

A cream soup thickened with a beurre manié.

3 pints vegetable stock	*Yield: 2½ pints*
2 large cucumbers	*8 portions*
2 teaspoons chopped shallot	
pinch mace	
pinch nutmeg	
salt and pepper	
1 oz. butter	
¾ oz. flour	

½ clove garlic
2 egg yolks or 2 teaspoons arrowroot } add when serving
⅛ pint cream (single)

garnish: 2 tablespoons cucumber balls or dice
2 tablespoons peas } cooked in stock
2 tablespoons sliced French beans

Peel the cucumber and cut it into ½-inch slices.

Bring the stock to boiling-point and add the cucumber and shallot. Cook until tender.

Pass the soup through a sieve, blending first in a liquidizer if available.

Beat the butter and flour thoroughly together to form a beurre manié, and whisk it into the soup a little at a time over gentle heat. Season and flavour with the spices, and boil for 5 minutes.

To freeze: Cool rapidly, pack into 1- or 2-pint containers, seal, and freeze. The garnish may be blanched and frozen separately.

To serve: Partially thaw the soup in the refrigerator, or stand the container in hot water, before transferring it to a saucepan to heat through. Add the juice from $\frac{1}{2}$ a clove of garlic if desired.

Mix the arrowroot or egg yolks together with the cream and combine with a little boiling soup.

Return the mixture to the saucepan and continue heating until it has thickened. The soup must not be allowed to boil if egg yolks have been included. Add the garnish when serving.

PREPARATION OF PURÉE AND CREAM SOUPS FROM FROZEN INGREDIENTS

$\frac{3}{4}$ lb. frozen vegetable purée (for example, artichokes, spinach, cauliflower)
1 small onion
1 pint stock (may use frozen concentrated stock)
bouquet garni
salt
pepper
$\frac{1}{2}$ oz. cornflour
$\frac{1}{4}$ pint stock (or milk for cream soups)
2–3 tablespoons cream ⎱
egg yolk if available ⎰ for cream soups

Yield: 1$\frac{1}{2}$ pints
4 portions

Make up the concentrated stock to 1 pint with boiling water, and combine with the frozen purée in a saucepan.

Heat gently until thawed.

Add the onion, bouquet garni, and seasoning; bring to the boil and simmer for 30 minutes.

Blend together the cornflour and stock (or milk).

Remove the soup from the heat, take out the onion and bouquet garni, and stir in the blended cornflour.

Return to the heat and continue to boil for about 5 minutes.

214

Remove from the heat, adjust seasoning, and stir in cream and egg yolks to cream soups.
Serve.

Alternative method of thickening for cream soups

(This method is particularly suitable for the preparation of soups from purées of green vegetables.)

Substitute ½ pint of béchamel or white sauce of coating consistency for the blended cornflour and stock (or milk).

If using a frozen sauce, thaw and reheat over hot water while the purée and stock are being simmered together.

COLD SOUPS

Note on thawing and serving cold soups

Soups to be served cold should be thawed in a refrigerator; this may take up to 18 hours. If time does not permit this method of thawing, transfer to a saucepan, reheat to boiling-point, then promptly chill in a bowl of cold water and ice cubes. Place in a refrigerator until served.

VICHYSSOISE

8 leeks
4 oz. onions
8 tablespoons butter
1 lb. potatoes
4 sticks of celery
4 sprigs parsley
salt
pepper
3 pints chicken stock
1 pint single cream—add after thawing.

Yield: 4 pints
8 portions

garnish: chopped chives

Melt the butter in a stewpan.
Slice the leeks and onions and cook gently in the butter without colouring until tender.
Add the stock, peeled and sliced potatoes, sticks of celery, seasonings, and parsley.

215

Cook until tender, for about 40 minutes.
Sieve the soup, blending first in a liquidizer if available.

To freeze: Cool rapidly, remove any fat from the surface. Pour into 1- or 2-pint containers. Seal and freeze.

To serve: Thaw the soup. Add the single cream, allowing ½ pint to each pint of soup, and keep chilled before serving. Garnish with chopped chives.

ICED CUCUMBER SOUP

2 large cucumbers
2 shallots
2 sprigs mint
3 pints chicken stock
1 tablespoon arrowroot
¼ pint single cream
salt and pepper
Angostura bitters
green colouring

Yield: 2½ pints
8 portions

garnish: diced blanched cucumber, chopped mint

Peel the cucumber and cut into small pieces.
Chop the shallot and cook in the stock for 15 minutes.
Add the cucumber and mint and cook until tender.
Remove the mint and sieve the soup, blending it first in a liquidizer if available.
Mix the arrowroot with the cream and add to the sieved soup.
Bring to the boil and simmer for 1 minute.
Colour lightly.

To freeze: Cool rapidly and pack into 1- or 2-pint containers.

To serve: Thaw the soup (see note on page 215). Adjust seasoning and consistency—mix thoroughly if there is a tendency to separate. Serve garnished with cucumber dice and chopped mint.

FISH SOUPS

SHRIMP BISQUE

4 tablespoons butter
4 tablespoons chopped celery
4 oz. chopped mushrooms

Yield: 2½ pints
8 portions

4 slices onion
4 slices carrot
salt and pepper
½ bay leaf
pinch mace
1 tablespoon peppercorns
2 tablespoons lemon juice
2 pints chicken stock
6 oz. shrimps
2 tablespoons white wine
½ pint double cream—add when serving

Melt the butter in a stewpan. Add the celery, mushrooms, onion, and carrot, and cook over gentle heat for 10 minutes.

Add the stock, lemon juice, herbs, and seasoning, and simmer for 20 minutes.

Strain through a sieve. Add wine and shrimps to the liquid, and simmer for 5 minutes.

To freeze: Cool rapidly, remove any fat from the surface. Pack into 1- or 2-pint containers. Seal and freeze.

To serve: Partially thaw in a refrigerator or over hot water. Transfer to a saucepan and reheat gently to boiling-point. Remove from the heat, stir in the cream then reheat, but do not boil. Serve immediately.

MISCELLANEOUS SOUPS

KIDNEY SOUP

½ lb. ox kidney
2 tablespoons cornflour
1–2 oz. butter
½ turnip
1 onion
1 carrot
3 pints beef stock
bouquet garni
seasoning

Yield: 2½ pints
8 portions

Remove skin and fat from the kidney. Wash and cut into pieces. Dip in seasoned cornflour, retain any cornflour left over to use for thickening at the end.

217

Prepare the vegetables and cut roughly into cubes.
Fry the kidney, and vegetables, briskly in the butter to develop a good brown colour, but do not let the kidney become hard.
Drain off any excess fat.
Add the stock, bouquet garni, and seasoning.
Bring to the boil and simmer gently for 1 hour.
Pour through a large sieve to remove the kidney and vegetables.
Retain the kidney and chop very finely.
Blend any remaining cornflour with a little water and add to the soup. Return it to the boil for a further 5 minutes.
Add all or part of the chopped kidney, according to taste.

To freeze: Rapidly chill, remove any fat solidifying on the surface. Pack into 1- or 2-pint containers. Seal and freeze.

To serve: Transfer to a saucepan, and over gentle heat thaw and gradually bring back to boiling-point. Simmer a few minutes, and adjust seasoning if necessary.

MINESTRONE SOUP

1½ oz. pork fat or dripping   *Yield: 2½ pints*
3 rashers streaky bacon   *8–10 portions*
2 oz. onion
1 leek
4 oz. carrot
4 oz. turnip
8 oz. potato
2 oz. celery
3 pints chicken, beef, or vegetable stock, or water
pinch dried sage
8 oz. cabbage
2 tomatoes
2 oz. green beans
2 oz. peas
2 oz. rice (uncooked weight)
1 tablespoon tomato purée—or more to taste
1 tablespoon chopped parsley } add when serving
grated Parmesan

Melt the dripping.
Remove the rind from the bacon rashers and finely dice the flesh.

218

Prepare the onion, carrot, turnip, potato, celery, and leek (white portion only) and dice finely.

Fry the prepared vegetables and bacon in the fat without colouring, for 10 minutes.

Add the stock and the pinch of sage, bring to the boil, and simmer for 1½ hours.

Skin the tomatoes, remove the seeds and dice the flesh.

Add the tomato flesh to the soup, together with the finely shredded cabbage, peas and beans and rice. Add tomato purée to flavour if liked.

Cook for 20 minutes.

To freeze: Cool rapidly, remove surface fat. Pack and seal.

To serve: Thaw in a covered pan over gentle heat. Bring to the boil. Add parsley and grated cheese and serve.

SOUP GARNISHES

TOASTED CROÛTONS

Toast slices of bread ⅜ inch thick.
Remove the crusts.
Cut into cubes of equal dimensions.
Place on baking trays and dry out in oven at 350°F (175°C), Regulo 4.

To freeze: Pack in convenient amounts. Seal and freeze.

To serve: Thaw in freezer wrappings at room temperature. Reheat in oven if desired.

PARMESAN CROÛTONS

5 slices bread ¼ inch thick
4 tablespoons grated Parmesan
1 egg yolk
2 tablespoons butter
½ teaspoon salt
pinch cayenne

Oven: 350°F (175°C), Regulo 4
Time: 20 minutes

Mix together the cheese, butter, yolk and seasonings.
Toast the bread on one side. Remove crusts.
Spread the untoasted side with the Parmesan mixture.

Cut into ¼-inch cubes.
Place on baking sheet and cook in oven until brown.

To freeze: Pack in convenient quantities. Seal and freeze.

To serve: Thaw in pack at room temperature. Serve. Reheat in oven if desired warm.

CHOUX PUFFS

6 oz. choux pastry (page 377)   *Makes 80*
1 oz. grated Parmesan   *Oven: 400°F (205°C), Regulo 6*
pinch cayenne   *Time: 20 minutes*
¼ teaspoon salt

Beat the cheese and seasonings into choux pastry.
Place the pastry in a piping bag with a ¼-inch plain nozzle.
Pipe in small pieces on to a well-greased baking tray.
Bake in a fairly hot oven. When firm to touch, make a small slit in the sides and dry out thoroughly in a cooler oven.

To freeze: Cool thoroughly. Pack in convenient amounts.

To serve: Thaw in freezer wrappings at room temperature, and serve. Or remove from the wrappings and spread out on a baking tray, thaw and reheat in a fairly hot oven for 10 minutes.

Sauces

HAVING a selection of basic sauces in the freezer puts the housewife in the same position as the hotel chef: able to produce elaborate sauces to order. The time taken for the sauce to thaw out is thus the only limiting factor, and if using a double boiler this need only be a matter of a few minutes. It is very tedious and time consuming to make the small quantity of a particular sauce sometimes called for in a recipe. This inconvenience may easily be overcome by having available in the freezer, sauces frozen in the small quantities often required, for example ⅛ pint, ¼ pint, and ½ pint.

Most basic sauces freeze well, and even for those which do not it is possible to take special measures before freezing to ensure a satisfactory result after thawing.

Basic sauces suitable for freezing are as follows:

(a) White sauces—béchamel and velouté

(b) Brown sauces

(c) Tomato sauces

(d) Egg yolk and butter sauces

(e) Flavoured butters

(f) Purée or fruit sauces

Many variations can be achieved from this short list by the addition of other ingredients, and these are given later in this section.

WHITE SAUCES

White sauces can be made in two consistencies, one suitable for serving with food and called a *pouring sauce,* and one for coating foods or for combining with other ingredients, called a *coating*

221

sauce. The thickness of the sauce is regulated by the amount of fat and flour mixture added. A thick sauce can be frozen and diluted after thawing, but unless flavoured liquid is used for this some flavour will be lost.

For a *pouring sauce* 1 oz. butter and 1 oz. flour are used to thicken 1 pint liquid.

For a *coating sauce* 2 oz. butter and 2 oz. flour are used to thicken 1 pint liquid.

These sauces can be frozen after cooling; frequent stirring is necessary as the sauce cools to prevent the formation of a skin on the surface. On thawing, the sauce will require whisking over a gentle heat to regain its smooth and glossy appearance.

The following treatments prior to freezing will help to retain the glossy appearance on thawing:

Either blend the completed and slightly cooled sauce in a liquidizer for 30 seconds, then bump the container of sauce on the work bench during packing to remove air bubbles;

or reserve one-quarter of the flour when thickening the liquid and complete the sauce using the reduced amount of flour. Mix the reserved flour to a thin paste with a little cold liquid, add, and mix well into the completed and slightly cooled sauce before freezing.

Béchamel Sauce

Milk flavoured with vegetables and herbs is used to prepare a béchamel sauce.

1 pint milk
2 oz. carrot
2 oz. onion
1 oz. turnip
pinch mixed herbs
½ bay leaf
1 blade of mace
2 cloves
10 peppercorns
1–2 oz. butter or margarine ⎫ depending upon the thickness
1–2 oz. flour ⎬ required

Peel and slice the vegetables. Add the vegetables and herbs to the milk, and heat gently for 30 minutes, until the milk is well flavoured. Strain the milk.

222

In a clean pan melt the margarine, add the flour, and mix well over low heat without colouring for 1 minute. Remove from the heat and add the strained milk gradually, beating well to avoid lumps forming. Return to the heat and bring to boiling-point, stirring continuously, and boil for 1–2 minutes until the sauce is glossy and smooth. Adjust seasoning.

To freeze: Cool rapidly, stir frequently to prevent the formation of a skin on the surface.

Treat as suggested on page 222 to ensure that the sauce retains a good gloss on thawing.

Pack in ¼-pint, ½-pint, or 1-pint containers.

To serve: Thaw in a refrigerator for 8–12 hours; or quickly over hot water or low heat.

Beat well during reheating, bring to the boil, and use as required.

VELOUTÉ SAUCE

1 pint white stock, well flavoured, from chicken, veal, or fish
1–2 oz. flour
1–2 oz. margarine or butter

For the preparation and freezing of a velouté sauce, follow the same directions as for the béchamel sauce, given above, substituting the white stock in place of the infused milk.

ADAPTATIONS OF A BASIC BÉCHAMEL OR VELOUTÉ SAUCE

Except where one of the two basic white sauces is specified, the following simple or more elaborate sauces may be made by adding the listed ingredients to ½ pint of hot sauce, either béchamel, or velouté. Partially thaw the sauce in a refrigerator for about 6 hours, or more quickly in a double boiler. Bring back to boiling-point before making the additions.

ANCHOVY SAUCE

Serve: with fish dishes.

2 teaspoons anchovy essence.

SAUCE AURORE

Serve: with meat, fish, eggs and vegetables.

4 tablespoons reduced tomato purée (page 230)
1 tablespoon softened butter
salt and pepper

} Beat into a
béchamel sauce

CAPER SAUCE

Serve: with boiled turbot, cod, or mutton.

1 teaspoon chopped capers
½ tablespoon vinegar

SAUCE CARDINAL

Serve: with fish.

2 tablespoons lobster butter (page 234)

CHEESE SAUCE

Serve: with fish, vegetables, eggs.

2 oz. grated cheese
salt and pepper
pinch of nutmeg

} Add to a béchamel sauce. Do
not reboil after adding the
cheese

SAUCE CHIVRY

Serve: with egg, fish, vegetables, poached chicken.

¼ pint white wine
1½ tablespoons mixed dried
chervil and parsley (or
tarragon) or use 3 table-
spoons of the fresh herbs if
available
1½ tablespoons chopped onion
or shallot

} Place all ingredients in a pan
and reduce to 1 tablespoon
liquid. Strain into a velouté
sauce. Mix well and reheat.

SAUCE DUXELLE

Serve: with white meat, cauliflower, white fish.

⅛ pint white wine
1 finely chopped shallot
1 tablespoon duxelles (page
245) or 2 oz. finely chopped
mushrooms or mushroom
stalks cooked in butter until
dry and not greasy

} Cook the shallot in wine and
reduce to 1 tablespoon. Add
with the duxelle to a velouté
sauce. Mix well and reheat.

PARSLEY SAUCE
Serve: with fish.

1 tablespoon finely chopped parsley

SHRIMP SAUCE
Serve: with fish.

½ oz. butter
½–1 oz. shrimp trimmings
1 oz. shrimp tails, chopped

Pound together the shrimp trimmings and butter and sieve; add to the hot sauce together with the chopped shrimp tails.

SAUCE SOUBISE
Serve: with eggs, cauliflower, fish.

6 oz. chopped onions
½ oz. butter
salt and pepper
pinch nutmeg
⅛ pint cream

Cook the onions in the butter, adding a pinch of salt, until tender but not coloured. Add the hot sauce to the onion. Sieve or blend in a liquidizer. Simmer, add cream, nutmeg, and seasoning.

SAUCE SUPRÊME
Serve: with vegetables, white meat, eggs, fish.

2–3 tablespoons double cream
1–2 egg yolks
1 tablespoon butter
salt and pepper
lemon juice

Stir into a slightly cooled velouté sauce.

These sauces may all be further enriched by the addition of up to 1 tablespoon of butter, in small pieces, just before serving.

BROWN SAUCE

A well-flavoured brown sauce is a very useful asset in the freezer, as it can easily be enriched, or varied, and served with so many dishes. Preparation of this sauce in large quantities is particularly recommended, since it is not easy to brown successfully small amounts of flour without risk of burning.

BROWN SAUCE

4 oz. dripping (freshly clarified)
4 oz. onion
4 oz. carrot
3 oz. celery
2 oz. diced blanched bacon or cooked ham
3 oz. flour
4 pints brown stock
2 tablespoons tomato paste
4 oz. chopped mushroom stalks
bouquet garni including   4 sprigs parsley
                                1 bay leaf
                                ½ teaspoon dried thyme

Slice the onion, carrot and celery.
Melt the dripping in a pan and fry the sliced vegetables and diced bacon (or ham) for 10 minutes.
Strain the vegetables and bacon from the fat.
Blend the flour into the fat and cook over a low heat, stirring frequently until the flour colours to a deep golden brown.
Remove the pan from the heat and pour in the hot stock.
Add the fried vegetables, bacon, tomato paste, chopped mushroom stalks and bouquet garni.
Simmer slowly for 2 hours.
Adjust the seasoning.
Strain.

To freeze: Cool rapidly. Remove any surface fat. Pack into ¼-pint, ½-pint, or 1-pint containers. Seal and freeze.

To Serve: Thaw slowly in a refrigerator, or more rapidly over hot water or gentle heat. Beat over gentle heat to regain the smooth texture of the sauce.

ADAPTATIONS OF A BROWN SAUCE

Thaw and reheat ½ pint of basic brown sauce over gentle heat. Bring to the boil and use it to prepare the following more elaborate sauces by making the correct additions.

SAUCE BIGARDE

Serve: with wild duck and game.

juice and finely shredded rind of 1 orange
1 teaspoon lemon juice
¼ pint red wine
1 teaspoon redcurrant jelly

} Retain half the shredded orange rind and blanch. Simmer remaining ingredients in the brown sauce for 10 minutes. Strain, then add the blanched orange rind to the sauce before serving.

CURRY SAUCE

Serve: with lamb, chicken, beef, rice and egg dishes.

6 oz. finely chopped onion
1 tablespoon butter
1 tablespoon curry powder—more or less according to taste
seasoning
1 tablespoon softened butter
2 teaspoons lemon juice

} Cook the onion in the butter until tender, about 15 minutes. Add curry powder and cook for 1 minute. Add to the sauce. Add seasoning and heighten the flavour with lemon juice. Beat in softened butter before serving.

SAUCE DIABLE

Serve: with grilled meat, chicken, and rechauffé dishes.

1 oz. onion or shallot chopped and cooked in butter
¼ pint white wine
black pepper, cayenne pepper
1 tablespoon Worcestershire sauce
1 teaspoon chopped parsley or fresh herbs
1 tablespoon softened butter

} Add the wine to the cooked shallot or onion and reduce the liquid to 3 tablespoons. Add to the sauce. Bring to the boil and simmer for 2 minutes. Add the peppers and Worcestershire sauce. Beat in the butter and fresh herbs on serving.

SAUCE DUXELLE

Serve: with small cuts of meat and fish.

Substitute ½ pint of brown sauce for the velouté sauce on page 224. Add 1 tablespoon tomato paste to the finished sauce.

SAUCE ITALIENNE

Serve: with pasta dishes and vegetables.

227

4 chopped shallots 4 chopped mushrooms 4 oz. tomatoes 1 tablespoon oil 1 sprig thyme (or $\frac{1}{8}$ teaspoon dried thyme) $\frac{1}{8}$ pint dry white wine $\frac{1}{8}$ pint stock	Remove the skin and seeds from the tomatoes and chop the flesh coarsely. Fry mushrooms, shallots, and tomatoes in a little oil. Add wine, herbs and stock, and boil until the volume is reduced by a half. Strain into the brown sauce. Bring to the boil and simmer 10 minutes.

SAUCE MADÈRE AND SAUCE AU PORTO

The wine used gives its name to the sauce.

Serve: with fillet beef, veal escalop, and in vol-au-vents to combine meats.

$\frac{1}{8}$ pint Madeira or port seasoning 1 tablespoon softened butter	Reduce the wine to 1 tablespoon and add to the sauce; simmer 5 minutes. Add up to 2 tablespoons wine (not reduced) if necessary to improve flavour. Add seasoning. Beat in butter before serving.

SAUCE PIQUANTE

Serve: with pork, beef, rechauffé dishes of meat and fish.

1 oz. finely chopped onion 2 tablespoons vinegar 1 tablespoon gherkins coarsely chopped 1 tablespoon capers cut in half salt and pepper	Cook the onion in the vinegar until soft and the liquid is reduced to 1 tablespoon. Add to brown sauce together with gherkins, capers and seasoning and simmer 5 minutes before serving.

SAUCE POIVRADE

Serve: with game.

10 peppercorns $\frac{1}{8}$ pint vinegar $\frac{1}{8}$ pint red wine	Crush peppercorns and boil in the vinegar. Reduce volume by a half. Add to the brown sauce together with the red wine. Simmer 10–15 minutes. Strain through muslin, reheat, and serve.

SAUCE ROBERT

Serve: with grilled meats, especially pork.

228

1 oz. finely chopped onion ¼ oz. butter ¼ pint white wine 2 tablespoons creamed French mustard 1 tablespoon chopped parsley	Cook onions in butter until tender. Add wine and reduce liquid to 2 tablespoons. Add to the sauce and simmer 10 minutes. Add the creamed French mustard. Adjust seasoning. Beat in parsley before serving.

TOMATO SAUCES

The following sauces may be made in quantity when the tomatoes are cheap and in good supply.

TOMATO SAUCE

8 rashers streaky bacon   *Yield: 2 pints*
4 oz. carrot
4 oz. onion
4 oz. celery
4 tablespoons butter
3 oz. flour
1 pint brown stock
½ teaspoon salt
¼ teaspoon sugar
bouquet garni including   8 sprigs parsley
                         1 bay leaf
                         ½ teaspoon thyme
4 lb. tomatoes or tinned equivalent
2 tablespoons tomato paste

Blanch and dice the bacon.
Peel and chop the vegetables.
Fry the vegetables and bacon in the butter for 10 minutes.
Mix the flour into the fried mixture and continue cooking for 5 minutes without colouring.
Remove from the heat, then add the stock.
Coarsely chop the tomatoes and add to the pan.
Add the seasonings, sugar, and bouquet garni. Simmer for 2 hours.

Pass the sauce through a sieve.
Add the tomato paste, and adjust the seasoning.

To freeze: Cool rapidly, remove any surface fat. Pack into ¼-pint,
½-pint, or 1-pint containers. Seal and freeze.

CONCENTRATED TOMATO PURÉE

This purée is useful where a full and fresh tomato flavour is required,
for example on pizza or in savouries, but it is not of the same
strength as the tomato paste supplied commercially in tubes and
tins.

2 oz. onion   *Yield: ¾ pint, or 15 × 1 oz. cubes*
1 tablespoon butter
1 teaspoon flour
2 lb. tomatoes
pinch sugar
bouquet garni including   2 sprigs parsley or ¼ teaspoon dried
        parsley
        ¼ bay leaf
        ¼ teaspoon thyme
        pinch basil
        3–4 coriander seeds or pinch powdered
        coriander
small piece orange peel
¼ teaspoon salt
1 tablespoon tomato paste

Skin and remove the seeds from the tomatoes.
Chop the flesh coarsely.
Finely chop the onion and cook it in the butter until tender.
Add the flour and continue to cook for a further 5 minutes without
colouring.
Add the tomatoes, sugar, bouquet garni, orange peel, and seasonings
to the pan, and slowly bring to a simmering-point. Simmer for 30
minutes or until the pulp is cooked and thick.
Remove the bouquet garni and peel. Sieve the sauce.
Add the tomato paste and adjust the seasoning.

To freeze: Cool rapidly. Spoon the purée into 1–2 oz. ice-cube
makers and freeze. Remove the cubes from the trays and pack.

EGG YOLK AND BUTTER SAUCES

The sauces in this group, of which hollandaise, maltaise, and béarnaise are the most widely known and used, are delicious, but require some care in making and thawing. With all egg sauces, if the heat is applied too strongly, then the egg will cook and harden where it comes in contact with the surface of the saucepan, giving the sauce a 'scrambled' texture. It is safer to make the sauce in a double pan (this should be of stainless steel or enamel, otherwise the sauce will become discoloured), or in a china basin over hot water. The aim is to cook the yolks without scrambling them, but at the same time cause them to absorb butter. Each yolk will absorb 2–3 oz. butter. For the best results the butter should be well softened without being liquid, and added in small pieces. Each addition should be well beaten in before the next is made. This should give a thickened and stable sauce, when lukewarm, which is the temperature at which it is served.

Cautionary note

Frozen egg and butter sauces require so much care in thawing and reheating that no saving of time is made when serving them. Preparation and freezing of these sauces may be considered worthwhile if there are a number of egg yolks to be used up or when eggs are very cheap. Alternatively these sauces may be prepared starting with previously frozen egg yolks.

HOLLANDAISE SAUCE

Serve: with broccoli, fish and cauliflower.

4 egg yolks
1 tablespoon lemon juice
½ teaspoon salt
¼ lb. butter
(2 tablespoons cream)

Beat the yolks and lemon juice, and add the salt. Heat gently over hot water and whisk until slightly thickened.
Soften the butter in a warm place and add gradually in small knobs, checking that each knob of butter is fully incorporated into the sauce before adding more. Do not overheat. (If the sauce should curdle, add 1 tablespoon cold water and beat well. If this fails to restore the consistency, put 1 teaspoon lemon juice and 1 tablespoon of the sauce

231

into a bowl and beat until creamy. Beat in the remaining sauce 1 tablespoon at a time, until all is thick and creamy again.)

To freeze: Cool, pack in ¼-pint or ½-pint containers.

To serve: Thaw gradually in a refrigerator for 18–24 hours. Beat a little of the sauce (about a quarter of the amount used) in a basin over hot water until creamy. Add remaining sauce in tablespoonfuls, beating between each addition.

ADAPTATIONS OF A HOLLANDAISE SAUCE

The following additions may be made to ½ pint of hollandaise sauce just before serving.

EGG WHITE AND CREAM

Serve: with fish, soufflés, and delicately flavoured vegetables.

2 stiffly beaten egg whites
or ¼ pint whipped cream } Fold in to the completed sauce.

The addition of egg white and cream serves to lighten and give volume to the sauce.

Hollandaise sauce with whipped cream is known as *Sauce Mousseline.*

HERBS

Serve: with such dishes as poached egg and boiled fish.

Minced parsley, chives, tarragon.
The addition of herbs gives extra flavour to the sauce.

PURÉES

2 tablespoons of vegetable purée, for example asparagus, artichoke
       or chopped pounded shellfish
       or duxelles (page 245)

ORANGE JUICE AND RIND

Serve: with cauliflower, asparagus and broccoli spears.

3 tablespoons orange juice
grated rind of 1 orange
Hollandaise sauce with orange juice and rind is known as *Sauce Maltaise.*

BÉARNAISE SAUCE

Serve: with grilled meat and grilled fish.

1 oz. chopped shallot
1 teaspoon dried tarragon
1 teaspoon dried chervil
6 peppercorns
½ bay leaf
⅛ pint dry white wine
⅛ pint vinegar (preferably wine vinegar)
pinch salt
3 egg yolks
6 oz. butter

Place the shallot, herbs, and seasonings, with the wine and vinegar, in a small pan, and boil until the liquid is reduced to 2 tablespoons. Strain.

Add the liquid to the beaten yolks. Beat the yolks over hot water until they begin to thicken.

Add the softened butter in small knobs, beating well between each addition to fully incorporate the butter, as for hollandaise sauce.

To freeze: Pack in ¼-pint or ½-pint containers.

To serve: Thaw gradually in a refrigerator for 18–24 hours. Place a small amount of fully thawed sauce in a basin over hot water and beat until smooth. Add the remaining sauce a tablespoonful at a time, beating well between additions.

ADAPTATIONS OF A BÉARNAISE SAUCE

The following additions may be made to ½ pint béarnaise sauce before serving.

SAUCE CHORON

Serve: with grilled meat, fish, chicken and eggs.

2 tablespoons of tomato purée (page 230)

FLAVOURED BUTTERS
SAVOURY BUTTERS

These are made by combining well-creamed butter with other ingredients and seasonings to prepare the following sauces.

The quantities listed are for 4 oz. butter.

ANCHOVY BUTTER
16 anchovy fillets, well pounded and sieved, or anchovy essence
cayenne pepper
colouring if necessary

LOBSTER BUTTER
3–4 oz. cooked, sieved lobster coral

SHRIMP BUTTER
3–4 oz. well-pounded cooked sieved shrimps

MAÎTRE D'HÔTEL BUTTER
juice of 1 lemon
4 teaspoons chopped parsley
seasoning

DEVILLED BUTTER
2 teaspoons curry powder
4 teaspoons chutney
few drops lemon juice
Worcestershire sauce to taste
salt and pepper

Preparation of Savoury Butters:
Shape between damp pieces of greaseproof paper.
Form into a long roll of about ¾ inch diameter and cut off into ½-inch slices with a wet knife,
or form into a solid block about ½ inch thick and cut into squares or diamond shapes,
or form into balls about ¾ inch diameter.
Spread out on greaseproof paper to freeze.
When hard, pack in plastic bags or waxed drums or cartons.
Serve direct from freezer on grilled meat or fish.

HARD SAUCES
These are basically a well-beaten mixture of butter and sugar to which a flavouring has been added.

234

4 oz. unsalted butter
$4\frac{1}{2}$ oz. sugar (soft brown, castor or icing)
suitable flavourings: 1 tablespoon of juice, and the grated rind of
                                 lemon or orange
                                 brandy or liqueur
                                 vanilla sugar
                                 2 tablespoons fruit purée

Shape between wet greaseproof paper as for savoury butters.

FRUIT SAUCES

Hot and cold fruit sauces may be simply prepared from frozen-fruit purées, for example, strawberry, raspberry, apricot, blackcurrant.

Served cold they are a suitable accompaniment to ice cream, meringues, mousses and creams.

Served hot they present an attractive alternative to custard or jam sauces when served with steamed or baked sponges.

Fruit Sauce—served cold

fruit purée
icing sugar
30–40 per cent syrup
liqueur if desired

Thaw the purée overnight in a domestic refrigerator.
Sweeten if necessary with a little icing sugar, and dilute with a little syrup if too stiff.
Add a little liqueur if desired.
Keep the sauce chilled until served.

Fruit Sauce—served hot

fruit purée
30–40 per cent syrup
castor sugar
arrowroot

Put the frozen purée with a little syrup into a thick-based pan to thaw, sweeten if necessary with castor sugar.

Simmer gently for 2–3 minutes.
Remove from the heat and stir in a little blended arrowroot (allow
1 teaspoon arrowroot to ½ pint sauce).
Return to the boil before serving.

APPLE SAUCE

Serve: with duck or pork.

There is little advantage in freezing the completed apple sauce
when it can be simply prepared from frozen purée, as follows:

½ pint apple purée
little water or syrup
strip of lemon peel
sugar to taste
knob of butter or margarine

Put a little water or syrup and a strip of lemon into a pan, and add
the frozen apple purée.
Heat gently until the purée has thawed, add sugar to taste.
Simmer for about 5 minutes.
Add a knob of butter or margarine before serving.

ORANGE SAUCE

Serve: with duck.

5 tablespoons sugar
2 tablespoons water
1 tablespoon white vinegar
juice 2 oranges
segments from 1 orange
grated rind of 1 orange
1 teaspoon arrowroot

Mix the arrowroot with a little orange juice.
Dissolve the sugar in the water, bring to boiling-point, and boil until
the sugar solution becomes a deep golden-brown. Watch carefully,
remove from heat and immediately add the vinegar and the remaining orange juice; dissolve the caramel. Add the orange rind, segments,
and blended arrowroot.
Bring to the boil and boil for 2 minutes.

236

To freeze: Cool rapidly and pack.

To serve: Thaw over gentle heat and gradually return to boiling-point. Serve with roast duck.

CRANBERRY SAUCE

Serve: with turkey.

1 lb. cranberries
¾ pint water
¾ lb. sugar

Thoroughly rinse the cranberries.
Dissolve the sugar in the water over gentle heat.
Add the cranberries, and cook gently for 15–20 minutes.

To freeze: Cool and pack in ¼-pint or ½-pint containers.

To serve: Thaw at room temperature, 2–4 hours.

MISCELLANEOUS SAUCES

The following sauces freeze well.

BOLOGNESE SAUCE

Serve: with pasta.

6 oz. onion   *Yield: 8 portions*
1½ tablespoons cooking oil
1½ lb. minced chuck steak (lean)
3 tablespoons tomato purée (page 230)
1 tin tomatoes (net weight 21 oz.)
1½ teaspoons salt
pinch pepper
3 teaspoons sugar
½ teaspoon dried basil
⅛–¼ pint red wine
1 clove garlic—add when serving

Chop the onion and fry in the oil until golden-brown and tender.
Add the minced meat and continue frying until meat loses its redness.
Add the tinned tomatoes, purée, seasoning, sugar, herbs and wine.
Simmer for 1 hour.

237

To freeze: Cool rapidly. Pack in cartons holding ¼ pint (1 portion) or
½ pint (2 portions).

To serve: Thaw as required, over a gentle heat. Add garlic to taste.
Allow to heat through thoroughly and cook for 30 minutes.
Serve with pasta.

CHOCOLATE SAUCE

Serve: with ice creams, pancakes, light sponge puddings, etc.

3⅜ oz. unsweetened chocolate (nominal 4 oz. block)
3 oz. castor sugar
2 teaspoons cocoa
½ pint water
2 egg yolks
2 oz. butter (optional)
rum

Blend the cocoa with a little of the water in a saucepan, add the rest
of the water.
Break the chocolate into squares and add to the saucepan together
with the sugar, heat gently until the chocolate has dissolved.
Then bring to the boil, and boil gently for about 10 minutes until the
sauce appears syrupy.
Remove from the heat and beat in the egg yolks, and butter if desired.
Stir in a little rum to taste.

To freeze: Cool the sauce by standing the pan in cold water, transfer
to a carton or tub, and freeze.

To serve: Thaw at room temperature for 6 hours if required cold, or
over gentle heat if required warm.

Vegetables

THE freezing of cooked vegetables is not normally recommended, as they are said to acquire a 'warmed-up' flavour and lose some of the colour and fragrance normally associated with fresh vegetables. Blanched vegetables in contrast seem to retain the qualities of the fresh vegetable.

There are certain occasions, however, when freezing cooked vegetables offers many benefits. Too frequently an otherwise interesting and appetizing meal is spoilt by dull vegetables, presumably because there is little time to spare for preparing the vegetables, after preparing the main dish. The inclusion in the menu of special vegetable dishes which have been prepared and frozen in advance is thus a way in which this situation may be remedied, without adding to the last-minute preparations. Pre-cooked vegetables are found to retain their quality very much better when covered in a sauce which at the same time increases the appeal of the vegetables.

Potatoes, so often the 'Cinderella' of the vegetables, can be very successfully frozen in a variety of ways, for example in the form of duchesse potatoes, potato croquettes, and chips. These more appetizing forms can be available from the freezer in less time than it would take to prepare simple boiled potatoes.

Some vegetables may be frozen to provide excellent supper dishes with a minimum of effort.

As with the selection of vegetables for blanching, choose only those of high quality when preparing vegetable dishes for freezing. Good results cannot be expected from stale or stored vegetables. The texture of potatoes is found to vary considerably, so it is always worth-while to prepare a sample batch before preparing any of the potato variations in quantity.

239

SPICED RED CABBAGE

2 lb. red cabbage   *Yield: 8–12 portions*
4 oz. shallots
4 oz. butter
¼ teaspoon each of powdered nutmeg, allspice, cinnamon
¼ teaspoon each of thyme, and caraway seeds
¼ pint red wine
1 tablespoon vinegar
1 lb. cooking apples
1 tablespoon brown sugar
1 clove of garlic—add when reheating

Shred the cabbage fairly finely.
Chop the shallots.
Heat the butter in a pan until foaming, add the cabbage and shallots, and cook in the covered pan with frequent tossing for 10 minutes.
Add the spices, herbs, wine and vinegar.
Cook over a low heat for 20 minutes or until the cabbage is crisp but tender.
Peel, core, and slice the apples, and arrange over the top of the cabbage.
Sprinkle with the sugar.
Cover and cook for 10 minutes.
Mix in the apple and sugar with the red cabbage.

To freeze: Cool rapidly. Pack in suitable amounts in polythene bags or waxed cartons.

To serve: Place unthawed cabbage in a covered saucepan over low heat. Stir occasionally until thawed. Add the juice from the clove of garlic. Increase the heat for a further 15–20 minutes. Serve piping hot.

DUCHESSE POTATO MIXTURE

2 lb. potatoes   *Yield: 20–30 shapes*
4 oz. butter
2 eggs or 4 egg yolks
salt and pepper
pinch grated nutmeg

Steam or boil the potatoes until soft.
Drain and place in a warm oven to dry off.

240

Put through a potato ricer or press through a sieve.

Place the sieved potato into a pan and stir over heat to complete the drying.

Add the butter and beat in well.

Season with the salt, pepper and nutmeg.

Beat in the eggs or egg yolks.

If the consistency is too stiff for piping, a little hot milk may be added to soften the mixture.

Place the duchesse mixture into a piping bag with a ½-inch star pipe.

Pipe out the mixture on to baking sheets lined with lightly oiled greaseproof paper.

To freeze: Place in the freezer until the shapes are hard. Pack in suitable quantities.

To serve: Transfer from the freezer wrappings on to baking sheets. Brush with egg wash if desired. Place in a cold oven set at 400°F, (205°C), Regulo 6, and bake for 20 minutes until heated through and browned.

POTATO CROQUETTES

These may be frozen after shaping or after frying.

duchesse potato mixture (previous recipe)   *Deep fat, 360°F (180°C)*
beaten egg
bread raspings

2 lb. duchesse potato mixture gives approximately 12–18 croquettes, and requires 2 beaten eggs for egg and crumbing.

Prepare a stiff duchesse mixture suitable for moulding.

Divide the mixture into even-sized portions of 1½–2 oz. weight, and shape into cylinders, balls, or pear shapes.

Brush each shape with the beaten egg, then toss in bread raspings.

Toss lightly to remove excess crumbs.

Freeze at this stage

or place the freshly made croquettes in deep fat at 360°F (180°C) and cook for 5 minutes. Drain on absorbent paper. Cool rapidly.

To freeze: Place *shaped* and *fried* croquettes on to baking sheets and freeze until hard. They are now easily packed and remain separate.

241

To serve: Allow *shaped* croquettes to thaw at room temperature for 2–3 hours. Heat deep fat or cooking oil to 360°F (180°C), and cook until golden-brown. Drain on absorbent paper.

Place *fried* croquettes in a fairly hot oven for 20 minutes.

Serve croquettes on a dish mat and garnish with parsley.

VARIATIONS

POTATO CROQUETTES CHEVREUSE

To 1 lb. duchesse mixture add 1 teaspoon dried or 2 teaspoons fresh chopped chervil.

POTATO CROQUETTES À LA LYONNAISE

To 1 lb. duchesse mixture add 2 oz. finely chopped onion (fried in butter until tender).

POTATO CROQUETTES À LA PARMESANE

To 1 lb. duchesse mixture add:　　1 oz. grated Parmesan
$1\frac{1}{2}$ teaspoons salt
pinch cayenne pepper

POTATO CAKES

These may be made with duchesse mixture, or from the following more homely recipe.

2 lb. potatoes　　　　　　　　　　　　　　*Yield: 16–20 cakes*
1 tablespoon chopped parsley
salt and pepper
1 egg
seasoned flour
beaten egg and bread raspings for coating are optional
mixture of oil and butter for frying

Cook the potatoes by boiling or steaming.
Push through a potato ricer or mash well.
Add the seasoning, egg, and chopped parsley, and beat until smooth.
Form into a cylinder and roll in flour.
Cut into even-sized pieces, and shape into cakes 3 inches in diameter in the flour.

The cakes may then be coated with beaten egg and tossed in bread raspings for a firmer finish.

To freeze: Freeze uncovered on trays. Pack in suitable amounts when frozen.

To serve: Heat butter, or a mixture of butter and oil, to foaming-point. Place frozen cakes into the pan and cook for 5 minutes on each side, using moderate heat. The cakes should be golden-brown and hot in the centre in this time.

BAKED STUFFED POTATOES

4 large potatoes	*Yield: 4 or 8 portions*
1 oz. margarine	*Oven: 425°F (220°C), Regulo 7*
salt and pepper	*Time: 1½ hours to bake the potatoes*

8 rashers streaky bacon
2 well-beaten eggs } or 4 oz. grated cheese
little grated Parmesan—add when serving

Scrub the potatoes and make a slit through the skin all around the centre. Bake in a fairly hot oven until the centre is quite cooked, about 1½ hours. Grill the bacon until it is crisp, and when it is cold place between pieces of greaseproof paper and 'crumble' it with a rolling pin.

Beat the eggs.

Cut the potatoes in half along the slit where the skin was cut, and scoop out the potato.

Press the potato through a nylon sieve (or a potato ricer if available). Mix in a bowl with the margarine, seasoning, and either the crumbled bacon and well-beaten eggs, or the grated cheese.

Refill the potato cases.

To freeze: Cool. Place a piece of cellophane over each half potato, and pack in a suitable container. Seal and freeze.

To Serve: Place in an ovenware dish, remove the pieces of cellophane from the potatoes. Cover the dish with foil and place in an oven at 375°F (190°C), Regulo 5, for about 40 minutes.

Remove the foil, sprinkle the surface with a little grated Parmesan, and lightly brown under a hot grill.

Serve with a watercress garnish.

243

COOKING FOR THE FREEZER

CHIPPED POTATOES

The sales of commercially prepared frozen chips are surprisingly high, so it would seem, that although they are so widely enjoyed, many people find them tedious to make at home. One method of overcoming this is to prepare chips in quantity and freeze them in amounts appropriate to the family's requirements.

Choose good-quality potatoes—old stored potatoes do not make good chips.

Peel and cut into regular-shaped chips.

Rinse in cold water and dry thoroughly.

(Blanch chips 1 minute in boiling water, if desired, to give chips which are lighter in colour.)

Fry in deep fat at 360°F (180°C) until tender, but not coloured. Lift out. Reheat the fat to 380°F (195°C) and fry the chips until lightly browned.

Drain on absorbent paper.

To freeze: Cool the chips quickly and pack in appropriate quantities.

To serve: Spread out the chips on a baking tray and place in a fairly hot oven for 5–10 minutes.

The following alternative methods may also be found useful:

BLANCHED CHIPS

Freezing chips which have been blanched only may be found a useful method, for example when high-quality potatoes are available and time does not permit the preparation and frying of chips in the quantities required.

Cut the chips as before and rinse in cold water.

Blanch 1 minute in boiling water; cool under cold running water, and dry.

To freeze: Freeze before packaging so that the chips remain separate. If they are frozen after packaging, they form a solid block which is difficult to separate in the fat bath.

To serve: Transfer the frozen, blanched chips to the fat bath at 380°F (195°C) until tender. (Adding frozen chips to the fat bath will considerably lower the temperature of the fat; thus it is important to keep the heat beneath the fat turned up high during this period.) Lift out. Reheat the fat bath to 380°F (195°C) and return the chips until lightly browned. Drain on absorbent paper.

244

CHIPS FROZEN AFTER FIRST FRYING

Excellent results are obtained with chips frozen after the first frying at 360°F (180°C). They may subsequently be thawed and browned when returned to the fat bath for a few minutes at 380°F (195°C).

However, this method may be thought the least convenient, as it involves the use of the fat bath before and after freezing.

DUXELLES

This is a useful purée of mushrooms to have in store for flavouring sauces, and as a stuffing for tomatoes.

1 lb. mushroom stalks and trimmings
2 oz. onion, or 4 tablespoons chopped parsley
4 shallots
salt and pepper
ground nutmeg
2 oz. butter

Wash and dry the stalks. Chop finely and remove any moisture.
Chop the onion and shallot.
Fry the onion in the butter, add the chopped shallot, and cook until tender.
Add the mushrooms, seasoning, and nutmeg, and cook over a low heat until all the moisture has evaporated.
(4 tablespoons of parsley can be added, and the onion omitted, depending upon the flavour required.)

To freeze: Pack in 2 oz. or 4 oz. quantities, as these are most useful.

To serve: There is no need to thaw duxelles before use—it will thaw quickly when mixed with other ingredients.

TOMATOES STUFFED WITH DUXELLES

Served as a savoury, or to accompany dishes cooked à la Provençale.

12 small tomatoes
8 oz. duxelles
3 oz. chopped lean ham

2 teaspoons tomato paste
2 tablespoons finely chopped parsley if not incorporated in duxelle mixture

Slice the tops from the tomatoes to form lids, or if preferred the lids may be cut from the bottoms of the tomatoes, as the top portion surrounding the stalk provides a flatter base on which to stand the tomatoes for baking.
Remove the seeds and inner pulp.
Sprinkle with salt and invert to drain out the liquid.
Mix together the cooked ham, duxelle, tomato paste, and parsley, and use to fill the tomato cases.
Replace the tops.

To freeze: Arrange on small trays or aluminium foil patty tins. Pack, seal, and freeze.

To serve: Thaw for 2 hours at room temperature. Place in an oven at 300°F (150°C), Regulo 2, for about 40 minutes to heat through. Do not overcook.

Courgettes à la Provençale

1½ lb. courgettes   *Yield: 8 portions*
¾ lb. onions
1 tin tomatoes (net weight 21 oz.)
seasoned flour
2 oz. grated Parmesan
3 tablespoons cooking oil

Wash or wipe the courgettes, but do not peel. Cut into slices 1–1½ inches thick.
Sprinkle with salt, leave 5 minutes, then dry on kitchen paper. Toss in seasoned flour.
Heat 2 tablespoons cooking oil in a pan. Add the courgette slices and fry until lightly coloured. Drain.
Slice the onions. Fry in the pan until just tender, adding more oil if necessary. Season well.
Slice the tomatoes.

To freeze: Take a casserole which can be spared from everyday use, or prepare a casserole mould. Arrange the courgettes, onions, and tomatoes in layers in the dish. Season well between layers, and add a little lemon juice. Finish with a layer of tomatoes. Place a piece of cellophane or other freezer wrapping on the surface. Freeze until solid (overnight) and wrap closely and seal.

To serve: Remove the freezer wrappings; if using a casserole mould, return to the original casserole. Place in a cold oven set at 375°F (190°C), Regulo 5, to thaw and complete cooking. This will take at least 1½ hours. When the mixture is thawed sprinkle the surface with the grated cheese. The surface may be browned under the grill before serving.

AUBERGINES À LA PROVENÇALE

This dish may be made substituting aubergines for courgettes.

Wipe and cut the aubergines into slices 1–1½ inches thick.
Sprinkle with salt and leave for up to 30 minutes.
Dry and complete the preparations exactly as for courgettes.

STUFFED PEPPERS

Stuffed peppers provide a convenient and appetizing supper dish. The filling may include any selection of a wide variety of ingredients; bacon, boiled ham, tomato, mushroom, cheese; usually combined with a small quantity of breadcrumbs, rice or corn, and moistened by a little sauce such as a béchamel.

Recipes are given for two suitable fillings each sufficient for 8 medium-sized red or green peppers.

To prepare and freeze peppers

Wipe the peppers and cut off a slice including the stalk to form a lid; carefully remove the seeds and white pithy veins from inside.
Blanch in boiling salted water for 2–3 minutes, cool under running water, and drain well.
Fill with the prepared filling. Replace the lid.
Place in a suitable ovenware dish or deep aluminium foil tray, and in each case cover with foil. Seal and freeze.

To serve: Partially thaw in a refrigerator for about 6 hours if time permits. Place the covered dish in an oven at 350°F (175°C), Regulo 4, for about 45 minutes. Remove the aluminium foil cover after about 20 minutes, and place a knob of butter on each pepper.

SPANISH RICE FILLING

8 oz. Patna rice (uncooked weight)
2 oz. oil
4 oz. lean bacon
¾ lb. onions
6 oz. mushrooms
½ lb. tomatoes
salt and pepper

Cook the rice in boiling salted water until it is barely cooked--the centre of the grain will still feel slightly hard.
Drain and rinse under cold water, and dry.
Chop the onion and bacon finely.
Heat the oil and add the onion and bacon. Fry over a low heat for 10 minutes.
Wash and slice the mushrooms, peel and slice the tomatoes. Add to the pan and continue cooking until tender.
Combine the mixture thoroughly with the rice and seasoning, pack into the peppers, and replace the tops.

CORN AND HAM FILLING

12 oz. tinned corn
8 oz. celery
1 oz. butter
8 oz. boiled ham—cut thick
¼–½ pint béchamel sauce
salt and pepper

Dice the celery and cook in butter for ten minutes until tender.
Cut the ham into ¼-inch cubes.
Mix the celery and ham together with the drained corn and béchamel sauce.
Add salt and pepper to taste.
Fill the prepared peppers.

CREAMED SPINACH

This can be served as a vegetable, or as a base for dishes 'à la Florentine', or in some savoury dishes.

2 lb. spinach
2 oz. butter

salt and pepper
6 tablespoons béchamel sauce, or a mixture with double cream in equal portions, reduced to coating consistency.

Thoroughly wash the spinach and shake until dry.
Shred the leaves, then chop fairly finely.
Melt the butter in a pan, add the spinach and seasoning, and cover with a lid.
Cook over moderate heat for 10 minutes, or until tender, shaking the pan frequently. Remove the lid and quickly evaporate off any excess liquid.
Place the hot spinach and hot béchamel sauce into a liquidizer and blend until smooth, or sieve the mixture. Season well.

To freeze: Cool rapidly. Pack in suitable amounts in containers lined with polythene to prevent staining.

To serve: Thaw in a covered pan set over low heat. Stir frequently. Heat through until seen to be just on boiling-point. Adjust seasoning and serve.

BAKED STUFFED MARROW RINGS

1 medium-sized marrow   *Yield: 8 portions*
¾ lb. cooked minced meat
6 oz. fresh breadcrumbs
1 large onion
1 oz. butter
1 tablespoon flour
¼–½ pint stock
salt and pepper
tomato paste, if desired
grated cheese—add when serving

Peel the marrow and cut into slices 1½–2 inches thick. Remove the seeds and pith from the centre.
Blanch in boiling water 2–3 minutes, or in steam for 4 minutes. Cool.
Finely chop the onion and fry gently in butter until soft.
Sprinkle in the flour, and stir in the stock. Bring to the boil.
Add the meat, breadcrumbs, seasoning, and a little tomato paste if desired.

To freeze: Lay the slices of marrow in a greased ovenware or aluminium foil dish and fill the centres with meat stuffing. Cover, seal and freeze.

To serve: Cover the stuffed marrow slices with buttered paper and bake at 375°F (190°C), Regulo 5, for about 1–1¼ hours. Uncover and sprinkle the marrow slices with grated cheese, and continue to bake for a further 10–15 minutes.
Serve with a tomato sauce or a good brown sauce.

Fish

WITH two exceptions mentioned below, the cooking of fish prior to freezing is not really recommended. The fish after cooking is rather fragile, due to the small content of connective tissue present; also there is a tendency for cooked fish to weep during reheating, which is liable to spoil the appearance of the dish.

In any case, raw fillets, cutlets, and even whole fish, require a short time to cook, and for most purposes the raw fish may be cooked directly from the frozen state. Thawing is only necessary when the fish is to be stuffed or given a coating prior to cooking. The time taken to prepare a fish dish is more often dictated by the time necessary to prepare a suitable sauce. Thus a supply of frozen raw fish together with a variety of suitable sauces will prove as convenient as a selection of completed fish dishes, and will be more economical of freezer space.

The two instances when freezing the prepared or pre-cooked dish is recommended:

(*a*) Recipes which use cooked fish—'Rechauffé Fish Dishes'. These fish dishes involve a long preparation and may be made in larger quantities as easily as in smaller ones. They may be made when the fish is cheap and of good quality, and thus provide a selection of quickly prepared meals in the freezer.

(*b*) Dishes made from fish which is in season for a short time only, usually the shellfish, and particularly where these involve a long or elaborate preparation.

RECHAUFFÉ FISH DISHES

White fish is generally used; cod, fresh haddock, or hake are all suitable, although some smoked fish may be used where the flavour is preferred.

251

Place the fish on a greased baking tin, pour over a little milk and cover with greased paper. Bake in an oven at 350°F (175°C), Regulo 4, until the fish becomes opaque, about 15 minutes. Any liquor which remains may be used in the sauce. Allowance for wastage of bones and skin should be made, about 4 oz. per lb. of raw fish.

FISH CAKES

1 lb. cooked potatoes	*Yield: 24 cakes of*
1 lb. cooked fish	*3-inch diameter*
2 oz. melted margarine	
2 eggs	
salt and pepper	
2 tablespoons chopped parsley, or 2 oz. chopped shrimps	

1 egg
bread raspings ⎬ optional
mixture of oil and butter for frying

Force the potatoes through a potato ricer or through a nylon sieve to ensure a smooth finish.

Remove the skin and bones from the fish and flake well, using two forks.

Mix the potatoes, fish, melted margarine, eggs, flavourings, and seasonings, and beat well until a smooth mixture is obtained.

Roll the mixture into a long cylinder and cut into twenty-four slices.

Shape the fish cakes with a palette knife until the edges are sharp.

The cakes can be coated with egg and bread raspings to give a crisper finish to the cakes.

To freeze: Place on baking sheets and freeze. When solid, pack in suitable quantities.

To serve: Melt 1 tablespoon oil and 1 oz. butter in a frying pan. Place the frozen cakes in the foaming fat and cook for 5 minutes on each side over a moderate heat. By this time they should be well browned on the outside and fully thawed and hot inside.

FISH PIE

1 lb. cooked potato	*Yield: 6 portions*
1 lb. cooked fish	*1 × 1-pint pie dish or*
2 oz. butter	*2 × ¾-pint pie dishes*

⅛ pint milk
2 tablespoons chopped parsley
salt and pepper
Sauces—½ pint béchamei (page 222)
            or sauce piquante (page 228)
            or sauce Italienne (page 227)

Put the cooked potato through a potato ricer or a nylon sieve.
Beat in the warmed milk, butter, parsley and seasoning.
Bone and skin the fish. Flake coarsely. Mix with the sauce.
Place the fish and sauce mixture at the base of the greased pie dish,
or suitable foil container.
Spread the potato mixture over the fish, and decorate with a fork.

To freeze: Freeze until firm and pack in moisture-vapour-proof film
or aluminium foil. Overwrap if necessary.

To serve: Place the pie dish or foil container into a cold oven. Set
the oven at 400°F (205°C) Regulo 6, and allow to cook for 1½ hours.
By this time the dish is heated through and the topping is browned.
Alternatively the pie may be fully thawed overnight in a refrigerator,
and placed in an oven at 425°F (220°C), Regulo 7, for 20–30 minutes
to heat through and brown the topping.

FISH TURNOVER, USING FLAKY PASTRY

8 oz. flaky pastry (page 371)   *Yield: 2 envelope shapes,*
8 oz. cooked fish   *each 3 portions*
1 oz. butter
4 tomatoes (skinned)
1 teaspoon curry powder
salt and pepper

Roll the pastry slightly. Cut into two.
Roll each piece into a square with 12-inch sides.
Remove skin and bones from the fish and flake coarsely.
Melt the butter and mix together in a bowl with the curry powder and
the fish. Season with the salt and pepper.
Place equal amounts of fish on each piece of pastry and cover with
slices of tomato.
Fold in the corners of the pastry to overlap in the centre and give an
envelope shape. Seal the edges.

To freeze: Wrap in aluminium foil or other suitable freezer wrapping. Seal and freeze.

To serve: Unwrap and place the frozen turnovers on a baking tray and bake at 475°F (245°C), Regulo 9. After 20 minutes, when the pastry should be risen, lower the heat to 400°F (205°C), Regulo 6. Continue cooking for 20–30 minutes.

FISH CROQUETTES

May be frozen after shaping or after frying.

3 oz. flour	*Yield 12–16 croquettes*
3 oz. margarine	*Fat bath 360°F (180°C)*
½ pint milk	*for 5 minutes*
1 blade mace	
1 bay leaf	
1 lb. cooked fish	

1 oz. chopped onion cooked in butter,
    or 8–12 anchovy fillets, pounded or finely chopped
salt and pepper
1 egg
beaten egg and breadcrumbs (to finish)

Pour a little of the milk over the fish when baking it in the oven. Infuse the rest of the milk with the mace and bay leaf. Strain the infused milk and the milk from the fish after baking.

Melt the margarine, add the flour. Continue cooking for a few minutes. Add the milk to the flour mixture and beat until smooth. The mixture should form a firm paste, just leaving the sides of the pan clean.

Cool slightly and beat in the egg a little at a time.

Remove the bones and skin from the fish and flake finely.

Add to the mixture in the pan together with the flavouring of anchovy or onion. Adjust the seasoning.

Spread the mixture on to a plate cover and allow to cool.

Divide into equal portions, shape into balls, using a little flour if necessary, and then into small cylinder shapes.

Remove excess flour. Dip each croquette into beaten egg, then roll in bread raspings.

Freeze at this stage

or fry in deep fat for 5 minutes. Drain on absorbent paper. Cool the croquettes thoroughly if they are to be frozen.

To freeze: Both *cooked* and *uncooked* croquettes keep their shape better if frozen before packaging. Freeze 2–3 hours until hard, pack in layers separated by cellophane. Seal.

To serve: Thaw *uncooked* croquettes for 2 hours at room temperature. Fry in deep fat at 360°F (180°C) for 5 minutes. Drain on absorbent paper and serve.

Place *cooked* croquettes while still frozen on a baking sheet, and re-heat for about 30 minutes in a fairly hot oven. Serve.

KEDGEREE

1 lb. cooked fish	*Yield: 8 portions*

8 oz. Patna rice (uncooked weight)
2 oz. butter
½ teaspoon salt
pepper and pinch of cayenne pepper
2 tablespoons béchamel or 2 tablespoons whipped cream
1 tablespoon chopped parsley

garnish: 2 hardboiled eggs
         lemon wedges

Cook the rice in fast-boiling salted water until tender—about 20 minutes.
Drain and rinse under cold running water. Dry well.
Skin and bone the fish, and flake coarsely.
Melt the butter.
Carefully mix together the fish, rice, melted butter, béchamel sauce, and chopped parsley. Add the seasoning.

To freeze: Pack in suitable amounts in foil containers. Cover with sheet foil.

To serve: Allow to thaw in a refrigerator overnight if to be used for breakfast. Place in a covered ovenware dish, or leave in the original covered foil containers. Put in cold oven, set at 400°F (205°C), Regulo 6, for 20–30 minutes. At the same time, hard boil the egg for the garnish. Serve garnished with egg slices and lemon wedges. Alternatively, place the dish straight from the freezer into a cold oven set at 400°F (205°C), Regulo 6, for 45 minutes. Stir the contents with a fork from time to time to speed up thawing and heating through.

SALMON MOUSSE

1 lb. tinned salmon or tuna fish   *Yield: 2 × 1 pint moulds*
½ pint mayonnaise   *or soufflé dishes*
¾ oz. gelatine   *each serving 4–6 portions*
⅛–¼ pint water
½ pint double cream
     or mixture of equal parts double and single cream
     or ½ pint evaporated milk
salt and pepper
ground mace
¼ pint aspic jelly

garnish: small pieces of tomato pith and cucumber peel

Pour a very thin layer of aspic into each mould and allow it to set.
Pour in a second very thin layer of aspic and arrange the garnish in it.
Alternatively prepare the soufflé dishes by securing a band of aluminium foil around the outside so that it stands up 2 inches above the rim of the dish.
Place the gelatine with the water in a basin standing in a saucepan of hot water and heat gently until dissolved.
Remove any skin and bone from the fish and mash it finely, or preferably blend it in a liquidizer with some of the mayonnaise.
Whisk the evaporated milk or cream mixture until thick (if using double cream alone it should only be half-whipped or it will become too stiff).
Fold in the evaporated milk (or cream) together with the rest of the mayonnaise, seasoning, and mace, to the fish mixture.
Slightly cool the gelatine and carefully stir into the fish mixture.
When it begins to thicken pour it into the prepared moulds or soufflé dishes.

To freeze: Place in the freezer without wrappings until hard. Cover the surface with cellophane or foil. Seal in a polythene bag.

To serve: Thaw out in a domestic refrigerator about 18 hours. Turn out the mousse set in moulds on to a dish and serve with salad.
Alternatively, pour a thin layer of aspic on to the surface of the mousse set in a soufflé dish and garnish with pieces of tomato pith and slices of cucumber. When set remove the band of aluminium foil.

256

9. Preparation of fruit for freezing; slicing peaches directly into container of syrup

10a. Blackcurrant Flan, page 308

10b. Fruit Gâteau, page 314

SOUSED HERRING

8 herrings

liquid to cover (amount depends upon shape of container):
  1 part lemon juice
  1 part white wine
  2 parts vinegar
2 bay leaves
1 blade mace
2 cloves
½ teaspoon powdered allspice or 12 allspice
1 teaspoon salt

garnish: green salad

Yield: 8 portions

Oven: 275°F (135°C), Regulo 1, for 3 hours

Remove the gut, head and backbone of the herrings. Wash and dry the fish.

Roll from head end to tail in neat rolls. Place in a fireproof dish.

Mix the liquids, herbs, and seasoning, and pour over the rolled herring to cover.

Cover the dish and cook.

Cool in the liquid.

To freeze: Drain from the liquid and pack in suitable quantities.

To serve: Allow to thaw at room temperature for 3 hours, or in a refrigerator for 8 hours.

Serve with green salad.

SHELLFISH

Opinion varies as to the quality of these fish when they are frozen raw; however excellent results have been obtained from freezing pre-cooked shellfish dishes.

SCALLOPS (*COQUILLES ST. JACQUES*)

Most methods of cooking scallops, such as poaching in wine, cooking gently in butter, and frying, are successful. One example of each method has been given.

COQUILLES ST. JACQUES À LA PARISIENNE

This is a classical way of preparing these fish.

8 scallops

257

4 oz. mushrooms
½ pint white wine
2 oz. onion
2 oz. shallots
bouquet garni including   2 sprigs parsley
                        ¼ bay leaf
                        ¼ teaspoon thyme
4 oz. butter
2 oz. flour
2 oz. grated cheese (Cheddar or Gruyère)
juice 1 lemon
salt and pepper
little milk

Yield: 8 portions as fish course
4 portions as main course

Clean the scallops and wash well.

Place the scallops in a pan with the wine, sliced onion, chopped shallot and a bouquet garni.

Bring to simmering-point and simmer for 5 minutes.

Drain the scallops from the liquid and keep hot.

Melt 2 oz. butter in a pan, add the lemon juice.

Peel or wash the mushrooms, slice thinly, and cook in the butter until tender. Remove from the pan.

Melt another 2 oz. butter in the pan, add the flour, and stir over heat until cooked, about 2 minutes.

Combine the liquid from the scallops and mushrooms and add sufficient milk to make up to ¾ pint.

Gradually add the liquid to the roux (butter and flour mixture) in the pan and beat well to give a smooth sauce. Place over heat and simmer for 5 minutes until the sauce thickens. Season well.

Cut the scallops into quarters, add to the mushrooms and half the sauce.

Divide between 8 scallop shells or dishes.

Reheat the remaining sauce to boiling-point. Remove from the heat and beat in the cheese. Adjust the seasoning.

Coat the scallop mixture with the cheese sauce.

To freeze: Cool rapidly. Place the shells on to trays and freeze. When frozen pack in moisture-vapour-proof film.

To serve: Remove shells from the freezer wrappings. Place in a cold oven set at 400°F (205°C), Regulo 6, and cook for 25–30 minutes. Sprinkle with browned breadcrumbs, add a few small knobs of butter, and place under the grill until lightly browned.

Coquilles St. Jacques à la Newburg

8 scallops
4 oz. butter
2 teaspoons lemon juice
2 teaspoons flour
½ pint single cream
4 egg yolks
4 tablespoons sherry
salt
cayenne pepper

Yield: 8 portions as fish course
4 portions as main course

Clean, wash and dry the scallops.

Melt 2 oz. butter, and when foaming add the scallops. Cook for 3 minutes.

Add the lemon juice and continue cooking for 1 minute; lift out the scallops and retain the liquor which remains.

Melt the remaining 2 oz. butter in a pan, add the flour, and mix in well.

Cook for 2 minutes.

Blend the yolks with the cream and add to the pan.

Add the sherry and liquor remaining from cooking the scallops. Cook over gentle heat until the sauce is thickened.

Add the scallops cut into quarters and season with the salt and cayenne.

To freeze: Cool rapidly and divide the mixture between the shells.

To serve: Thaw out in a refrigerator 6–8 hours. Stand the shells in an ovenware dish which can be loosely covered with foil. If the shells are deep a little water may be poured into the casserole to surround the shells. Reheat in an oven at 325°F (165°C), Regulo 3, for 30–40 minutes.

Alternative method of freezing and thawing: Transfer the completed scallops and sauce to a suitable container. Seal and freeze.

To serve: Partially thaw in a refrigerator and transfer the scallops and sauce to a basin. Reheat over a saucepan of gently boiling water. Return to the shells and serve.

Scallops Fried

8 scallops
1 large egg

Yield: 4 portions

dried breadcrumbs
salt and pepper
pinch cayenne
2 oz. butter

Clean, wash and dry the scallops; season well. Dip in the egg and coat thoroughly. Coat with breadcrumbs.
Melt the butter in a frying pan and heat until foaming.
Fry the scallops 1 minute on each side.

To freeze: Drain the scallops on kitchen paper. Cool rapidly. Pack, separating the layers with cellophane.

To serve: Unwrap and place on a baking sheet in an oven at 400°F (205°C), Regulo 6, for 20 minutes. Serve with sauce béarnaise.

CRAB

DRESSED CRAB

For each medium-sized cooked crab allow:
2 oz. fresh breadcrumbs
salt, pepper
1 teaspoon French mustard
chopped parsley ⎫
sieved egg yolk ⎬ add when serving
thin wedges of lemon ⎭

Clean and prepare the crab according to the directions on page 175.
Then with the handle of a teaspoon scrape out the brown creamy meat from the big shell and place in a basin.
Add any brown meat from the underside of the body.
Gently tap the inside edge of the shell, break it along the natural line to make a wider opening. Wash well, dry and rub with a little oil.
Mix the brown meat with the breadcrumbs and seasonings.
Place neatly in the centre of the shell.
Cut the body of the crab in half, and remove all the white meat from the leg sockets.
Crack the claws and legs and remove the white meat. Care should be taken that no shell is included in the white meat.
Flake the white meat finely.
Place the white meat in the shell on each side of the brown meat.

To freeze: Pack in a suitable wrapping for the freezer.

To serve: Thaw 6–8 hours in a domestic refrigerator. Decorate with sieved egg yolk and chopped parsley along lines dividing white and brown meat. Serve with wedges of lemon and a sharp mayonnaise, tartare sauce, or vinaigrette, and brown bread and butter.

LOBSTER

This fish also freezes successfully in different ways. In most towns lobsters are available only in a cooked state, but if treated carefully they give a good result in recipes calling for a fresh lobster.

(For killing and preparing a live lobster see notes on page 175.)

LOBSTER THERMIDOR

4 small cooked lobsters   *Yield: 8 portions*
6 oz. butter
2 tablespoons oil
salt and pepper
4 shallots
½ pint white wine
2 teaspoons dried chervil or 2 tablespoons chopped fresh chervil
1 teaspoon chopped dried tarragon
1 pint béchamel sauce
1 tablespoon French mustard
4 oz. grated Parmesan ⎫
browned breadcrumbs ⎬ add when serving
               ⎭

Cut the lobsters in halves. Clean, remove coral.
Remove the meat from the shell and claws. Cut into dice. Wash and dry the shells.
Melt 3 oz. butter with the oil until foaming. Add the lobster meat and heat through. Season with salt and pepper.
Strain the lobster meat from the fat and keep hot.
Chop the shallots finely and add to the pan. Cook until tender.
Add the wine and boil to reduce the liquid to ¼ pint.
Add the reduced wine to the béchamel sauce, together with the herbs, mustard and seasoning. Bring to boiling-point and remove from the heat. Beat in the remaining 3 oz. butter gradually.
Put a layer of sauce inside each half shell, divide the lobster meat equally between the shells and coat with the remaining sauce.

To freeze: Cool rapidly. Freeze and then pack each half-lobster individually. Pack together in quantities as required.

To serve: Remove freezer wrappings and place the filled half-shells on a baking sheet. Place in a cold oven set at 400°F (205°C), Regulo 6, and allow to heat through for 35 minutes.

Sprinkle with Parmesan cheese and browned breadcrumbs if liked, and place under a hot grill to brown.

Lobster à l'Armoricaine

This classic dish demands live lobster, but is very good with pre-cooked fish, even if this is not strictly correct.

4 cooked lobsters *Yield: 8 portions*
4 tablespoons oil
2 oz. chopped onion
2 oz. chopped shallot
$\frac{1}{8}$ pint brandy
4 tomatoes
1 teaspoon tomato paste
$\frac{1}{4}$ pint fish stock
$\frac{1}{4}$ pint dry white wine
2 tablespoons chopped parsley
1 teaspoon dried tarragon or 1 tablespoon chopped fresh tarragon
4 oz. butter

Place the lobster with its back uppermost. Remove the claws.
Cut the lobster into half, dividing the head from the tail.
Cut the tail into segments across the joints, leaving it on the shell.
Crack the claws and remove the meat from them.
Cut the head in half lengthways, remove the stomach sac and intestine and discard.
Remove the coral and place in a basin or mortar.
Heat the oil in a pan, add the lobster pieces, meat side down, and the meat from the claws, and toss in the oil for several minutes until the pieces are pinkish brown in colour. Remove from the pan.
Place the chopped onion in the pan and cook gently until tender. Add the shallot and mix well together.
Return the lobster to the pan, arranging it in a layer over the onion, and return to a gentle heat.

Pour over the brandy and ignite. Shake the pan gently until the flame subsides.

Skin the tomatoes, remove the seeds, and chop coarsely.

Add to the pan with the tomato paste, stock, wine and herbs.

Bring almost to the boil and allow to simmer for 20 minutes.

Remove the lobster pieces from the pan and take the flesh from the shells if desired—this dish may be served with the meat still on the shells. Continue heating the liquids in the pan over gentle heat until they are reduced and thickened.

To freeze: Return the lobster to the thickened liquids and cool together. Pack in suitable containers. Retain the head shell for serving if required.

Pound the coral with the butter, then sieve. Pack and freeze separately.

To serve: Thaw the coral butter until it is well softened. Gently thaw and reheat the lobster mixture in a double boiler. Remove from the heat and gradually work in the coral butter. Return to a gentle heat sufficient to warm up the sauce again, but prevent it from simmering, otherwise the sauce may curdle. Serve in the shell, or in a ring of steamed rice. Decorate with fresh herbs.

LOBSTER NEWBURG

8 oz. cooked lobster meat (from a good 1½ lb. lobster)
2 oz. butter
1 dessertspoon flour
¼ pint sherry
4 egg yolks
½ pint single cream
salt and pepper
little grated nutmeg
4 pieces hot buttered toast—add when serving

Yield: 4 portions as fish course

Cut the meat from the tail and claws into 1-inch cubes.

Melt the butter, add the lobster, and heat gently for about 5 minutes. Lift out lobster and mix in the flour.

Remove from the heat, stir in the sherry, nutmeg, and seasoning, then heat for a further 2–3 minutes. Replace the lobster.

Combine the yolks with the cream, and add to the pan.

Stir until the mixture thickens, but do not boil.

263

To freeze: Cool and pour into a suitable container. Seal and freeze.

To serve: Remove to a domestic refrigerator for 2–3 hours until partially thawed. Reheat in a double boiler, stirring occasionally and taking care to prevent the sauce from boiling. Adjust seasoning and serve immediately on hot buttered toast.

PAELLA

The following is an excellent party dish which may be prepared and frozen.

There is scope to vary the ingredients according to taste and availability.

½ lb. conger eel or skate   *Yield: 12 portions*
1 stick celery
1 shallot
salt and pepper
bouquet garni
2½ pints water
½ raw chicken
¼ lb. streaky bacon
½ lb. Dublin Bay prawns
½ lb. shrimps
oil
1 onion
4 tomatoes
¼ teaspoon saffron
12 oz. Patna rice (uncooked weight)
peas   ⎫
diced carrots ⎬ optional
               ⎭
2 oz. blanched almonds
2 oz. dried apricots (soaked overnight)
½ pint cooked mussels or cockles

Coarsely chop the celery and shallot, wash and trim the eel (or skate) and place altogether in a large saucepan with the water, seasoning, and bouquet garni.
Simmer for 30 minutes.
Lift out the eel and strain the liquor into a jug.
Heat a little of the oil in a pan, divide the chicken into suitable portions and cook for 10–15 minutes. Cool slightly and cut into bite-size pieces.

Gently fry the picked prawns and shrimps for 3–5 minutes and remove.

Finally, cut the bacon into ½-inch pieces, fry for 2 minutes. Lift out.

Skin and remove the seeds from the tomato and chop coarsely.

Chop the onion and cook in the oil until lightly browned.

Add the rice, tomato, and any vegetables used, toss in the hot oil for a few minutes.

Make up the liquor retained from cooking the eel to 2¼ pints.

Dissolve the saffron in a little of the hot liquor. Add with remaining liquor to the pan containing the rice and other vegetables.

Cook fairly rapidly for about 12 minutes.

Add the eel or skate, chicken, bacon, prawns, shrimps, cooked mussels, apricots coarsely chopped, and the almonds.

Mix in gently and continue to cook for about 5 minutes.

The paella may be served fairly dry or rather moist, according to taste.

If part is to be frozen, a moist finish is more suitable.

To freeze: Freeze the paella after all the ingredients have been combined with the rice, and while the mixture is still very moist. Pack in suitable quantities and freeze.

To serve: Transfer to a greased ovenware dish, cover with greaseproof paper. Place in a cold oven set at 375°F (190°C), Regulo 5, for at least 1 hour. Stir to hasten thawing and reheating.

CRÊPES À LA MARINIÈRE

The following dish makes an excellent supper dish or a party dish.

It may be assembled entirely from pre-frozen ingredients; it can also be frozen after completion. The shellfish and other ingredients can be varied according to taste and availability. Single pancakes may also be stuffed with the mixture and served individually.

7 pancakes of 8-inch diameter (page 401)   *Yield: 8 portions*
8 oz. creamed spinach (page 248)
½ lb. kipper fillet—cut very thinly
   or finely mashed smoked haddock
½ pint cheese sauce (page 224)
½ lb. peeled shrimps
lemon juice—add when serving

Thaw the creamed spinach mixture, and if necessary the cheese sauce.

Gently fry the shrimps for a few minutes in hot butter. Drain and add them to the cheese sauce.

Place the first pancake on an ovenware plate and spread on the first filling of shrimps in cheese sauce.

Cover with a second pancake, spread on a layer of creamed spinach. Add a layer of thinly sliced kipper fillet or smoked haddock (baked in a little milk and finely mashed).

Cover with a third pancake.

Continue layers of pancakes and fillings.

Freeze when completed sealed in aluminium foil or other suitable wrapping.

or pour a little lemon juice over the surface, cover lightly with aluminium foil, and place in a hot oven for about 15–20 minutes and serve.

To serve frozen crêpes: Loosely wrap in foil, thaw and reheat in a fairly hot oven for 30–40 minutes.

USING FROZEN RAW FISH

As stated at the beginning of this section, a variety of fish dishes can be quickly prepared from frozen fillets or cutlets. Most proprietary packs recommend cooking from the frozen state and this is satisfactory for most purposes, but partial or complete thawing is necessary before fish can be skinned and rolled, coated, or stuffed. A selection of the more homely or classical dishes can be prepared quickly and simply, using a suitable sauce from the freezer and adding an appropriate garnish. Thaw the sauce to be served over a gentle heat at the same time as the fish is being cooked.

PREPARATION OF FILLETS OF PLAICE AND SOLE

For steaming and baking

Thaw first in order to remove the skin if still present, and roll up the fillet with the skin side inside.

Grease a plate to fit the top of a saucepan for steamed fish and an ovenware dish for baked fish, and lay in the rolled fillets. (If using still-frozen fillets, lay these on the plate or in the dish so that they barely overlap.)

Add a little liquid to the fish to be steamed and cover with another plate.

Steam 10–15 minutes.

Place a knob of butter on the fish to be baked and cover with greaseproof paper. Place in the oven at 375°F (190°C), Regulo 5, for 15–20 minutes.

Drain any curdy liquid which results from the steamed or baked fish into the sauce being used.

For frying

Thaw fillets sufficiently to enable a coating of flour, egg and breadcrumbs, or batter to adhere to the surface.

Fry in shallow fat, using a mixture of butter and oil heated to foaming point, or in a deep-fat bath at 360°F (180°C).

The following sauces may be served as a coating sauce for steamed or baked fillets:

Aurore (page 223)   Parsley (page 225)
Cheese (page 224)   Soubise (page 225)
Duxelle (page 224)   Suprême (page 225)

Suitable pouring sauces, served as an accompaniment to fried fillets are:

Hollandaise (page 231)   Shrimp (page 225)
Cardinal (page 224)   Suprême (page 225)
Parsley (page 225)

PREPARATION OF CUTLETS AND STEAKS

Cutlets and thick steaks are more usually baked, poached, or grilled. It is usually simpler in all cases, except where stuffing is to be used, to remove the skin and bone after cooking. For baking prepare the cutlets and steaks in the same way as for fillets.

For poaching choose a saucepan with a wide base and use sufficient fish stock to half cover the cutlets. Half the quantity of stock may be substituted by white wine.

For grilling, place the fish on a well-oiled grid and cook evenly on both sides.

The cooking time should be extended slightly for each method used if the fish is to be cooked directly from the frozen state.

The following sauces may be served as an accompaniment to cutlets and steaks of halibut, turbot, hake, and cod:

Aurore (page 223)   Hollandaise (page 231)
Caper (page 224)   Béarnaise (page 233)
Soubise (page 225)   Choron (page 233)
Piquante (page 228)

Meat Dishes

ALTHOUGH most varieties of meat prepared and cooked by conventional methods may be satisfactorily frozen, it seems a wise policy to select and freeze those cuts, and use those methods, for which freezing has been found especially successful, or of particular advantage. However, in order to indicate any special techniques or steps which should be taken, examples have also been included of all the usual methods by which meat dishes are prepared.

ROAST MEATS

Roast joints of beef, veal, lamb, and pork may all be frozen after cooking for serving hot or cold at a later date. The former use is not particularly recommended, as thawing and reheating of only a small joint takes at least an hour in a moderate oven, and even when kept wrapped in foil the texture is likely to be rather dry.

Likewise meats served cold must be allowed considerable time to thaw before they can be carved. To avoid delay it is possible to slice the meat before freezing, and to pack the slices between double thicknesses of cellophane, although the result is not likely to be as good as with meat freshly sliced. The use of frozen joints can be a useful form of advance preparation for a buffet supper or summer picnic, when sufficient time for thawing out before carving can be estimated and allowed.

The meat should be left in the freezer wrappings while thawing.

Approximate thawing times for whole roast joints of meat

In a domestic refrigerator	at least 8 hours per lb.
At room temperature	about 4 hours per lb.

Because of the difficulty of judging the exact amount of time to allow for a joint of meat to thaw, it is good practice to begin thawing overnight in a domestic refrigerator. The following day the meat may be left to continue thawing in the refrigerator if the rate of thawing is proving satisfactory, or where necessary the rate can be accelerated by bringing out the meat to room temperature.

In an emergency, although not recommended for general practice, the meat may be thawed very rapidly when stood before a fan.

Left-over meat

Freezing is recommended as a means of dealing with meat left over from the week-end joint which proves too large for the family's appetite.

(*a*) Trim off any fat and remove any bone present.

(*b*) Slice the meat, and pack between double thicknesses of cellophane if required to be served cold, or place in aluminium foil containers and cover with gravy or sauce if it is to be served hot.

Frozen left-over meat may be used for the preparation of re-chauffé meat dishes.

GRILLED AND FRIED MEATS

Steaks, chops, and escalopes are usually only cooked prior to freezing when a time-consuming or more elaborate preparation other than simple grilling or shallow frying is intended. Unless given the protection of a special coating or a sauce, these cuts of meat become rather dry during frozen storage. However, if they are to be included with other ingredients in a sauce, or coated with egg and breadcrumbs, or a batter, and fried in deep fat, cooking prior to freezing offers many advantages.

For the hostess, frozen fried foods, provided they are properly coated, are a great asset, as they normally require only to be thawed and reheated in a hot oven to give excellent results. The need for the hostess to disappear to the kitchen to complete last-minute frying preparations is no longer necessary.

VEAL ESCALOPES

The price of veal, particularly the best cuts, varies during the year, thus such cuts are well worth preparing and freezing.

8 pieces of veal fillet, each 3–4 oz. in weight.   *Yield: 8 portions*
2 eggs
4 oz. fresh breadcrumbs
4 oz. butter
salt and pepper

Trim and flatten the fillets between pieces of greaseproof paper.
Beat the eggs.
Coat the fillets in the egg.
Season the breadcrumbs and toss the escalopes in them until completely coated. Shake away the excess crumbs.
Heat the butter in a pan (use only half the butter if the pan will hold only 4 escalopes) until it is at foaming-point. Cook the escalopes over moderate heat for 1 minute on each side. Drain well on absorbent paper and chill rapidly.

To freeze: Separate the escalopes with a double thickness of cellophane. Pack in suitable quantities. Seal and freeze.

To serve: Arrange the frozen escalopes on a baking tray. Reheat for 25 minutes in a fairly hot oven and serve.

Veal Escalopes in Cream Sauce with Mushrooms

8 escalopes, each weighing 3–4 oz.   *Yield: 8 portions*
salt and pepper
3 medium-sized onions
2–3 oz. butter
1 tablespoon flour
$\frac{1}{4}$ pint chicken stock
$\frac{1}{2}$ pint single cream
2 egg yolks
$\frac{1}{2}$ lb. button mushrooms

Trim the escalopes and season with salt and pepper.
Finely chop the onions and cook in half of the butter until soft.
Blend in the flour gradually, then add the stock and cream.
Season and bring just to boiling-point, simmer for 20 minutes.
Blend the onion sauce in a liquidizer or pass through a nylon or hair sieve.
Slice and cook the mushrooms in the remainder of the butter, and lift out.

270

Fry the escalopes for about 1 minute on each side, and place in a casserole or casserole mould (page 99) with the mushrooms in a layer over each escalope.

Bring the onion sauce just to boiling-point, remove from the heat and beat in the egg yolks. Pour the sauce over the veal and mushrooms.

To freeze: Cool the veal, mushrooms and sauce in the casserole or casserole mould, cover with aluminium foil or other suitable freezer wrapping, seal and freeze.

To serve: Partially thaw the veal and sauce in a refrigerator first if time permits. Then stand the casserole in a tin of cold water and reheat at 350°F (175°C), Regulo 4, for ¾–1 hour. The egg-based sauce must not be allowed to boil.

Cutlets and Noisettes of Lamb

Allow 2 cutlets or noisettes per serving, 3 pieces of best end of neck, each with 5–6 bones, are required for 8 portions. Ask the butcher to chine the piece of best end of neck, but to leave it in the piece.

Preparation of Cutlets

Have the rib bones chopped about 3½ inches from the thick portion of meat (noix).

Remove the chine bone (split backbone) and the flap of meat. Cut the cutlets ¾ inch thick and trim, leaving only one rib bone in each cutlet. Bare the last 2 inch of bone and trim excess fat from around the meat.

Preparation of Noisettes

Remove the chine bone, and the rib bones.

Trim away excess fat. Season the meat with salt and pepper and herbs if liked. Form into a roll with the thick end of meat surrounded by the flap, and tie around the roll at intervals of 1½ inches.

Slice into portions of equal thickness, cutting between the ties so that each noisette is held in shape by string during cooking.

Finish of Cutlets and Noisettes

Per lb. of cutlets or noisettes:
2 beaten eggs
8 oz. fresh breadcrumbs
salt and pepper

4 oz. butter
4 tablespoons oil

Season the cutlets.

Brush the cutlets or noisettes with the beaten egg until thoroughly coated.

Place in the seasoned breadcrumbs and toss until the egg is completely covered. Shake off any surplus crumbs.

Heat half the butter and oil in the pan until foaming. Fry the noisettes or cutlets in the fat over a moderate heat for 2 minutes on each side. They should be browned and set on the outside.

Add more oil and butter when necessary.

Drain well on absorbent paper.

To freeze: Cool rapidly. Pack using a double thickness of cellophane to separate the pieces. Seal and freeze.

To serve: Place the frozen noisettes or cutlets on to a baking sheet. Cook for 30 minutes at 400°F (205°C), Regulo 6.

MEAT STEWS AND CASSEROLES

Depending upon the selection of meat and other ingredients used in the preparation of a stew or casserole, so a simple or sophisticated dish may be produced. Most meat stews and casserole dishes freeze well and can be reheated on top heat or in the oven within an hour with little attention necessary other than occasional stirring and an adjustment to seasoning when serving. The presence of such a dish in the freezer is a valuable asset in the event of unexpected guests.

VEAL STEW

2 lb. pie veal
½ lb. button onions (approx. 16)
½ lb. button mushrooms (approx. 16)
1 pint white stock
salt and pepper
bouquet garni
⅛ pint white wine
3 oz. margarine
2½ oz. flour

Yield: 8 portions
Stew: 1½ hours

272

11a. Lemon Savarin, page 398

11b. Pineapple Sorbet, page 330

12a. Cornish Pasties, page 281
12b. Meat Balls in Tomato Sauce, page 278
12c. Lamb Noisettes with Spaghetti, page 271

½ pint milk
2 yolks of egg
3 tablespoons cream   } add on serving
1 tablespoon lemon juice

garnish: 16 bacon rolls; 8 triangular croûtes; parsley sprigs

Wipe, trim and cut meat into cubes.

Peel the onions, and slice if large.

Wash and trim the mushrooms.

Place the veal, onions, seasoning and the bouquet garni into the boiling stock. Simmer for 1 hour.

Add the mushrooms and wine and continue simmering for 30 minutes, when the veal should be tender. Strain the liquid from the meat and vegetables.

Melt the margarine in a saucepan, stir in the flour, and cook together for 3 minutes.

Gradually stir in the veal stock and then the milk, and bring the sauce to the boil, stirring continuously.

Add to the veal and vegetables.

To freeze: Cool rapidly, and pack in suitable amounts. Seal and freeze.

To serve: Partially thaw in a refrigerator if time permits before returning to the saucepan; or place the frozen mixture in a covered pan over gentle heat to thaw. Stir occasionally to prevent the mixture sticking. Bring to boiling-point. Mix the yolks and cream together and pour on about ¼ pint hot sauce from the veal stew and mix well together. Return the sauce plus yolks and cream to the pan, and stir into the stew, and reheat gently without boiling. Add the lemon juice and adjust seasoning. Serve on a hot entrée dish, arranging some mushrooms and onions at each end of the dish. Garnish with bacon rolls, croûtes, and parsley sprigs.

VEAL FRICASSÉE

This is a very popular dish and freezes well.

Use the recipe given for veal stew, using all stock in place of stock and milk, so bringing the total quantity of stock used to 1½ pints, and use the following method:

Melt the margarine or butter, and fry the prepared veal cubes for 5 minutes.

Add the onion and continue frying for a further 5 minutes without browning. Remove the veal and onion from the pan.

Stir the flour into the fat and cook over gentle heat until the flour becomes pale fawn in colour.

Stir in the stock.

Return the veal and onions to the sauce with the bouquet garni, seasoning, and white wine, and simmer for 1 hour. Add the mushrooms and continue simmering for 30 minutes.

Follow the directions for freezing, thawing, and finishing with egg yolks and cream, given in the recipe for veal stew.

VEAL SAUTÉ MARENGO

2 lb. pie veal   *Yield: 8 portions*
6 oz. onion   *Stew: 1 hour*
3 tablespoons oil
1½ teaspoons flour
½ pint white wine
1 tin tomatoes (net weight 14 oz.)
   or 1 lb. tomatoes, skinned and seeded
salt and pepper
¼ teaspoon dried thyme
¼ teaspoon dried tarragon
4 oz. mushrooms
1 tablespoon oil for cooking mushrooms
garnish: heart-shaped croûtes fried in butter or oil; chopped parsley;
        cooked button onions.

Wipe, trim, and cut the meat into large cubes about 2 inches square. Peel and finely chop the onions.

Heat the oil; put in the veal and fry until golden-brown, together with the onions.

Remove the meat and onions.

Mix the flour with the oil and fry over gentle heat until a deep golden-brown colour is developed. Add the wine and mix to a smooth sauce.

Return the meat and onions to the pan, and add the tomatoes, salt and pepper, and herbs. Simmer for 1 hour.

Toss the mushrooms in the tablespoon of oil over gentle heat (leave button mushrooms whole, but slice or quarter large mushrooms).

To freeze: Rapidly cool the veal in the sauce and pack in suitable amounts. Cool the mushrooms and pack in amounts corresponding to the veal packs.

To serve: Partially thaw the veal in a refrigerator if time permits before returning to the saucepan, or thaw in a covered pan over gentle heat. When the stew has completely thawed, add the mushrooms and thoroughly reheat. Arrange in a hot entrée dish with the button mushrooms neatly placed at each end. Garnish with the croûtes and button onions, and sprinkle with the parsley.

CURRIED VEAL

The recipe is given for veal, although chicken, rabbit, or beef would be equally suitable.

3 lb. pie veal	*Yield: 8 generous portions*
2–3 oz. butter	*Stew: 1½ hours*

6 oz. onion
1 medium dessert apple
3 teaspoons curry powder
3 teaspoons curry paste
1½ oz. flour
1½ pints stock
3 tablespoons coconut if liked
3 oz. sultanas
salt
paprika
lemon juice—add when serving

garnish: lemon slices and chopped parsley

Remove any skin from the meat and cut into 1-inch cubes.
Fry lightly in the butter, and lift out from the pan.
Finely chop the onion and apple, and fry together until lightly browned.
Sprinkle in the curry powder and flour, and add the curry paste.
Gradually stir in the stock, add the coconut, sultanas, salt and paprika.
Bring to boiling-point, return the meat to the pan, and continue to simmer for about 1½ hours.

To freeze: Cool rapidly, remove any fat and pack in suitable containers. Seal and freeze.

To serve: Transfer to a pan over gentle heat and gradually bring back to boiling-point. Stir to prevent mixture sticking. Adjust seasoning and consistency.

Add a little lemon juice before serving. Garnish with lemon slices and chopped parsley.

Serve with boiled rice and mango chutney.

RAGOÛT OF BEEF

3 lb. braising steak
oil for frying
18 button onions
8 oz. button mushrooms
3 oz. streaky bacon
2 teaspoons flour
½ pint burgundy
½ pint stock
bouquet garni, including   bay leaf
                             thyme
                             celery stalk
seasoning
garnish: cooked diced carrots and peas

Yield: 8-10 portions
Oven: 325°F (165°C) Regulo 3
Time: about 2 hours

Cut meat into 2-inch squares.

Heat a little oil in a heavy pan, and quickly brown the meat; lift out into a casserole.

Cut the mushrooms into quarters, and the bacon into 1-inch pieces. Gently toss mushrooms, bacon and onions in the pan, and allow to brown.

Lift out and keep separate from the meat.

Mix the flour into the fat and fry slowly until a good brown colour develops.

Gradually stir in the stock, then add the wine, bouquet garni, and seasoning, heat to boiling-point and pour over the meat contained in the casserole.

Cook gently in the oven for about 1 hour.

Add the mushrooms, bacon, and onion, and continue to cook for a further 1 hour.

Remove the bouquet garni, cool, skim off any surplus fat.

To freeze: Place in a clean casserole or suitable freezer container. Seal and freeze.

To serve: Place in a cold oven set at 375°F (190°C), Regulo 5, for 1½ hours or until completely heated through.

Adjust seasoning and consistency.

Garnish with cooked diced carrots and peas.

FLEMISH BEEF

3 lb. chuck steak

$\frac{1}{8}$ pint vinegar

$\frac{1}{4}$ pint oil

1 oz. finely chopped onion

1 teaspoon chopped parsley

thyme

salt and pepper

} for marinade

2 oz. butter

6 medium-sized onions

1 oz. flour

$\frac{3}{4}$ pint beer

$\frac{1}{4}$ pint stock

Yield: 8-10 portions

Oven: 350°F (175°C), Regulo 4

Time: 1$\frac{1}{2}$–2 hours

Trim off any fat and skin from the meat and leave in large pieces of about 2 × 3 inches; put into the marinade for at least an hour—it is better if it can be left overnight.

Thinly slice the onions and fry in butter in a heavy pan until lightly coloured; lift out.

Briskly fry the meat and lift out.

Blend the flour and seasoning into the fat, then gradually add the stock and beer; bring to the boil.

Place the meat and onions in layers in a casserole, and pour over the thickened stock and beer; cook in a moderate oven for 1$\frac{1}{2}$ hours.

To freeze: Cool rapidly, remove surplus fat, transfer carefully to a clean casserole, or pack in suitable container. Seal and freeze.

To serve: Place in a cold oven set at 375°F (190°C), Regulo 5, for 1$\frac{1}{2}$ hours. Adjust seasoning and consistency.

BEEF GOULASH

3 lb. chuck steak

2 oz. butter

1$\frac{1}{2}$ lb. onions

salt, pepper

paprika

3 oz. tin tomato paste

$\frac{3}{4}$–1 pint stock

8 oz. sour cream

2 teaspoons flour

} add when serving

Yield: 8-10 portions

Stew: about 1$\frac{1}{2}$ hours

cooked rice or noodles (about 12 oz. uncooked weight)—add when serving.

Wipe the meat and cut into 2-inch squares.
Dice the onions and toss in butter over gentle heat until they are soft.
Remove the onions and brown the meat over a high heat.
Stir in the salt, pepper, paprika, tomato paste, and the onions.
Add the stock, and cook over a low heat for about 1½ hours.

To freeze: Cool rapidly, skim off surplus fat, pack in suitable containers.
Seal and freeze.

To serve: Thaw over gentle heat and bring up to boiling-point, stir in the sour cream mixed with the flour. Reboil and continue to cook gently for a further 10 minutes.
Serve in a dish bordered with rice or noodles.

Meat Balls

1 lb. minced chuck steak
   or any lean minced meat
½ lb. sausage meat
½ lb. minced pie veal

or 2 lb. minced steak

Yield: 36 balls
8 portions

4 oz. bread
2 oz. onion
4 tablespoons chopped parsley
2 teaspoons salt
½ teaspoon pepper
4 eggs
4 tablespoons oil
1 pint tomato sauce (page 229)

Crumb the bread; or soak in water for 5 minutes, squeeze dry and beat well.
Mix the bread with the meat or meats.
Finely chop the onion and add to the meat together with the parsley and seasonings.
Beat the eggs and thoroughly mix in with the other ingredients.
Divide into 36 pieces and shape into balls on a flavoured board.
Heat the oil in a pan and fry the meat balls, allowing sufficient space for them to be shaken and retain their round shape until set. When set and brown, remove from the pan.

Place the meat balls in the heated tomato sauce and simmer on top heat or in the oven for 30 minutes.

To freeze: Cool rapidly, pack in suitable amounts. Seal and freeze.

To serve: Place unthawed meat balls in sauce over a gentle heat until thawed, simmer 20 minutes, or place in cold oven set at 400°F (205°C), Regulo 6, and allow to heat through for 1½ hours.

SPICED PORK

8 pork chops, loin or spare rib,	*Yield: 8 portions*
or 2 lb. lean belly rashers	*Oven: 400°F (205°C), Regulo 6*
1 oz. butter	*Time: 30 minutes*

2 oz. onion
2 oz. celery
2 tablespoons brown sugar
1 tablespoon mustard
2 teaspoons salt
1 teaspoon paprika
1 tablespoon tomato purée
2 tablespoons Worcestershire sauce
2 tablespoons vinegar
⅛ pint lemon juice
¼ pint water

Cook the chops (or rashers) over moderate heat until they are well browned and the excess fat has melted out. Place into a casserole. Chop the onions and the celery and fry in the butter until tender. Add the dry ingredients, then the liquids and seasoning. Pour over the pork chops and cook in a covered casserole for 30 minutes.

To freeze: Cool rapidly, remove any surface fat. Pack in suitable amounts. Seal and freeze.

To serve: Thaw in the refrigerator overnight and reheat in a fairly hot oven for 20 minutes, or place directly from the freezer into a cold oven set at 400°F (205°C), Regulo 6, for about 1 hour.

PORK IN ORANGE SAUCE

2 pieces loin of pork, each weighing	*Yield: 8 portions*
about 1½ lb. after removing bones	*Oven: 425°F (220°C),*

salt and pepper

2 onions

½ pint orange juice (frozen or fresh)

2 tablespoons vinegar

2 tablespoons redcurrant jelly

garnish: orange segments

Regulo 7, for 30
minutes, reduced to
350°F (175°C),
Regulo 4, for a further
45 minutes

Wipe the meat, remove the bones and skin if necessary. Form into a roll and secure at intervals with string.

Rub the meat with salt and pepper, and place in a fairly hot oven.

Thinly slice the onions, and add to the pan containing the meat after it has been in the oven about 15 minutes, so that they are lightly browned in the fat from the pork.

After a further 20 minutes reduce the oven temperature and pour off any excess fat.

Gradually stir in the orange juice and vinegar mixed together, and continue to cook for at least another 45 minutes.

Stir in the redcurrant jelly.

To freeze: Cool the meat and sauce rapidly, keeping it covered. Place the two pieces of loin, together with the orange sauce, in an aluminium foil container, or other suitable freezer container. (When smaller servings are required, the two loins may be packed separately and the sauce divided equally between them.) Seal and freeze.

To serve: Return to a cold oven set at 350°F (175°C), Regulo 4, for about 1 hour. Serve the meat on a dish with the sauce poured over and around, and garnished with segments of fresh oranges.

BRAISED KIDNEYS IN RED WINE

16 lamb's kidneys—allow 2 per person

8 oz. mushrooms

2 oz. flour

2 oz. butter

½ pint red wine, or wine and stock

seasoning

cooked rice ⎫
chopped parsley ⎭ add when serving

Yield: 8 portions
Stew gently for 30 minutes

Slice the mushrooms and cook gently in a little butter; lift out from the pan.

Halve the kidneys, remove the skin and tubules. Wash and toss in seasoned flour, retaining any flour left-over.

Gently cook the kidneys in butter for a few minutes—do not let them become hard.

Add the wine, or wine and stock (it should be sufficient to just cover the kidneys) and the mushrooms.

Simmer for about 30 minutes.

Blend the remaining flour with a little stock, and use this to adjust the consistency of the wine sauce.

To freeze: Cool rapidly, remove any surplus fat, seal in suitable containers, and freeze.

To serve: Partially thaw in a refrigerator or over hot water. Return to a pan. Gradually heat the kidney mixture to boiling-point over a low heat, simmer a few minutes. Adjust seasoning and consistency. Serve in a dish bordered with rice and garnished with parsley.

MEAT DISHES WITH PASTRY

THE addition of pastry to any meat dish is likely to enhance its popularity. The prepared meat dish may be frozen before or after baking, whichever is more suitable for its future use. If it is ultimately to be served hot, there is little point in baking before freezing.

Another alternative which may be found useful is to put a raw pastry topping of shortcrust or flaky pastry over a cooked, or partly cooked, meat filling before freezing. In this way the time taken for the pastry to cook is the same as that required to thaw and reheat the meat filling.

USING SHORTCRUST PASTRY

CORNISH PASTIES

Cornish pasties may be frozen before or after baking

1 lb. shortcrust pastry (page 367)	*Yield: 8 pasties*
12 oz. sliced shoulder mutton or beefsteak	*Oven: 425°F (220°C),*
6 oz. potato	*Regulo 7, for 10*
2 small onions	*minutes and 350°F*

3 tablespoons stock or water   (175°C), Regulo 4,
seasoning   for 30–40 minutes
egg wash

Prepare the pastry.

Cut the mutton and potato into small dice (if using beefsteak this should be minced). Finely chop the onion.

Mix together the meat, potato, onion and seasoning, and add sufficient stock to moisten the ingredients.

Divide the pastry into 8 portions and roll each into a circle about 5–6 inches in diameter (cut round a basin or cake tin of the appropriate size to obtain a neat edge). See Fig. 18.

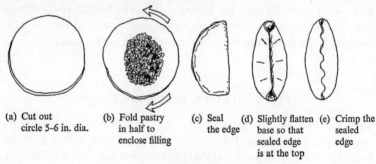

(a) Cut out circle 5–6 in. dia. (b) Fold pastry in half to enclose filling (c) Seal the edge (d) Slightly flatten base so that sealed edge is at the top (e) Crimp the sealed edge

FIGURE 18
SHAPING CORNISH PASTIES

Pile the meat filling in the centre of each circle of pastry.

Damp the edges of the pastry and fold them together to make a half-circle.

Seal around the edges.

Slightly flatten the base of the pastry so that the sealed edge is on the top. It is customary to crimp the edge between the thumb and first finger of one hand and the first finger of the other hand to give a fluted appearance.

Freeze at this stage.

or brush the pasties with egg wash.

Bake in a hot oven until the pastry is set, then reduce the temperature for a further 30 minutes.

To freeze: Freeze *unbaked pasties* after shaping, then pack and seal. Cool *baked pasties*, wrap and place in a box where possible for added protection, seal and freeze.

To serve: Brush *unbaked pasties* with egg wash, and while still frozen transfer to an oven at 425°F (220°C), Regulo 7. Bake for 10–15 minutes, before reducing the temperature to 350°F (175°C), Regulo 4, for a further 40 minutes.

Thaw *baked pasties* in a refrigerator for about 12 hours if to be served cold, or thaw and reheat in a fairly hot oven for about 20 minutes if required hot.

USING FLAKY PASTRY

Meat Pies

These can be frozen before or after baking, giving equally good all-round results. However, the pastry tends to rise better and to become more tender if baked before freezing, and the pie takes less time to prepare for the table after freezing. But it takes up more freezer space and requires more protection.

Steak and Kidney Pie

6 oz. flaky pastry (page 371)	*Yield: 6 portions. 1½ pint*
1½ lb. braising steak	*pie dish Oven: 475°F (245°C)*
6 oz. kidney	*Regulo 9, until pastry is set.*
3 oz. chopped onion	*Lower to 350°F (175°C),*
1 tablespoon flour	*Regulo 4, for a further 2 hours*
1 teaspoon salt	
½ teaspoon pepper	
½ pint brown stock	

Roll the pastry until it is 2 inches larger than the pie dish all round. Allow to relax for 15 minutes.

Wipe the meat and cut into 1½-inch cubes, or thin slices approximately 1½ inches by 2½ inches.

Prepare the kidney and cut into small pieces.

To give an even distribution, place the kidney pieces in the beef slices and roll up. Roll in the seasoned flour before placing in the pie dish. (A quicker method is to toss the cubed beef and kidney in the flour before placing in the pie dish.)

Add ¼ pint stock (half the total amount) and the chopped onion.

Wet the edge of the pie dish.

Cut off a 1-inch strip all the way round the edges of the pastry and place on to the dampened edge of the pie dish. Dampen the surface.

Roll the pastry for the top a little more if necessary and place on top of the pie. Seal the layers of pastry, trim, and flake the edges of the pastry with a knife.

Freeze at this stage.

or brush with egg wash.

Make leaves and a small rosette from the trimmings of pastry. Pierce a hole in the centre of the pie and arrange the leaves and rosette.

Bake in a very hot oven until the pastry is set; lower the temperature and continue to cook for 2 hours.

To freeze: Freeze the *unbaked pie* until the pastry is hard; do not pierce a hole in the pastry. Seal in sheet freezer wrappings.

Remove the *baked pie* from the oven and add the remaining stock. Cool rapidly, pack and freeze.

To serve: Place the *unbaked pie* in an oven at 475°F (245°C), Regulo 9, for 25 minutes until the pastry is risen and set. Make a small steam vent after the pastry has thawed. Lower the heat to 375°F (190°C), Regulo 5, and continue to cook for 2 hours until the meat is tender. If the filling appears dry, add more stock before serving.

Thaw and reheat the *baked pie* in a fairly hot oven for about 1 hour.

USING RAISED PASTRY

MEAT PIES

These can be made and frozen raw or baked. As they are usually served cold, there is more advantage in cooking these pies before freezing, as they only require thawing before use. Pork, veal and ham, or spiced beef fillings may be used with equal success.

VEAL AND HAM PIE

1 lb. raised pastry (page 380)	*Yield: 2 pies 7 inches diam-*
1 lb. pie veal	*eter each giving 4–6 portions;*
¾ lb. raw lean ham	*or 4 pies 5 inches diameter*
½ teaspoon salt	*each giving 3 portions*
pinch pepper	*Oven: 450°F (230°C), Regulo 8,*
pinch ground mace	*for 15 minutes; reduce*
pinch nutmeg	*to 350°F (175°C), Regulo 4,*
¼ teaspoon mixed dried herbs	*for 2 hours*
grated rind of 1 lemon	
¼ pint stock	

284

egg wash

¼ pint jellied stock (1½ teaspoons gelatine)

Make the pastry and while still warm divide into the number of pies to be made.

Subdivide the portions into one-third for the lids and two-thirds for the bases. Keep the pastry for the lids in a warm place.

Mould the pastry for the base either around a greased cake tin, inside a loose-bottomed cake tin, or inside a greaseproof-paper circle, and allow to set. See Fig. 19.

(a) Mould 2/3rds. of pastry around a greased tin or jar

(b) Make lid from remaining 1/3rd. of pastry and seal to inside edge of pastry base about 1 in. from top

(c) With scissors make ½ in. cuts through the sealed edge

(d) Turn alternate sections down to produce the battlement effect

FIGURE 19

SHAPING AND FINISHING VEAL AND HAM PIE

Wipe, trim, and finely dice the veal.

Blanch the ham in boiling water for 5 minutes. Cool, trim, and dice.

Mix the ham, veal, herbs seasonings, lemon rind, and stock.

Pack the meats into the set pastry. Dampen the inside edge of the pastry.

Roll the lid from the remaining one-third of pastry to fit inside the top of the pie.

Seal the edges of the lid to the dampened inside edge of the base.

Decorate the edge without breaking the seal.

Make leaves and a rosette from the pastry trimmings.

Freeze at this stage

or pierce a hole in the centre of the lid and decorate with the leaves and rosette.

Pies which have been made inside a cake tin may be cooked inside the same tin, otherwise fasten a circle of paper around the pastry to hold the pie in a good shape while it is cooking.

Brush with egg wash.

Place in a very hot oven for 15 minutes, then reduce the heat and continue cooking the pie for about 2 hours.

Paper may be placed over the pie if it is browning too much.

Cool. When cold remove the rosette and fill the pie with jellied stock which is just at setting point.

Thoroughly cool the pie before wrapping for the freezer.

To freeze: Freeze *unbaked pies* after shaping. If a cake tin has been used for shaping, this may be removed when the pie is frozen. If it has been moulded inside a circle of greaseproof paper, this may be allowed to remain, and suitable freezer wrappings used in addition.

Thoroughly cool *baked pies* before wrapping and sealing.

To serve: Support *unbaked pies* inside cake tin or circle of grease-proof paper and bake at 450°F (230°C), Regulo 8, for 25 minutes. Make a small steam vent in the top after the pastry has thawed. Lower the heat to 350°F (175°C), Regulo 4, and continue baking for about 2½ hours. Cool the pie and fill with jellied stock as before.

Thaw *baked pies* in a domestic refrigerator about 18 hours. If desired instead of adding the jellied stock before freezing it may be withheld at this stage and added on thawing. Adding warm stock to the still-frozen pie hastens thawing in the centre while the stock sets quickly on coming into contact with the cold filling.

MEAT LOAVES

Meat loaves may be frozen before or after baking. If it is intended to serve the loaf hot, there is little advantage in baking first, as the re-heating will take at least half the complete baking time.

SAVOURY MEAT LOAF

2 lb. raw minced beef	*Yield: 2 loaves 8½ × 4½ inches.*
½ lb. minced fat pork	*Oven: 350°F (175°C), Regulo 4*
8 oz. fresh white breadcrumbs	*Time: 1½ hours*

2 small onions
2 teaspoons chopped parsley
½ green pepper if liked
1 teaspoon mixed herbs
1–2 tablespoons Worcestershire sauce
salt and pepper
1 dessertspoon French mustard
4 eggs

Finely chop the onions and green pepper.
In a large bowl combine thoroughly all the ingredients with the exception of the eggs.
Beat the eggs and add to the mixture binding it together.
Grease the bread tins for loaves baked prior to freezing, or line with aluminium foil for loaves to be frozen raw, and pack the mixture into them.

Freeze at this stage
or cover the tins with foil and bake for about 1½ hours.

To freeze: Freeze *unbaked loaves* in the tins overnight; afterwards remove from the tins complete with the aluminium foil lining. Seal in suitable freezer wrappings.
Cool *baked loaves*, remove from their tins, wrap in suitable freezer wrappings, seal and freeze.

To serve: Return *unbaked loaves* to the tins and cover with foil. Bake at 350°F (175°C), Regulo 4, for 2 hours. Serve hot with tomato sauce (page 229).
Thaw *baked loaves* in their wrappings in the refrigerator for 12–18 hours.
Serve cold with green salad.

RECHAUFFÉ DISHES

MEAT CROQUETTES

Meat croquettes may be frozen before or after frying

1 lb. cooked beef	*Yield: 8 portions*
¼ lb. lean ham or bacon	*12–16 croquettes*
2 oz. onion	

287

salt and pepper
2 oz. margarine
2 oz. flour
½ pint brown stock
2 egg yolks (optional)
2 beaten eggs
bread raspings

serve: with tomato sauce

Blanch the ham in boiling water for 5 minutes. Rinse in cold water and drain.
Mince the ham and beef.
Peel and finely chop the onion.
Melt the margarine and gently fry the onion in it until tender. Strain the onion from the fat.
Add the flour to the fat in the pan and cook gently until it is a golden-brown colour.
Add the stock, beat well, and return to the heat, stirring continuously.
Cook for 2 minutes; the mixture should just leave the sides of the pan.
Add the beef, ham, and onion to the pan. Season well.
Beat in the egg yolks.
Spread the mixture over a large plate to cool.
When cold divide into 16 portions, mould in flour into smooth cylinder shapes. Coat with beaten egg. Toss in the bread raspings until the egg is evenly coated, then reshape the croquettes.

Freeze at this stage

or fry in deep fat at 360°F (180°C), until golden-brown in colour. Drain well on kitchen paper.

To freeze: Both *shaped* and *fried croquettes* keep their shape better if frozen before packing. Pack and seal when hard.

To serve: Thaw *shaped croquettes* in their wrappings for about 2 hours at room temperature. Fry in deep fat at 360°F (180°C) until golden-brown.
This takes about 5 minutes.
Place *fried croquettes* on a baking tray in a fairly hot oven for 20–30 minutes. Serve with tomato sauce or other suitable sauce.

Poultry and Game

ROAST POULTRY

THE comments made in the previous section on the freezing of joints of meat apply equally to the freezing of roast poultry. There is no saving of time or improvement in quality when birds are roasted before freezing.

However, where for reasons of convenience it is decided to roast poultry prior to freezing, the following two methods are recommended in order to get the best results:

ROAST CHICKEN to be served cold

Weight about 3 lb.   *Yield: 4–6 portions*

Thoroughly clean the bird, removing any pin feathers. Wash and dry the cavity, and place a knob of butter inside.
Cover the breast with slices of streaky bacon.
Place in a baking tin and cover with well-buttered greaseproof paper.
Either roast at 425°F (220°C), Regulo 7, for 10 minutes, then 350°F (170°C), Regulo 4, for about 1½ hours, or at the lower cooking temperature for the whole of the cooking time—about 1¾ hours.

To freeze: Drain the chicken, cool rapidly, and pack in aluminium foil or another freezer wrapping which fits snugly to the bird, excluding as much air as possible. Seal and freeze.

To serve: Thaw in a refrigerator about 18–24 hours, or at room temperature for about 9 hours. (If foil was used as a packaging material, replace this with polythene, as foil slows down the rate of thawing.)
Serve with salad.

ROAST CHICKEN to be served hot

Weight about 3 lb. *Yield: 4–6 portions*

Prepare the bird as above, fill with forcemeat stuffing (page 404) and cook as above.

To freeze: Drain well, cool and remove the stuffing; pack this separately in a foil container. Pack as above, seal and freeze.

To serve: Partially thaw the chicken and stuffing overnight in the refrigerator until sufficiently pliable for the stuffing to be returned to the cavity of the bird. Place in a fairly hot oven for 30 minutes or until completely heated through.

FRIED CHICKEN JOINTS

These are very useful to have to hand in the freezer, as they can quickly be transformed into the main course of any meal with a minimum of effort.

2¼–3½ lb. chicken *Yield: 4 portions*
1 beaten egg
4 oz. bread raspings
salt and pepper
2 tablespoons oil
2 oz. butter

Clean and joint the chicken and skin the portions. (Directions for jointing birds are included on page 166).
Dry thoroughly. Brush with the beaten egg.
Season the bread raspings generously with salt and pepper. Dip the egg-coated joints into the bread raspings and toss until evenly covered. Remove the excess raspings with gentle shaking.
Heat the butter and oil until foaming. Cook the joints in the fat over moderate heat for 5 minutes on each side, when they should be golden brown and crisp on the outside.
Drain thoroughly on absorbent paper. Cool rapidly.

To freeze: Separate the joints with layers of cellophane, and pack in suitable quantities. Seal and freeze.

To serve: Place frozen joints on to a baking tin and place in a fairly hot oven for 50 minutes. Test to see the flesh is cooked. Serve with a sauce, or as Chicken Maryland, with corn fritters and fried bananas.

CORN FRITTERS

These may be prepared, frozen and reheated at the same time as the chicken joints.

8 oz. cooked corn (tinned corn is very suitable)   *Yield: 8 portions*
2 eggs
3–4 oz. fresh breadcrumbs
salt and pepper
1 teaspoon baking powder
oil and butter for frying

Mix the baking powder with the breadcrumbs and add the egg yolks, seasoning, and coarsely chopped corn.
Whisk the egg whites until stiff and fold into the other ingredients.
Put sufficient oil and butter in the pan to give a depth of $\frac{1}{2}$ inch, heat until foaming.
Add the fritter mixture in tablespoonfuls to the hot fat, turn when half cooked.
Drain well on absorbent paper.

To freeze: Cool thoroughly and pack, placing a layer of cellophane between layers. Seal and freeze.

To serve: Spread out the frozen fritters on a baking tray and place in a fairly hot oven for about 10–15 minutes.

CHICKEN BREASTS À LA KIEV

This delightful party dish can be made without being extravagant when one possesses a freezer, as the remaining parts of the chickens can be turned into casserole dishes, fried joints, or frozen raw.

8 chicken breasts from birds $3\frac{1}{2}$ lb. weight   *Yield: 8 portions*
    (smaller birds give rather small rolls)
8 oz. butter
2 tablespoons chopped chives   }   *or* 2 tablespoons chopped parsley
$\frac{1}{4}$ teaspoon tarragon   }   and 2 tablespoons lemon juice
salt and pepper
1 tablespoon seasoned flour
2 beaten eggs
6 oz. fresh breadcrumbs

Skin the chicken breasts and cut one slice of meat from each side of the breastbone, extending the slice as far as possible in all directions. Flatten the meat to be approximately ¼ inch thick. Place with the skin side next to the table.

Cream the butter, add the herbs and seasonings.

Divide the mixture into 8 pieces and mould into cylinders. Place in a refrigerator for about 30 minutes to become firm.

Place one piece of butter on to each flattened breast. Fold one end of the meat over the butter, then fold the sides inwards to the centre, and finish by rolling up. This should enclose the butter.

A drumstick may be enclosed in one end of the roll.

Dip the rolls into seasoned flour, then coat with the beaten egg.

Toss in the breadcrumbs until completely and evenly coated.

To freeze: Separate the rolls with a double thickness of cellophane. Pack in suitable quantities.

To serve: Thaw for 1–2 hours at room temperature. Place the rolls into deep fat at 340°F (170°C) and cook for 15 minutes. They should be golden brown outside, and the butter inside should be liquid when they are cut. Drain on absorbent paper and serve immediately.

Poulet Sauté Chasseur

This dish may be prepared using veal in place of the chicken joints.

8 chicken joints	*Yield: 8 portions*
2 oz. onion	*Oven: 375°F (190°C)*
seasoned flour	*Regulo 5, for*
2 teaspoons tomato purée	*30 minutes*
1 pint brown stock	
½ pint white wine	
salt and pepper	
3 oz. butter	
2 tablespoons oil	
1 tablespoon flour	
1 clove garlic ⎫ add after thawing	
½ lb. mushrooms ⎭	

Clean and trim the chicken joints. Dip into seasoned flour.

Chop the onion finely.

Heat the butter and oil until foaming, and fry the joints until well browned all over. Remove the joints from the pan and place into a casserole.

Place the onion in the pan and fry until browned and tender. Stir in the tablespoon of flour and cook for 2 minutes.

Pour in the stock, wine, and tomato purée.

Bring to boiling-point, season, then pour over the chicken joints.

Bake in a fairly hot oven for 30 minutes.

To freeze: Cool rapidly. Remove any surface fat. Pack in suitable amounts.

To serve: Place the frozen joints and sauce into a covered pan and thaw out over low heat. This should take about 30 minutes.

Add the whole clove of garlic when beginning to thaw. Add the sliced mushrooms and simmer for 15 minutes. Remove the clove of garlic before serving.

CHICKEN À LA KING

1 lb. cooked diced chicken   *Yield: 8 portions*
½ lb. mushrooms
1 large green pepper
3 oz. butter
1½ oz. flour
salt
paprika
1 pint milk and chicken stock mixed
sippets of toast—when serving.

Finely slice or quarter the mushrooms and cook in a little of the butter, then lift them out.

Finely chop the pepper and cook in the remainder of the butter for about 5 minutes.

Blend in the flour, salt, paprika, and gradually stir in the mixture of chicken stock and milk.

Bring to the boil stirring all the time until the sauce thickens.

Add the mushrooms and diced chicken. Simmer for 5 minutes.

To freeze: Thoroughly cool and pack in suitable containers. Seal and freeze.

To serve: Transfer to the top of a double boiler for 30 minutes to thaw and thoroughly reheat.

Serve with sippets of toast.

CHICKEN IN CIDER

2 medium chickens
3–4 oz. butter
18–24 button onions
4 oz. streaky bacon
$\frac{1}{2}$–$\frac{3}{4}$ pint cider
bouquet garni
seasoning
4 large dessert apples—add when serving

Yield 8 portions
Oven: 375°F (190°C)
Regulo 5, for 1 hour.

Peel the onions, and cook gently in a little butter, and lift out of the pan.

Joint the chicken (page 166), toss in seasoned flour and lightly brown in the butter.

Blanch the bacon, cut into 1-inch pieces and add to the pan.

Arrange the chicken portions in one or two casseroles (as is convenient) with the onions and bacon. Pour over the cider; add seasoning and bouquet garni.

Cover the casserole with a tightly fitting lid.

Cook in a moderate oven for about 1 hour.

To freeze: Rapidly cool and freeze in a casserole or suitable container.

To serve: Peel and core the apples, cut into quarters or eighths and cook in butter for 2–3 minutes.

Transfer the covered casserole to a cold oven set at 375°F (190°C), Regulo 5, for about 1$\frac{1}{2}$ hours, or until thoroughly heated through; add the apple slices after the first hour.

Adjust seasoning and serve.

COQ AU VIN

2 chickens, each 2$\frac{1}{2}$–3$\frac{1}{2}$ lb. weight
8 oz. lean bacon
12 oz. onions
2 oz. butter
2 tablespoons oil
2 tablespoons brandy
$\frac{1}{2}$ teaspoon salt
$\frac{1}{4}$ teaspoon pepper
1 tablespoon tomato purée
1 pint red wine

Yield: 8 portions
Oven: 375°F (190°C)
Regulo 5, for 1 hour.

good pinch nutmeg
bouquet garni including $\frac{1}{8}$ teaspoon thyme
$\frac{1}{2}$ bay leaf
$\frac{1}{8}$ teaspoon majoram
2 teaspoon sugar
beurre manié: 1 oz. flour, 1 oz. butter
12 oz. button mushrooms
1 clove of garlic—add when reheating

Clean and joint the chickens (page 166). Trim the portions.
Cut the bacon into strips $\frac{1}{4}$ inch wide and 1 inch long. Simmer in water for 10 minutes. Rinse in cold water and drain.
Peel and finely chop the onions.
Melt the butter and oil, place in the bacon strips and fry until browned. Remove from the pan.
Fry the onions in the fat until browned and tender. Remove from the pan.
Fry the chicken joints until browned all over. Return the bacon and onion to the pan; add seasoning. Cover and cook over low heat for 10 minutes.
Add the brandy and ignite the contents of the pan. Gently rotate the pan until the flame subsides.
Add the wine, tomato purée, sugar, nutmeg, and bouquet garni.
Transfer to a casserole and cook in the oven for 1 hour.
Take out the chicken pieces.
Mix together the 1 oz. flour with 1 oz. of butter for the beurre manié. Heat the liquid in a pan. Whisk in the beurre manié a little at a time; allow to cook between additions. The sauce should be just thick enough to coat the chicken joints.
Gently fry the mushrooms in a little butter and place with the chicken joints in a suitable container. Coat with the sauce.

To freeze: Cool rapidly. Remove any surface fat. Seal in moisture-vapour-proof wrappings and freeze.

To serve: Place the frozen chicken into a casserole with a cover.
Place in a cold oven set at 400°F (205°C), Regulo 6, for 1 hour.
Add a clove of garlic when the mixture is beginning to thaw, but remove before serving.

CHICKEN TETRAZZINI

2 lb. cooked chicken meat   *Yield: 8 portions*
$\frac{3}{4}$ lb. spaghetti

4 oz. butter
1 lb. button mushrooms
2 oz. flour
1 pint chicken stock
½ pint single cream
salt and pepper
good pinch of mace
4 tablespoons sherry
4 oz. grated cheese—Swiss or Cheddar (1 oz. added just before serving).

Slightly undercook the spaghetti in fast-boiling salted water. Drain. Wash or skin the mushrooms, slice thinly, and fry gently in 2 oz. of the butter.
Cut the cooked chicken meat into thin strips (julienne).
Melt the remaining 2 oz. butter in a heavy pan and stir in the flour. Remove from the heat and gradually work in the stock.
Bring back to the boil, stir in the cream and cook over a low heat for about 10 minutes.
Add salt, pepper and mace.
Stir in the sherry and all but 1 oz. of the cheese.
Use part of the sauce to moisten the chicken; the rest is mixed with the spaghetti and mushrooms.

To freeze: Choose casseroles or containers which can be used in the oven for reheating, and put first a layer of spaghetti mixture, and then place the chicken on top. Cool, seal and freeze.

To serve: Place in a cold oven set at 375°F (190°C) Regulo 5, for up to 1½ hours. Sprinkle over the remaining 1 oz. of cheese, and lightly brown under a hot grill before serving.

GAME

BRAISED GROUSE

brace of grouse
¼ lb. streaky bacon
1 medium onion
1–2 oz. butter
1 carrot
½ turnip

Yield: 4 portions
Stew gently for 1½ hours

1 pint stock
bouquet garni
salt and pepper
1 oz. flour (blended with water) ⎫
12 button mushrooms   ⎪
12 button onions   ⎬ add after thawing
2 tablespoons sherry   ⎭

Finely chop the onion and bacon, fry in a little melted butter, and lift out.
Dice the carrot and turnip, fry until lightly browned, and lift out.
Wipe the birds with a damp cloth and remove any pin feathers. Truss as for roasting.
Add the birds to the pan and brown them well on all sides.
Add stock, bouquet garni, vegetables, bacon, and seasoning.
Bring up to boiling-point and simmer for about 1½ hours.

To freeze: Transfer the birds to a casserole or suitable container.
Strain the gravy from the pan over the bird. Cool rapidly, remove any fat from the surface. Seal and freeze.

To serve: Thaw and reheat in a covered casserole in a fairly hot oven for about 1½ hours, or in a saucepan on top heat.
Toss the button onions and mushrooms in a little butter for 5–10 minutes.
Blend the flour with a little cold water and stir into the gravy, and bring to the boil.
Add the mushrooms, onion, and a little sherry, and continue to simmer for about 15 minutes.
Remove string from the birds and serve in a dish with the onions and mushrooms at each end, and the gravy poured over and around the birds.

BROWN STEW OF RABBIT

1 rabbit	*Yield: 4 portions*
1 oz. butter or margarine	*Stew gently for 2 hours*
1½ oz. seasoned flour	
1 onion	
1 carrot	
1 small turnip	
salt and pepper	

bouquet garni
1 pint stock

garnish: bacon rolls

Wash and joint the rabbit, cutting the back into two or three pieces. Dip in seasoned flour and fry briskly in hot fat until lightly browned all over.

Prepare the vegetables; cut into cubes, and lightly brown them in the hot fat.

Place the rabbit, vegetables, seasoning, and bouquet garni in a pan and pour over the stock.

Bring to the boil and simmer for about 2 hours until tender.

To freeze: Cool rapidly remove any surplus fat, pack into a suitable container, and freeze.

To serve: Return to a saucepan and reheat, beginning over gentle heat and gradually bring the stew back to boiling-point. Adjust seasoning, consistency, and colour if necessary.

Garnish with grilled bacon rolls.

Hot and Cold Sweets

A GREAT variety of the usual hot and cold sweets may be stored in the freezer, as well as a whole new range of frozen desserts. Most of the sweets normally served hot can be frozen raw and placed in the oven straight from the freezer, taking very little longer to cook than the freshly made equivalent. Cold sweets are most conveniently frozen ready to serve, needing only to be thawed. This is satisfactorily done overnight in the refrigerator, although for quicker results the sweet may be thawed at room temperature. It is important that once thawed the sweet be kept in a refrigerator until served.

HOT AND COLD PUDDINGS USING PASTRY

Most pastry sweets can be frozen provided that the fruit or filling used will freeze and thaw satisfactorily. Custards do not freeze.

Weight of pastry

Where a weight of pastry is given in the list of ingredients, the weight always refers to the quantity of flour from which the pastry was prepared.

For example, 8 oz. shortcrust pastry implies that quantity of pastry produced from 8 oz. of flour.

USING SHORTCRUST PASTRY

FRUIT PIE

Fruit pies may be frozen before or after baking.

Suitable fruits include apple slices, apricots, plums, damsons, blackcurrants, redcurrants, gooseberries, blackberries, rhubarb, and

299

mixtures of these fruits. Fruit pies may be eaten hot or cold, but more usually hot.

Prevention of browning

To prevent the discoloration of apples and apricots used in fruit pies frozen before baking, sprinkle the fruit first with a solution of ascorbic acid.

For each 1 lb. of fruit use 500 milligrams ascorbic acid dissolved in ¼ pint cold water.

Size of pie dish	¾ pint	1 pint	1½ pint
Portions	3–4	5–6	6–8
Amount of shortcrust pastry (page 367)	4 oz.	6 oz.	8 oz.
Amount of flan pastry (page 368)	4 oz.	5 oz.	6 oz.
Amount of prepared fruit	14 oz.–1 lb.	1–1¼ lb.	1½ lb.

Allow: 3 oz. sugar to each 1 lb. fruit.

0–2 tablespoons water per lb. fruit depending upon the juiciness of the fruit.

Bake at 425°F (220°C), Regulo 7, for 15 minutes.
Reduce to 350°F (175°C), Regulo 4, for a further 30 minutes.

Prepare the quantity of pastry and fruit required.
Place half the fruit into the pie dish. Sprinkle with the sugar.
Place the remaining fruit on top, arranging it into a firm rounded shape.
Roll out the pastry to the shape of the pie dish, but 1 inch wider all round. Cut off the 1-inch strip.
Wet the rim of the pie dish and press the strip of pastry all around it. Dampen the surface with water.
Place the remaining pastry over the fruit and seal the edges against the pastry strip.
Trim the edge of the pie, flake the pastry with a knife, and decorate.

Freeze pies raw at this stage (do not make a steam vent)
or bake in a hot oven until the pastry is set. This takes about 15 minutes. Then lower the oven temperature or the position of the pies in the oven, and continue cooking until the fruit is tender, usually another 30 minutes.

To freeze: Freeze *raw pies* before packing, *baked pies* after cooling. They are most easily packed in sheet polythene or Saran, and sealed with freezer tape.

To serve: Bake *raw pies* at 425°F (220°C), Regulo 7, for 20 minutes.

Lower the temperature to 350°F (175°C), Regulo 4, and continue cooking for 35 minutes.

Thaw and reheat *baked pies* to be served hot in a moderate oven for 30–40 minutes.

Or if to be served cold, thaw at room temperature for 5–6 hours.

FRUIT PLATE PIE

These pies may be frozen before or after baking.

Fruits suitable for pies listed in the previous recipe, are also suitable for plate pies. These pies are eaten hot or cold, but more usually cold.

	6 inches	7 inches	8 inches
Size of plate, diameter			
Portions	3–4	6	8
Amount of shortcrust pastry (page 367)	6 oz.	8 oz.	10 oz.
Amount of flan pastry (page 368)	6 oz.	7 oz.	8 oz.
Amount of prepared fruit	12 oz.	1 lb.	1¼ lb.

Allow: 3 oz. sugar per 1 lb. fruit.

Bake at 425°F (220°C), Regulo 7, for 15 minutes.
Lower to 375°F (190°C), Regulo 5, until cooked.

Prepare the quantities of pastry and fruit required.

Cut the pastry into two, one piece slightly larger than the other.

Roll out the smaller portion to the size and shape of the plate.

Line the plate with the pastry.

Place half the fruit on to the pastry, leaving a 1-inch rim uncovered all around the edge. Sprinkle the sugar evenly across the fruit. Arrange the remaining fruit neatly and firmly on top in a rounded shape.

Damp the free edge of the pastry.

Roll the remaining portion of pastry into a piece large enough to cover the pie, and place in position.

Seal the edges of the pastry together, trim, flake with a knife, and decorate.

Freeze pies raw at this stage

or bake in a hot oven until the pastry is set.

Lower the position of the pie in the oven or reduce the temperature to 375°F (190°C), Regulo 5, and continue cooking until the fruit is tender and the base pastry cooked through.

Freeze and serve as for fruit pies.

MINCE PIES

Shortcrust, flan, flaky, or puff pastry may be used for preparing mince pies (pastry recipes, page 367–374). These may be made as plate pies as in the previous recipe, using mincemeat of the same weight in place of the fruit; or shaped in a flan case as a deep covered tart; or as small tartlets.

Preparation of tartlets

12 oz. shortcrust, flan, flaky, or
   puff pastry
1 lb. mincemeat

Yield: about 1½ dozen pies
Oven: 425°F (220°), Regulo 7,
for 15–20 minutes (shortcrust and flan
pastry); 450°F (230°C), Regulo 8, for
20 minutes (flaky and puff pastry)

Roll the pastry to ⅛ inch thickness.

Cut the bases with a crimped cutter 1 inch larger than the top rim of the patty pans. Cut the tops of the tartlets with a crimped cutter of the same size as the tops of the patty pans.

Line the pans with the larger circles of pastry.

Place a generous amount of mincemeat into each pan, damp the edge of the pastry circle. Place on the pastry top, press to seal the edges, and snip the centres to decorate.

It is more usual for flaky or puff pies to be made on a baking sheet. Roll the pastry to about ⅛ inch thickness.

Cut two circles of the same diameter for each tart, usually 3–4 inches, and with plain edges.

Place half the circles on to a baking sheet. Dampen the edges of each circle.

Place a mound of mincemeat in the centre. Cover with the remaining circles.

Seal the edges of the pastry together, leaving the cut edge free, and flake the cut edge with a knife. Snip the centres of the tops to decorate. A combination of pastries may also be used; for example, flan pastry base with a flaky pastry top.

To freeze: Freeze *unbaked tartlets* in patty pans, foil pans, or on sheets. Do not make a steam vent in the lid prior to freezing. When hard remove from the pans if the pans are needed. Pack and seal.

Cool, pack, and seal *baked tartlets*, then freeze. Note that puff-pastry tarts take twice as much storage space when cooked.

To serve: Bake *unbaked tartlets* as follows: shortcrust and flan pastry at 425°F (220°C), Regulo 7, for 20 minutes; flaky pastries at 450°F (230°C), Regulo 8, for 20 minutes.

Thaw *baked tartlets* for 2 hours at room temperature if being served cold, or thaw and reheat in a moderate oven for 10 minutes if required hot.

USING SUET PASTRY

STEAMED FRUIT PUDDING

These puddings may be frozen satisfactorily before or after steaming. A list of suitable fruits is included in the preparation of fruit pies on page 299.

Aluminium foil basins are excellent for the storage of raw and pre-cooked puddings.

Size of basin	1 pint	1½ pint	2 pint
Amount of suet pastry (page 370)	4 oz.	6 oz.	8 oz.
Amount of prepared fruit	6–8 oz.	8 oz.	12 oz.
Portions.	3–4	6	8

Allow 4 oz. sugar to each 1 lb. fruit.

Steam for 2 hours.

Grease a pudding basin evenly.

Divide the pastry into one-third for the top, two-thirds for the lining.

Roll the lining pastry to ½-inch thickness, place into the basin, and mould from the centre of the base outwards until the pastry is evenly spread to the top rim.

Put half the fruit into the basin, add the sugar, then the remaining fruit. The fruit should not more than three-quarters fill the basin.

Dampen the top rim of pastry.

Roll the remaining one-third of pastry into a round to fit on top of the fruit.

Place in position, dampen around the edge, and fold the lining pastry over it. Seal well.

Cover with greased greaseproof paper and aluminium foil.

Freeze puddings raw at this stage

or steam for 2 hours if to be cooked before freezing, and cool thoroughly.

To serve: Place frozen puddings into a steamer. Steam *raw mixtures* for 2½ hours and *cooked mixtures* for 45 minutes.

USING PUFF PASTRY

Gâteau Mille Feuilles

The preparation of this exciting sweet is neither lengthy nor difficult given a supply of puff pastry readily available in the freezer either in bulk form or preferably pre-shaped as required. Although the weight of pastry used in this sweet is only 4 oz., it will be found simpler to work with twice this quantity for easier rolling and shaping. Freezing the completed gâteau is not particularly recommended, as this takes up a lot of space in the freezer and the pastry does not keep very crisp.

4 oz. puff pastry *Yield: 6–8 portions*
8 oz. apricot jam *Oven: 450°F (230°C), Regulo 8*
$\frac{1}{2}$ pint double cream
2 oz. chopped walnuts
vanilla-flavoured glacé icing (page 350 quarter quantity)
small amount of coloured glacé icing.

If using frozen puff pastry, thaw about 3 hours at room temperature before rolling. Roll out $\frac{1}{16}$ inch thick, or as thinly as possible and cut into three circles 6–8 inches in diameter (or three rectangles, each 6 × 12 inches). *or* use frozen pre-shaped pastry if available.
Bake the pastry until well risen, set and dry throughout. Cool.
Whip the cream until stiff.
Layer the pastry pieces using the jam and two-thirds of the cream between the layers.
Spread the sides of the gâteau with the remaining cream and press with the chopped nuts to finish.
Run glacé icing over the top of the gâteau and feather ice with the coloured icing.

USING CHOUX PASTRY

Profiteroles and Chocolate Sauce

choux pastry cases—frozen after shaping *Allow 4–5 profiteroles*
   or after baking, page 377 *per portion*
$\frac{1}{2}$ pint whipped cream
chocolate sauce (page 238) $\frac{1}{2}$ pint is
   sufficient for 4 portions

Bake or thaw the cases as necessary.
Fill with vanilla-flavoured whipped cream.
Pile in individual dishes and coat with hot chocolate sauce.

FILLED FLANS

ALMOND CHERRY FLAN

This flan may be frozen before or after baking.

Pastry case

Use flan pastry (page 368)

Filling

1 lb. black cherries, stoned	*Yield: sufficient for 2 × 6-inch flan*
6 oz. ground almonds	*rings, each serving 4 portions*
9 oz. castor sugar	*Oven: 400°F (205°C), Regulo 6*
2 eggs	*Time: 30 minutes*
almond essence	
2 oz. shredded almonds	

Pile the cherries into the unbaked flan cases.

Mix the ground almonds and sugar. Add sufficient egg, beating well between additions, to give a soft paste.

Add a few drops of almond essence to give a good flavour.

Spread the almond paste evenly over the cherries. Spike shreds of almond into the paste to decorate.

Freeze the flan raw at this stage

or bake in a fairly hot oven until the pastry is cooked and the top golden-brown and crisp. This should take about 30 minutes.

To freeze: Freeze *unbaked flan* before packing, *baked flan* after packing.

To serve: Place *unbaked flan* into a cold oven set at 400°F (205°C), Regulo 6, and cook for 45 minutes. Cool and serve.

Thaw *baked flan* in wrappings at room temperature for 3 hours, and serve.

CHIFFON PIES

Pastry case

Use flan pastry (page 368).

or biscuit crust casing (page 369). This crust gives excellent results for flans frozen with filling ready to serve.

Chiffon Filling

This is a basic recipe for the preparation of a chiffon mixture and is suitable for all flavours except those from citrus fruits. An amended recipe for these is given below.

4 eggs
¾ pint milk
1 tablespoon gelatine
3 oz. castor sugar
¼ teaspoon salt
flavourings

Yield: sufficient for 3 × 6-inch flan cases,
each serving 4 portions

Prepare the flan cases; if using flan pastry, bake and cool. Chill or bake biscuit crust.

Separate the eggs.

Sprinkle the gelatine into the milk and leave to soak for 5 minutes. Then pour into a saucepan with half the sugar and the salt, and heat to scalding-point.

Beat the egg yolks, and gradually pour on the hot milk, beating all the time.

Return to the pan and stir over low heat until the custard thickens. Add the flavouring, see below.

Chill the custard until it begins to set.

Whisk the egg whites until they just stand in peaks; add the remaining sugar and rewhisk until stiff.

Fold the egg whites into the custard until evenly mixed.

Pour into the flan cases, chill, and decorate.

FLAVOURINGS

CHOCOLATE

2 oz. melted chocolate
1 tablespoon rum
decorate with chocolate vermicelli

COFFEE

4 teaspoons instant coffee
1 tablespoon hot water
decorate with whipped cream

LEMON AND ORANGE CHIFFON MIXTURE

4 eggs
1 tablespoon gelatine

Prepare the chiffon mixture
according to the basic method,

306

⅛ pint water
⅛ teaspoon salt
3 oz. castor sugar
¼ pint lemon juice
or 2 tablespoons lemon juice
    and ¼ pint orange juice
1 teaspoon grated lemon (orange) rind
add a further 3 oz. castor sugar to the egg whites

using the fruit juice and water
in place of the milk, and adding
the rind to the thickened
custard.

To freeze: Allow to freeze until hard. Pack and seal. It is useful to freeze the case on a foil container and use sheet wrapping.

To serve: Allow to thaw in a refrigerator for 8 hours or 2 hours at room temperature. The wrapping should not touch the surface of the pie, otherwise it will damage the finish.

BAKED CHEESE CAKE

2 oz. crushed petit beurre or
    digestive biscuits
1 lb. cottage or curd cheese
few drops lemon juice
1 teaspoon grated orange rind
1 tablespoon cornflour
2 tablespoons well-whipped cream
2 egg yolks
2 egg whites
4 oz. castor sugar

Yield: 8-inch cake tin;
8–12 portions
Oven: 350°F (175°C), Regulo 4
Time: about 1 hour

Butter the sides and place a circle of buttered greaseproof paper in the bottom of an 8-inch cake tin—select one with a removable base. Sprinkle thickly with biscuit crumbs.
Sieve the cheese into a large bowl.
Blend together the lemon juice, grated rind, and cornflour.
Add to the cheese.
Stir in the whipped cream.
Beat the egg yolks in a basin over hot water until they begin to thicken; cool and stir into the cheese mixture.
Whisk the egg whites until quite stiff; whisk in half the sugar; stir in the remaining sugar.
Fold the beaten egg whites into the cottage cheese mixture and transfer the completed mixture to the prepared tin.

Bake in a moderate oven for about 1 hour.
Allow to cool in the oven.

To freeze: Lift out the cheese cake from the tin and freeze until hard. Pack and seal, and for greater protection pack in a cardboard box.

To serve: Thaw in a domestic refrigerator for about 8 hours, or 3 hours at room temperature, and serve while still slightly chilled.

RHUBARB CREAM FLAN
Stoned apricots are also suitable

Pastry case

biscuit crust casing (page 369), using brown sugar.

Filling

12 oz. rhubarb
5 oz. castor sugar
3 eggs
6 oz. cream or curd cheese
¼ pint double cream

Yield: sufficient for 2 × 6 inch flans, or 1 × 9 inch flan serving 8 portions

Prepare the flan case using brown sugar in place of castor sugar in the recipe, and chill.
Prepare the rhubarb and cut into 1-inch pieces.
Place in a pan with a tightly fitting lid, add the sugar, and cook covered over a low heat until tender.
Separate the eggs. Beat the yolks and add to the rhubarb. Stir over low heat until the mixture thickens, then cool.
Mix the cheese and whipped cream. Fold into the fruit custard.
Whisk the whites until stiff, then fold into the custard cream.
Spoon into the flan cases and chill.

To freeze: Freeze unwrapped until hard, then pack and seal.

To serve: Thaw in wrappings for 8 hours in the refrigerator, or 2–3 hours at room temperature.

BLACKCURRANT FLAN
This flan may be frozen before or after baking.

8 oz. flour
1½ dessertspoons cinnamon
5 oz. butter

Yield: 2 × 6-inch flan cases each serving 4 portions
Oven: 400°F (205°C), Regulo 6
Time: 30 minutes

1½ oz. ground almonds
1½ oz. castor sugar
½ standard egg
1½ teaspoons lemon juice
1½ lb. blackcurrants
6 oz. sugar

Sift the flour and cinnamon.
Cut the butter into the flour, then rub in until fine crumbs are formed.
Mix in the ground almonds and the sugar.
Sprinkle with the lemon juice, beat in the egg and mix to a firm paste.
Divide this paste into two. Reserve a quarter from each half to use as lattice topping on the flans.
Roll the remaining pastry and line the flan rings or foil cases.
Prepare the blackcurrants.
Half fill each flan case with fruit, then sprinkle evenly with 3 oz. sugar.
Fill the flan cases with the remaining fruit.
Combine the trimmings with the reserved pastry. Cut into strips about ⅓ inch wide and arrange in lattice pattern on top.

Freeze the flan raw at this stage

or bake in a fairly hot oven. Before placing the flan in the oven brush the pastry lattice with water and sprinkle with castor sugar to produce a glaze.

To freeze: Freeze *unbaked flan* prior to packaging.
Thoroughly cool *baked flan*, wrap, seal, and place in a box for added protection.

To serve: Brush the lattice of *unbaked flan* with water, sprinkle with castor sugar. Place in a cold oven set at 400°F (205°C), Regulo 6, and cook for 45 minutes. Cool and serve.
Thaw *baked flan* in loosened freezer wrappings for about 3 hours at room temperature.

PUDDINGS WITH CRUMB TOPPING

FRUIT CRUMBLE

Size of dish	¾ pint	1 pint	1½ pints
Portions	3–4	5–6	6–8
Weight of prepared fruit	8 oz.	12 oz.	1 lb.
Weight of topping (made weight)	6 oz.	9 oz.	12 oz.
Allow 3 oz. per 1 lb. fruit.			

Topping

6 oz. flour
3 oz. castor or brown sugar
4 oz. margarine
suitable fruits—apple and apricots (if treated with antioxidant, page 300), plums and damsons

Sift the flour.
Add the margarine and cut into the flour.
Rub the margarine into the flour until the fat begins to oil and the mixture sticks together.
Add the sugar and mix in until evenly distributed.
Arrange the fruit in a greased pie dish, sprinkling the sugar half-way through the fruit.
Sprinkle the topping over the fruit and lightly press down.

Freeze the pie raw at this stage.

To freeze: As the pudding is served hot, it is more suitably frozen raw. Package after making, seal and freeze.

To serve: Place uncooked pudding in a cold oven set at 400°F (205°C), Regulo 6, and bake for 30 minutes. Lower to 375°F (190°C), Regulo 5, and continue cooking for 30–45 minutes, until the fruit is tender and the topping crisp.

PUDDINGS USING BATTERS

Sweet Pancakes

Preparation and freezing pancakes, page 401.   *Allow 2 pancakes*
Thaw pancakes in a fairly hot oven and fill as   *per portion*
follows.

Suitable fillings

Fill with ice cream, roll up and brush with apricot glaze (page 349); sprinkle surface with chopped blanched almonds;

or

fill with fruit purée, roll up, and arrange shapes of brandy butter or other liqueur-flavoured butter (page 234) on the surface;

or

fill with ice cream and chopped walnuts, coat with hot chocolate sauce (page 238).

FRUIT FRITTERS

½ pint yeast batter, page 395
or ½ pint pancake batter, page 401   *Fat bath 360°F (180°C)*
  substitute 1 oz. flour with 1 oz. fresh breadcrumbs
8 bananas or 6 cooking apples

Yield: 8 portions

Prepare the batter.
Prepare the fruits; peel, core, and slice apples into ¼-inch rings. Slice bananas into three length ways—if bananas are very large, cut each slice into two.
Using a carving fork, dip the fruit into the batter and then into the fat bath.
Do not use a frying basket; turn if necessary.
When lightly browned lift out on to absorbent paper.
If serving immediately, toss in castor sugar and serve very hot.
To freeze: Cool thoroughly, separate slices with a double thickness of cellophane, seal and freeze.
To serve: Spread out on a baking tray. Thaw and reheat in a fairly hot oven for about 10 minutes. Toss in sugar and serve.

PUDDINGS USING CAKE MIXTURES

STEAMED PUDDINGS

Steamed puddings are not usually considered the most interesting sweet and therefore they may not be thought worthy of freezer space. However, the presence of a good jam sponge in the freezer may well save the day in some households during the school holidays.

A recipe, with variations, for the preparation of a steamed pudding made from a rich cake mixture is included, as this requires the most time in preparation. Puddings made from plain cake mixtures, quick-mix puddings, and the traditional suet pudding, all freeze equally well, but recipes for them are not included as any favoured recipe may be used satisfactorily.

Use aluminium-foil basins for freezing and steaming.

USING RICH CAKE MIXTURE

STEAMED PUDDING

This pudding may be frozen before or after steaming.

The ingredients given in the basic steamed pudding recipe which follows yield about 2¼ lb. raw pudding mixture.

The following table serves to give some guidance on the weights of raw mixture for different sizes of basin, and the number of portions.

Size of basin or mould	2½-inch dariole mould	¾-pint basin	1-pint basin	1½-pint basin	2-pint basin
Number of portions	Allow 2 moulds per person	3	4	6	8
Weight of wet mixture	10 oz. sufficient for 6 moulds	8 oz.	10 oz.	1 lb.	1 lb.- 2 oz.

Basic recipe for steamed pudding

8 oz. margarine
8 oz. castor sugar
4 eggs
12 oz. flour
3 teaspoons baking powder
add one of the following flavourings:

Steam: 1½–2 hours (30 minutes only for dariole moulds)

LEMON

Add the grated rind of 1 lemon after beating in the egg.

JAM, SYRUP, OR MARMALADE

Add 2–4 oz. depending on the size of the basin used. Place in the basin before adding the pudding mixture.

SULTANA

Mix in 8 oz. cleaned sultanas with the flour.

CHOCOLATE

½ oz. vanilla sugar—substitute in place of ½ oz. castor sugar
9 oz. flour } substitute in place of the 12 oz. flour
1 oz. cocoa }
3⅜ oz. plain chocolate (nominal 4 oz. block). Melt and add after beating in the egg
3 tablespoons milk—add when flour partly folded in

General Method for Steamed Pudding

Allow the margarine to stand in a warm place overnight so that it will soften and cream easily.
Beat the margarine and sugar by hand or in a food mixer until light and fluffy.

Beat the eggs slightly and add gradually to the creamed margarine and sugar, beating well between additions.

Sift the flour together with the baking powder and fold into the creamed mixture.

Grease the basin or moulds with oil or butter.

Weigh in any topping, for example, jam, marmalade, or syrup.

Weigh in the appropriate quantity of raw pudding mixture.

Cover the basin with greased greaseproof paper and aluminium foil.

Freeze the puddings raw at this stage

or steam for 1½–2 hours if to be cooked before freezing, and cool thoroughly.

To serve: Place the frozen puddings into a steamer, and steam *raw puddings* for about 2½ hours, and *cooked puddings* for about 45 minutes.

Raw dariole moulds cooked from the frozen state in the steamer require 45–50 minutes (30 minutes in an oven at 375°F (190°C), Regulo 5).

EVE'S PUDDING

This pudding can be made using a rich cake mixture or the quick-mix cake mixture as a topping.

Suitable fruits include apples, and apricots (treated with an antioxidant, page 300, when the pudding is frozen raw), plums, damsons, gooseberries.

Size of dish	¾ pint	1 pint	1½ pint
Portions	3	4–5	6–8
Weight of rich cake mixture, page 344	6 oz.	8 oz.	10 oz.
Weight of quick-mix cake mixture, page 346	5 oz.	6 oz.	8 oz.
Weight of prepared fruit	6 oz.	8 oz.	12 oz.

Allow 3 oz. of sugar per 1 lb. fruit.

Grease the pie dish.

Arrange half the fruit in the pie dish, sprinkle with the sugar, and cover with the remaining fruit.

Spread the cake mixture over the top of the fruit to cover it evenly.

Freeze the pudding raw at this stage.

To freeze: Apple Eve is more suitably frozen raw, as this pudding is always served hot. Freeze until hard, then pack and seal.

313

To serve: Place *raw pudding* into a cold oven set at 400°F (205°C), Regulo 6, and bake for 30 minutes. Lower the heat to 375°F (190°C), Regulo 5, and continue baking for about 30 minutes or until the cake mixture is cooked. Serve.

PUDDINGS USING SPONGE OR GENOESE BASES

SPONGE FRUIT FLAN

8-inch sponge or genoese flan case (page 347 or 348)   *Yield:*
12 oz. fresh fruit or 1 lb. tin of fruit   *6 portions*
¼ pint fruit juice ⎫
1½ teaspoons gelatine ⎬ gelatine glaze, see notes on page 316
colouring if required ⎭ (sweets using gelatine)
    or ¼ pint 'Quick Jel' glazing made with fruit juice
apricot glaze (page 349, half quantity)

Make the gelatine glaze by dissolving the gelatine in the fruit juice in a basin over hot water.

Heat the apricot glaze and brush the top edges and the sides of the flan case to give a good finish. Allow to dry.

Drain the fruit and arrange in pattern in the flan case.

When the gelatine glaze is at setting-point, use it to brush or lightly coat the fruit.

Or make the 'Quick Jel' glaze and coat the fruit immediately with it.

To freeze: Freeze until hard, pack and seal.

To serve: Loosen packaging from pressing on the top surface and sides of the flan. Thaw at room temperature for 2–3 hours, covered by packaging or in an air-tight container.

FRUIT GÂTEAU

2 × 8-inch genoese or sponge   *Yield: 8 portions*
    sandwich cakes (page 348 or 347)
½ pint double cream
2 oz. chopped walnuts
½ pint orange jelly
11 oz. tin mandarin oranges ⎫ or any other combination of
13 oz. tin pineapple segments ⎬ fruits or single fruit
maraschino or glacé cherries, angelica to decorate

314

Make the orange jelly and cool.

Run a thin layer of jelly into an 8-inch sandwich tin and chill to set. Drain the fruits thoroughly. Arrange the orange segments and decoration in a pattern on the set jelly. Add sufficient jelly to coat the fruit.

Allow to set.

Place one sandwich cake, with the top surface downwards, on top of the jelly.

Whip the cream until stiff, retain about one-third of it in a small basin. Fold the drained pineapple segments into the remaining two thirds of the cream and spread over the second sandwich cake.

Dip the top of the first sandwich tin into hot water to loosen the jelly. Invert the tin and the jelly-decorated sandwich cake on top of the cream mixture.

Spread the sides of the gâteau with the remaining cream and smooth with a palette knife to give an even finish.

Press the chopped walnuts against the edge.

The gâteau is most easily assembled on a foil plate or rigid board on which it can be frozen.

To freeze: Freeze the gâteau until hard, then pack and seal.

To serve: Loosen the packaging from the top and sides of the gâteau. Allow to thaw for 4 hours at room temperature, covered, or in an air-tight container, or it may be placed in a refrigerator overnight. Serve chilled.

FRUIT ROLL

This may be prepared from a freshly baked Swiss roll or one which has been frozen unfilled. If preferred, 1 oz. chopped walnuts or glacé fruits may be folded in with the flour when preparing the Swiss roll mixture. Use the recipe for fatless sponge (page 347) or genoese mixture (page 348).

1 Swiss roll
¼ pint double cream
1 teaspoon vanilla sugar
¼ lb. fresh, tinned, frozen or glacé fruit,

Whip the cream and sugar until stiff.

Unroll the Swiss roll.

Spread with a layer of cream (retain one-third for the outside of the roll), then a layer of cleaned and chopped fruit.

315

Roll up.

Place the roll on a foil plate or rigid board.

Coat with the remaining cream. Fork mark to give an attractive finish if desired.

To freeze: Freeze until hard, then pack and seal.

To serve: Loosen the packaging from the surface of the cake. Thaw covered or in an air-tight container for 3 hours. Serve chilled. Cut off a thin slice from each end before serving so that cream mixture is showing.

SWEETS USING GELATINE

When gelatine-set jellies are frozen, the ice crystals formed disrupt the structure of the jelly. Whilst remaining set when thawed, the clear bright appearance is lost and the jelly has a granular and uneven texture. When used in thin layers to coat fruits and puddings, it is acceptable, but for jellies to be served as such, freezing is not recommended.

When gelatine is used as the setting agent for creams, soufflés, etc., the granular effect of the jelly is masked and the thawed results are excellent.

A selection of these dishes using gelatine is given.

FRUIT CREAMS

Suitable fruits include strawberries, raspberries, blackcurrants, apricots, peaches.

½ pint sweetened fruit purée—from tinned,   *Yield: 6–8 portions*
   frozen or fresh fruit
¼ pint double cream
¼ pint single cream
1 tablespoon gelatine
4 tablespoons fruit juice or water
2 teaspoons lemon juice for bland-flavoured fruits, e.g. apricots, strawberries

Stew fresh fruit first for a few minutes with a little sugar—1 lb. fruit yields approximately ½ pint purée.

Mix the gelatine with the fruit juice or water.

Place the basin in hot water over gentle heat until the gelatine dissolves. Mix the creams and whisk until thick. (The mixture gives a semi-stiff foam and never stiffens as much as double cream alone. Double cream used alone should only be half whipped for this sweet.)
Fold the purée, cream and lemon juice together.
Fold in the warm gelatine solution and mix in well.
Mould or spoon into individual wax cases.

To freeze: Freeze until hard, then pack and seal.

To serve: Thaw overnight or for at least 8 hours in a refrigerator, or alternatively thaw at room temperature for 4 hours. When ready for serving fruit creams should be soft throughout, but still chilled.

BAVAROISE

This is basically a gelatine-set mixture of custard and cream; or custard, cream and fruit purée, in equal proportions.

12 egg yolks	*Yield: sufficient for 1 charlotte*
8 oz. castor sugar	*russe 6-inch diameter, and 2 × 1-pint*
1½ pints milk	*moulds (4 portions each)*
1 vanilla pod	
3 tablespoons gelatine	
5 tablespoons water	
1 pint double cream	or equal amounts of each to make 1½ pints
½ pint single cream	

Place the vanilla pod with the milk in a pan and slowly heat to scalding-point.
Mix the gelatine and water in a basin and place in a saucepan of hot water over gentle heat until the gelatine is dissolved.
Beat the egg yolks and sugar until the foam is almost white and thick enough to leave a trail. Gradually pour on the hot milk and mix well.
Return to a pan over low heat and stir continuously until the custard thickens.
Add the gelatine mixture to the hot custard and mix in. Add the flavouring.
Cool rapidly, stirring frequently to prevent a skin forming on the surface.
Chill until just at setting-point.
Mix the creams and whisk until thick.
Beat the custard and the cream together and mould.

FLAVOUR VARIATIONS

COFFEE

Add 6 teaspoons instant coffee powder to 1½ pints custard.

CHOCOLATE

Add 4 oz. melted chocolate to 1½ pints custard.

FRUIT

Use half the basic recipe given above.
Combine ¾ pint fruit purée with the custard and cream before moulding.

.

MOULDING VARIATIONS

FANCY MOULD

Pour the mixture into a fancy mould to set.

STRIPED BAVAROISE

Two or more flavours and colours can be combined: chocolate and vanilla, strawberry and vanilla, coffee and vanilla.
Pour a layer of flavoured bavaroise about 1 inch deep into a fancy ring mould, jelly mould, or individual glass dishes.
Chill to set (a quick set is achieved by surrounding the mould with ice-cold water and ice-cubes).
When set cover with a layer of vanilla bavaroise. Alternate layers until the mould is filled.

CHARLOTTE RUSSE

2 dozen sponge finger biscuits   *Yield: 6-inch Charlotte Russe*
¼ pint flavoured jelly   *mould or cake tin. 6–8 portions*
pieces of fruit or glacé fruits for decoration
2 pint bavaroise (half the basic recipe), vanilla or fruit flavoured

Pour a thin layer of jelly into the base of the tin and set (see note on page 316—sweets using gelatine).
Arrange a bold decoration of the fruits and glacé fruits.
Add just sufficient jelly to cover the pieces.
Arrange sponge fingers around the side of the tin.
Once the mould has been decorated, fill with vanilla or fruit bavaroise and set.

To freeze: Freeze and seal most bavaroise mixtures in the mould;

318

although if preferred it is possible to unmould a charlotte russe on to a foil plate or rigid board when it is frozen hard before packing in suitable wrappings.

To serve: Unmould if necessary. Loosen wrappings so that they do not press on to the surface of the mixture. Allow to thaw at least 12 hours in a refrigerator or 4–6 hours at room temperature. Serve chilled.

COLD SOUFFLÉS

Basically these are similar mixtures to bavaroise, but lightened by the addition of whisked egg whites. Various flavours can be used, but the example given uses lemon.

LEMON SOUFFLÉ

4 eggs
4 oz. castor sugar
finely grated rind and juice of 2 lemons
1 tablespoon gelatine
4 tablespoons water
¼ pint double cream
¼ pint single cream
whipped cream
2 oz. chopped nuts } add when serving

Yield: soufflé case 7–8 inch diameter, or 2 × 5 inch diameter, 8 portions

Secure a band of foil around the side of the soufflé case, projecting 3–4 inches above the top rim.

Mix the gelatine with the water in a small basin and place in a saucepan of water over gentle heat until it dissolves.

Separate the eggs. Place the yolks, 3 oz. sugar, rind and juice of the lemons into a bowl. Whisk over hot water until the mixture is thick enough to leave a trail. Continue whisking away from the heat until the mixture is cool.

Mix the creams and whisk until they are of the same consistency as the egg-yolk mixture.

Whisk the egg whites until just stiff.

Whisk the remaining 1 oz. sugar into the egg whites.

Fold the creams and egg-yolk mixture together gently.

Fold in the warm gelatine, followed by the egg whites. Continue folding until the mixture is even.

319

Pour into the casing, allowing the mixture to come 2 inches over the rim of the dish. Allow to set.

VARIATIONS

Fruit

Use fresh, frozen, or tinned strawberries, raspberries, apricots, or peaches.

Replace lemon rind and juice by 2 tablespoons lemon juice and ¼ pint fruit purée.

Replace 4 tablespoons water by ⅛ pint fruit juice.

Omit the ¼ pint single cream.

If using fresh fruit, stew first in a little water to soften the fruit and produce a little fruit juice.

Strain fruits (stewed, frozen, and tinned) and use juice to dissolve the gelatine.

Purée fruits in a liquidizer, or pass through a sieve, and combine with the whisked double cream.

Chocolate

Replace the lemon rind and juice by 5 oz. chocolate and 2 tablespoons of brandy.

Use ½ oz. vanilla sugar in place of ½ oz. of sugar in the recipe.

Use 1 dessertspoon gelatine (note this is half the quantity used for the lemon soufflé).

Whisk the egg yolks and sugars over hot water as before.

Add the brandy.

Dissolve the gelatine in the water, add and melt the chocolate.

Combine the egg yolk and brandy mixture, together with the creams, gelatine, and melted chocolate, and finally the whisked egg whites as for the lemon soufflé.

To freeze: Freeze until hard. Fold the foil down over the top of the soufflé. Pack and seal.

To serve: Remove from the packaging. Lift the foil from the top surface and straighten it again around the sides. Place in a large polythene bag or air-tight container and thaw in the refrigerator for 8–12 hours, or at room temperature for 4–6 hours. Remove the foil. Press chopped nuts against the edge of the soufflé showing above the dish, and decorate the top edge with piped, whipped cream. Serve chilled.

Chilled Cheese Cake

1 lb. curd cheese
2 tablespoons gelatine
¾ pint tinned sweetened orange juice
2 tablespoons lemon juice
2 oz. sugar
1 teaspoon grated lemon rind
1 teaspoon grated orange rind
¼ teaspoon salt
¼ pint double cream
¼ oz. unsalted butter
2 oz. crushed digestive biscuit crumbs
2 tablespoons clear orange marmalade—add on serving

Yield: 8-inch loose-bottomed cake tin. 12 portions

Mix the gelatine with ¼ pint orange juice, stand in a bowl over hot water until the gelatine dissolves.

Add the sugar and salt and stir until dissolved. Mix in the remaining fruit juices, and rinds, then cool.

Sieve the curd cheese.

Half whip the cream, and fold into the cheese.

Fold in the combined fruit juices and gelatine.

Chill until just setting, then whisk until light.

Place a circle of greaseproof paper on the base of the sandwich tin.

Soften the butter, and grease the paper and sides of the cake tin.

Toss in the biscuit crumbs and coat the surface. Shake out loose crumbs.

Pour in the cheese mixture and allow to set.

To freeze: Freeze until hard. Remove the cheese cake from the tin. Wrap in foil, and seal.

To serve: Unwrap the foil from the cake. Remove the greaseproof paper and place on to a serving plate; spread a thin layer of marmalade over the top. Place inside a large polythene bag to thaw for at least 8–12 hours in a refrigerator, or 4–5 hours at room temperature. Serve chilled.

FRUIT COOKED IN WINE

Quite a few fruits, which do not freeze and thaw satisfactorily when raw, can be cooked in syrups in which they are then served, giving excellent results. Examples of pear and peach dishes are given.

321

FRUIT IN WINE SYRUP

All recipes using red or white wine are suitable. It will be found that due to the concentration of sugar, the syrup is still viscous at 0°F (−18°C); therefore a water-tight container is essential.

PEARS IN RED WINE
peaches also suitable

8 small pears or peaches   *Yield: 8 portions*
8 oz. sugar
¼ pint water
¼ pint burgundy
1 stick cinnamon
whipped cream—add when serving

Dissolve the sugar in the water and add the cinnamon stick.
Peel the pears, but leave the stalk and core intact.
Place the pears into the syrup and simmer in a covered pan for 15 minutes.
Add the burgundy and continue simmering for another 15 minutes with the pan uncovered. Turn the pears if necessary to ensure adequate cooking all round.
Remove the pears.
Place the pan over moderate heat and reduce the syrup until it is of a fairly thick consistency.
Pour the syrup over the pears and cool rapidly.

To freeze: When frozen together the pears tend to loose moisture on thawing and thin the syrup. Freeze in separate containers.

To serve: Thaw separately for at least 8 hours in a refrigerator. Pour syrup over the drained pears. Serve chilled with whipped cream.

POIRES ALMA

A more elaborate and sweeter dish:

8 ripe pears   *Yield: 8 portions*
½ pint port or port-type wine
½ pint water
strips of rind from 2 oranges
6 oz. castor sugar

2 tablespoons raspberry jam
2 tablespoons redcurrant jelly
¼ pint whipped double cream ⎫
chopped browned almonds ⎭ add when serving

Add the sugar to the port and water and heat to dissolve.
Add the orange rind.
Peel, core and halve the pears.
Poach in the wine syrup until tender.
Remove the pears.
Add the jam and jelly to the syrup and stir over heat until they dissolve.
Reduce the syrup over moderate heat until it is thick. Strain over the pears.
Cool rapidly.

To freeze: Freeze the pears and syrup separately—see previous recipe.

To serve: Thaw overnight in the refrigerator. Arrange the pears with the cut sides uppermost. Pour over the syrup. Keep chilled. Just before serving, pipe a rosette of whipped cream into the hollow of each pear and scatter with the chopped browned almonds.

PEARS IN WHITE WINE

This sweet is served hot.

8 pears   *Yield: 8 portions*
about ½ pint white wine   *Oven: 375°F (190°C), Regulo 5*
8 oz. sugar   *Time: 35–45 minutes*
2 tablespoons Kirsch ⎫
whipped cream ⎭ add on serving

Peel the pears, leaving the stalk and core intact.
Place in a casserole.
Cover with the white wine.
Sprinkle with the sugar.
Bake for 35–45 minutes.
Serve immediately or cool rapidly when to be frozen.

To freeze: Pack in suitable containers.

To serve: Place in a moderate oven for 45 minutes, or until completely warmed through. Just before serving pour over the Kirsch if used.
Serve with whipped cream.

STUFFED PEACHES

8 peaches
4 oz. ground almonds
1½ oz. softened butter
3 oz. icing sugar
grated rind of 1 small lemon
about 1 tablespoon orange juice
⅛ pint sherry
icing sugar

Yield: 8 portions
Oven: 400°F (205°C), Regulo 6
Time: 15–20 minutes

Mix the ground almonds with the 3 oz. icing sugar.
Work in the softened butter.
Add the lemon rind. Add sufficient orange juice to give a stiff paste.
Divide into eight pieces each about the size of a peach stone.
Dip the peaches into boiling water, peel, halve, and stone.
Place the almond stuffing into the hollow left by the stone, and rejoin the peach halves.
Place on an uncovered fireproof dish. Pour the sherry over the peaches. Dredge heavily with icing sugar.
Place in a fairly hot oven and cook for 15–20 minutes until the sugar forms a glaze.
Cool rapidly.

To freeze: Pack in suitable amounts or individually in foil patty tins.

To serve: Thaw at room temperature for about 2 hours. Place uncovered in a fairly hot oven for 10 minutes to heat through before serving. Serve with cream or a light custard-type sauce.

MILK PUDDINGS

Milk puddings are not usually the most popular item on the menu and rarely justify freezing. This is because many thickened with eggs, such as baked custards and caramel creams, do not freeze satisfactorily, and most of those thickened with cereals such as baked rice, tapioca, or sago puddings, or cornflour moulds, take very little time or skill to prepare.

However, rice, sago, and semolina often form the base of more elaborate cold sweets when creamed, or when both creamed and set with gelatine. The preparation of these bases is time consuming and requires care if a smooth result is to be obtained. Sufficient mixture for several bases or shapes may be prepared at the same time with

324

little extra work involved. The bases or moulds are best frozen before finishing. They may be thawed and finished with fruit, jam glaze, whipped cream and nuts, to produce a variety of different cold sweets when required.

CREAMED RICE

4–6 oz. Carolina rice   *Yield: 6–8 portions*
(use the larger quantity if a moulded shape is required)
2 pints milk
1 vanilla pod
¼ pint whipped cream
4–6 oz. sugar

Wash the rice, place with the milk, vanilla pod, and sugar, into a double saucepan, and cook for about 2 hours. The rice should be tender and the mixture fairly thick and creamy.
Remove the vanilla pod, allow to cool, stirring frequently to prevent the formation of a skin.
Fold in the whipped cream.

To freeze: Freeze in bulk, in shapes such as ring moulds, or Victoria sandwich tins, or in individual portions in dishes or waxed cartons. Seal in suitable freezer wrappings.

To serve: Thaw in wrappings overnight in a refrigerator, or for 3–4 hours at room temperature. Finish as required.

FINISHES

PEACH RUPERT

creamed rice—use half quantity   *Yield: 4 portions*
4 half peaches
apricot glaze (page 349, use half quantity)
whipped cream
chopped nuts

Divide the creamed rice into the individual dishes, and spread evenly across the top to give a smooth surface.
Carefully place half a peach on to the creamed rice in each dish with the rounded surface uppermost.
Glaze the fruit and surrounding rice with the hot apricot glaze.
Decorate with whipped cream and chopped nuts.

FRUIT CONDÉ

Fresh or tinned fruit may be used.

creamed rice—use half quantity *Yield: 7-inch Victoria sandwich tin;*
2 pears (or 2 peaches, or 2 bananas and   *4–6 portions*
some cherries)
apricot glaze (page 349, use half quantity)
whipped cream

Line the sandwich tin with a circle of lightly oiled greaseproof paper
and mould the rice in the sandwich tin before freezing.
Unmould after thawing.
Slice the tinned fruit. If using fresh fruit, blanch peaches and pears
in boiling syrup for 30 seconds, dip banana slices into lemon juice.
Arrange the fruit on top of the rice mould, slightly overlapping the
slices.
Stone and quarter the cherries and use to add colour when using
banana slices.
Glaze the fruit and sides of the mould with hot apricot glaze.
Decorate with whipped cream.

GELATINE-SET RICH CREAMED RICE

4 oz. Carolina rice   *Yield: 1 × 8-inch ring mould*
1 pint milk   *(6 portions) and 2 individual*
vanilla pod   *dishes*
4 egg yolks
3 oz. castor sugar
1 tablespoon gelatine
2–3 tablespoons cold water
⅛ pint double cream

Cook the rice in boiling water for 2–3 minutes, drain well, or wash
very thoroughly in cold water.
Place the rice, with ¾ pint milk and the vanilla pod, into a double
saucepan and cook slowly until tender.
Remove the vanilla pod.
Beat the yolks with the sugar until thick. Heat the remaining ¼ pint of
milk to scalding-point, then pour over the yolks and sugar and mix
in well.
Return the custard to a gentle heat and cook until thickened.

Soak the gelatine in the cold water and stir into the hot custard.
Mix the rice with the custard mixture and allow to cool.
Half whip the cream and fold into the cold mixture.
Pour into the moulds and set.

To freeze: Freeze in the moulds, unmould and pack in semi-rigid containers, or if the mould can be spared the rice may be left in the mould in the freezer.

To serve: Thaw overnight in the refrigerator or for 4–6 hours at room temperature. Dip into hot water to unmould. Finish as required.

FINISHES

INDIVIDUAL MOULDS, JELLY AND CHARLOTTE RUSSE MOULDS

Serve masked with sweetened fruit purée.

RING MOULDS

Fill the centre with fruit salad or suitable fruits in a heavy syrup; *or* fill the centre with whipped cream, sweetened with vanilla sugar and lightened by the addition of whisked egg white (allow 1 egg white to 1 pint of cream), decorate the border of rice with piped chestnut purée and cream;
or for RIZ IMPERATRICE
Soak 4 oz. diced candied fruits in 1 tablespoon of Kirsch and mix into the cooled rice mixture before folding in the cream. Pour into a ring mould and freeze in the usual way. Unmould and serve masked with a cold fruit sauce (page 235).

The following recipes are also suitable to be served as a sweet course:

Frozen Desserts

WITH the ownership of a freezer a large new range of frozen desserts and ice creams are within the scope of the housewife/hostess. Most are easily prepared, are different, and keep well.

However, they are much too cold to be served straight from the freezer. 3–4 hours' thawing in a refrigerator or, better still, 6–8 hours' thawing in the ice compartment of a domestic refrigerator at about 20°F (−7°C) will give an ice cream of the same consistency throughout. 20°F (−7°C) is the temperature maintained in the ice compartment of domestic refrigerators given a one-star rating, see page 64. The temperature found in the ice compartments of refrigerators given a two- or three-star rating would be too low to result in any significant thawing prior to serving. The difficulty when thawing is to get the centre soft without the surface becoming too thawed and runny.

Examples of several types of frozen desserts are given for water ices, ice creams, and ice-cream sweets.

WATER ICES

These are very cold to the palate and most delicious, giving a very true flavour of fresh fruit. A small portion is all that can be taken at a time.

They are made from a flavoured sugar syrup; the flavourings usually being of fruit, but liqueurs or wines may also be included.

If the syrup is frozen and used with no further beating, a coarse crystal formation results, and the water ices are known as *granite*. If a finer crystal is preferred, the syrup when frozen to a slush is beaten well until it becomes opaque. It is again allowed to freeze and is then rebeaten. Ices prepared in this way are usually called *sorbets*.

328

The water ices should be thawed either in a refrigerator for 3 hours, or an ice compartment at about 20°F (−7°C) for 6 hours. They will keep in an acceptable condition for two or three days in the ice compartment at this temperature.

The most usual flavours used in fruit water ices are strawberry, apricot, peach, melon, pineapple, lemon, grapefruit, and orange; suitable liqueurs include Kümmel, cherry brandy, Kirsch, rum, maraschino; champagne, sherry, or muscatel may also be used.

Ingredients and preparation of the basic syrup

1 lb. sugar
1½ pints water

Dissolve the sugar in the water, boil until the syrup is reduced to 1¼ pints.
Cool thoroughly and use this syrup as the basis for the preparation of the following sorbets:

Portions: About 16 portions of sorbet may be served for each pint of syrup used in the recipe.

APRICOT OR PEACH

1 pint apricot or peach purée
1 pint syrup
juice of 2 lemons

GRAPEFRUIT

1 pint syrup
1 pint grapefruit juice (from 4 large grapefruit)
juice of 1 lemon

LEMON

grated rind of 3 lemons (soaked in the syrup for 2 hours before straining)
1 pint syrup
juice of 4 lemons
juice of 2 oranges

MELON

1¼ pint syrup .
1¼ pint melon purée
⅛ pint lemon juice
1 dessertspoon brandy

ORANGE

1 pint syrup
¾ pint orange juice tinned, fresh or frozen.
¼ pint lemon juice
finely grated rind of 1 orange
finely grated rind of 1 lemon

PINEAPPLE

½ pint syrup
½ pint pineapple purée (a medium-sized pineapple should yield
½ pint of purée)
juice from ½ lemon

STRAWBERRY

1 pint purée (from 2 lb. fruit), fresh or frozen.
1 pint syrup
4 tablespoons lemon juice
4 tablespoons orange juice

Basic method for preparing a sorbet

Make and cool the syrup.
Purée the fresh, frozen, or drained tinned fruit in a liquidizer or
force through a sieve.
Thoroughly mix the cold syrup, fruit and flavourings.
Freeze until the mixture begins to ice and become slushy.
Spoon into a large chilled bowl and whisk with an electric food
mixer or a hand-operated beater until evenly mixed. The mixture
becomes opaque and increases slightly in volume.
Replace in the freezer for 2 hours, or until just beginning to ice again.
Spoon the freezing mixture out into a bowl a second time, and repeat
whisking.
Fill serving dishes, moulds, or waxed cases sufficient for individual
portions. Fruit skins, for example scooped-out melon or pineapple
shells, whole orange and grapefruit skins may be frozen separately
and used for serving the sorbets.
Pack, seal and freeze.

To serve: Partially thaw large moulds for 6–8 hours, and individual
moulds for 3–4 hours, in the ice compartment of a refrigerator at
20°F (−7°C). These ices liquefy rapidly at room temperature.

ICE CREAMS

There are many methods of making ice cream. Several require the use of an ice-cream maker which keeps the mixture in constant motion whilst freezing and so produces small ice crystals. These recipes have not been included, as they are generally available with the machines.

All the recipes given are suitable for making with no further beating necessary once they are frozen. Usually a custard made from hot milk, sugar, and egg yolks forms the first step in preparing ice creams, although for a smoother texture a sugar syrup may be combined with the egg yolks in place of the milk, as may be seen in the recipe for Fruit Ice Cream on page 333.

Ice creams made at home tend to have firmer consistencies than commercial ice creams, which are highly aerated. However, the variety of flavours which may be incorporated, combined with this unaccustomed texture, make them a welcome dessert even to the most jaded ice cream palate.

It is intended that the examples given should serve as a basis for future experiment; equally satisfactory results can be expected with all recipes of a similar type and prepared by a similar method.

VANILLA ICE CREAM

Yield: 1½–2 pints
12–16 portions

1 teaspoon gelatine
2 tablespoons water
8 oz. hot milk (½ pint minus 4 tablespoons)
2½ oz. castor sugar
½ oz. vanilla sugar
1 teaspoon flour
pinch salt
1 egg yolk
½ pint single cream
½ pint double cream

Mix the gelatine with the cold water in a basin and then add the hot milk.
Stir to dissolve the gelatine.
Mix together the sugars, flour and salt, and gradually beat the milk and gelatine into this mixture.
Pour into the top section of a double saucepan, place directly over a

low heat, and stir until the liquid has thickened. Place the pan into the outer water bath containing hot water, and continue cooking for 10 minutes. Add the yolk to the mixture in the double boiler and continue cooking for 1 minute.

Cool rapidly, then chill until just setting.

Whisk the mixture until light and frothy.

Mix the creams together and then whisk until they are thick.

Fold the custard and creams together and continue folding until well mixed.

Freeze in moulds, waxed containers, or ice trays.

To serve: Thaw in ice compartment of refrigerator at 20°F (−7°C) for 6–8 hours.

LEMON ICE CREAM

4 egg yolks

3 oz. icing sugar

juice of large lemon

¼ pint double cream

⅛ pint single cream

2 egg whites

2 tablespoons icing sugar

Yield: about 1 pint
6–8 portions

Whisk the yolks, the 3 oz. icing sugar, and lemon juice over a pan of hot water until the mixture thickens.

Remove from the heat. Whisk until cold.

Mix the creams and whisk until they are thick enough to leave a trail.

Beat the egg whites until they are stiff, but not dry.

Add the 2 tablespoons icing sugar and whisk until stiff once again.

Fold the yolk mixture and the creams together.

Fold in the egg white until evenly distributed.

Freeze in moulds, waxed containers, or ice trays.

To serve: For the best results, thaw in the ice compartment of a domestic refrigerator at 20°F (−7°C) for 6–8 hours.

CHOCOLATE ICE CREAM

1 pint milk

1 vanilla pod

5 egg yolks

4 oz. castor sugar

Yield: 2 pints
12 portions

8 oz. plain chocolate
⅛ pint water
⅛ pint double cream

Add the vanilla pod to the milk and infuse over gentle heat for 30 minutes.
Remove the pod, bring milk to scalding-point.
Beat the yolks and sugar until they are thick enough to leave a trail.
Pour in the hot milk gradually and stir well.
Return the pan to a low heat and stir until the custard thickens.
Strain the custard, cool rapidly with frequent stirring.
Melt the chocolate in the water and cool.
Half whip the cream.
Fold the chocolate into the custard.
Finally fold in the cream.
Freeze in moulds or waxed containers.

To serve: Thaw in the ice compartment of a domestic refrigerator at 20°F (−7°C), for 6–8 hours.

VARIATIONS TO THIS RECIPE

Omit the chocolate and ⅛ pint of water and make the following additions:

COFFEE

Infuse 3 oz. finely ground coffee in the milk and strain, *or* dissolve an equivalent amount of instant coffee (5–6 teaspoons).

NUTS

Select nuts according to taste (2 oz. per pint of milk). Blanch, then pound nuts in a mortar or on a board, and infuse in the milk. (Hazelnuts are first roasted before being pounded.)

VANILLA

To prepare vanilla ice, simply omit the chocolate and ⅛ pint of water.

FRUIT ICE CREAMS

RICH FRUIT ICE CREAMS

¼ pint water ⎫
8 oz. sugar ⎬ yield about ½ pint syrup
             ⎭

*Yield: 1½–2 pints
8–10 portions*

4 egg yolks
½ pint fruit purée (from 1 lb. fruit)
¼ pint double cream
¼ pint single cream

Dissolve the sugar in the water, and bring to the boil. Continue heating to 220°F (105°C). Cool slightly.
Beat the egg yolks in a bowl and pour in the syrup, beating continuously.
Pour the mixture into a double pan, return to the heat, and cook, beating all the time until the mixture resembles thick cream.
Strain into a bowl and whisk until cold, when the mixture becomes frothy and white.
Mix together the double and single creams and whisk until thick enough to leave a trail over the surface.
Fold the cream, egg mixture, and the fruit purée together.

To freeze: Pack in suitable containers.

To serve: Transfer to the ice compartment of a domestic refrigerator at 20°F (−7°C) for 6–8 hours before serving.

WHISKED FRUIT ICE CREAM

The above recipe for fruit ice cream may be slightly modified, to give a plainer ice cream, as follows:
Omit the egg yolks and single cream.
Prepare the syrup as before, cool, and combine with equal quantities of fruit purée and half-whipped double cream.
Chill the mixture 24 hours in an ice compartment of a domestic refrigerator at about 20°F (−7°C).
Remove to a bowl, whisk until double the volume.
Pack and freeze.

MIXED ICE-CREAM SWEETS

These may be prepared from home-made or commercial ice creams.

NEAPOLITAN ICES

Arrange layers of ice cream and water ices in any container to give the required shape, or line a mould with water ice, and fill with ice cream. Unmould to serve.

CASSATA

A very popular combination of ice cream, water ice, and cream, frozen together. It is usually frozen in a bombe mould, but a polythene bowl or basin with a lid is a good substitute.

½ quantity chocolate ice cream (p. 332)   *Yield: 10 portions*
¼ quantity strawberry water ice (p. 330)   *2-pint bowl container*
cream filling: ⅜ pint double cream
        1 tablespoon icing sugar
        1 tablespoon chopped candied peel
        1 tablespoon chopped angelica
        1 tablespoon coarsely chopped glacé cherries
        1 tablespoon sultanas or raisins
        1 teaspoon orange liqueur

Place the chopped candied fruits into the orange liqueur and stir well to coat the pieces.
Whip the cream and sugar until stiff.
Fold the candied fruits and liqueur into the cream.
Slightly thaw the chocolate ice cream and strawberry water ice.
Mould the chocolate ice cream in an even layer around the sides and base of the bowl. Chill if thawing too much.
Mould the strawberry water ice evenly on top of the chocolate ice cream, chilling if necessary.
Finally fill the centre with the cream mixture. Return the completed cassata to the freezer.

To serve: Thaw for 8 hours in an ice compartment of a domestic refrigerator at 20°F (−7°C). Dip the bowl in hot water for a second or two, invert on to the serving dish. Cut and serve in wedge-shaped portions.

ICE-CREAM CHARLOTTE

This can be made either using a scooped-out genoese cake, or using sponge fingers as when making charlotte russe.
The filling is made from alternate layers of ice cream and fruit water ice, or alternatively ice cream layered with fruits in syrup or whole fruit jam.

vanilla ice cream (page 331, use half quantity)
fruit water ice (page 329, use half quantity)
    or 8 oz. fruit cooked in syrup
    or 8 oz. whole fruit jam

1 × 8-inch genoese cake (page 348)
    or 1 × 6-inch charlotte mould lined with 2–3 dozen sponge fingers
apricot glaze (page 349, use quarter quantity)

Mark a circle on the top of the genoese cake ½ inch from the edge. Cut
through the marks and remove the centre circle ½ inch thick in one
piece. This acts as the lid to the sweet.

Scoop out the cake mixture, leaving the sides and base ½ inch thick.
(Alternatively line the sides and base of the charlotte russe mould
with trimmed sponge fingers.)

Place alternate layers of ice cream and fruit water ice (or fruit or jam)
inside the case, starting and finishing with a layer of ice cream.

Place the lid on to the genoese cake. Quickly brush all over with hot
apricot glace. Pack, seal, and freeze.

If the charlotte mould has been used, freeze the contents in it, then
unmould, pack and seal.

To serve: Thaw in a refrigerator for 4–6 hours, or in an ice compartment at 20°F (−7°C) for 8 hours, before serving.

ICE-CREAM ROLL

Plain or chocolate-flavoured Swiss rolls filled with various ice creams,
and fruit in syrup if desired.

1 sponge or genoese Swiss roll (page 347 or 348)
½ pint ice cream

Cook the Swiss roll, roll up unfilled and cool.
Partially thaw ice cream.
Open the roll gently, and spread with ice cream, and fruit if used, in a
generous layer.
Reroll the Swiss roll.
Wrap in suitable freezer wrappings and freeze.

To serve: Thaw in a domestic refrigerator for 3–4 hours, or in the ice
compartment of a domestic refrigerator at 20°F (−7°C) for 5 hours.
Serve in slices with fruit.

FROZEN DESSERTS

FROZEN APRICOT MOUSSE

4 oz. dried apricots   *Yield: 8 portions*
3 eggs

3 oz. castor sugar
¼ pint double cream
⅛ pint single cream

Soak the apricots in water, then cook in the same liquid until tender.
Strain the fruit, and blend in a liquidizer, or sieve.
Beat the eggs and sugar until they are thick enough to leave a trail.
Mix the creams and whisk until this mixture is thick enough to leave
a trail.
Fold the apricot purée into the egg mixture.
Finally fold in the cream.
Place into a mould, waxed container, or ice trays. Freeze quickly.

To serve: Thaw for at least 6 hours in a refrigerator.

BISCUIT TORTONI

2 egg whites   *Yield: 8 portions*
4 oz. icing sugar
½ pint double cream
½ pint single cream
¼ oz. vanilla sugar
4 oz. crushed ratafias or macaroons
⅛ pint Madeira or sherry

Mix the creams together. Beat until thick, then beat in half the icing
sugar and the vanilla sugar.
Whisk the egg whites until stiff, and beat in the remaining icing sugar.
Fold the whisked egg whites into the cream mixture.
Finally fold in the Madeira or sherry and crushed ratafias or maca-
roons.
Mould in individual portions in waxed cases, or dishes, or in a mould.
Freeze quickly.

To serve: Thaw for 3–4 hours in a refrigerator. Serve whilst still
chilled and slightly hard.

ICE-CREAM SUNDAES

Given a supply of ice cream in the freezer, whether commercial or
home-made, there is no end to the variety of ice-cream desserts which
can be made simply and quickly. A list of suitable additions which
can be made or served with a plain ice cream is given overleaf:

Defrosted fresh fruits—raspberries, peaches, strawberries, apricots
Fruit purées
Fruit sauces (page 235)
Apricot glaze (page 349)
Chocolate sauce (page 238)

Serve with:

Cats' tongues (page 364)
Meringues
Shortbread fingers (page 365)

Decorate with:

Coconut—fresh if available
Grated chocolate
Glacé fruits
Whipped cream
Roasted almonds
Whole walnuts

Cakes - Preparation and Finishing

THE freezer provides an excellent means of storing cakes which would otherwise only remain fresh and acceptable for a short time. Raw or cooked, plain or rich, decorated or undecorated, they are usually ready for eating within a couple of hours' thawing or cooking time. If a food mixer is also available, enough cakes for one month or more are easily and efficiently prepared in a single baking session. Large amounts of several mixtures can be made, and each one slightly varied to give a selection of cakes of different sizes and flavours.

Freezing raw cake mixtures

Some raw mixtures freeze well, and if there is insufficient time available during the preparation session to cook all the cakes, then it is a useful alternative. However, although raw cake mixtures may be stored satisfactorily in cartons, they must be at least partially thawed before they can be transferred to tins. If the tin can be spared from everyday use, it is easier to freeze the cake mixture in the tin in which it will ultimately be cooked. The sides and base of tins likely to rust should be lined first with greaseproof paper or foil. Where the tin cannot be spared from everyday use, the cake mixture may be frozen in the appropriate tin lined with aluminium foil; when the cake mixture is frozen it may be removed complete with the aluminium foil lining and packed and sealed. When required it may be returned to the original tin for baking.

Cakes made from frozen raw mixture may have a slightly smaller volume, as the raising action may not be quite as effective after frozen storage; but they are likely to have a better flavour. There is also a saving of space achieved by freezing the raw cake mixture rather than the completed cake. However, it is necessary to take into account the baking and cooling time for cooking a raw mixture, against only the thawing time required for pre-cooked cakes.

When all the factors have been considered, the decision whether to freeze the raw cake mixture or the completed cake remains one of personal preference and convenience. However, it is worth making the following distinctions between the different cake mixtures:

Plain cakes made by the rubbing in method are quick to prepare and there seems little point in freezing the raw mixture. But after baking they have a limited keeping time, and thus it is well worth while to freeze the baked product, which will be found moist and fresh when thawed several weeks or months later.

Rich cake mixtures take a longer time to prepare, but remain fresher for a longer period after baking under normal conditions of storage than do plain cakes. Thus they would be of equal use raw or cooked in the freezer.

Sponge cakes stale quickly and preferably should be eaten on the same day as baking—thus freezing the baked product is an advantage. The raw sponge mixture does not freeze satisfactorily.

Rich fruit cakes keep so well in an airtight tin and at the same time mature and mellow, that there seems little point in using valuable freezer space to store them.

Thawing and baking raw cake mixtures

Small cakes frozen in their tins or waxed paper cases, and mixtures frozen in sandwich tins, may be put directly from the freezer to the oven to bake. However, it is advisable to thaw larger cake mixtures for at least an hour at room temperature first.

Packing baked cakes

Cakes are fairly brittle after freezing and therefore, after sealing in a suitable packaging material, should be placed in cardboard boxes where possible for added protection. Iced cakes should be frozen prior to wrapping to prevent damage to the decoration. This will take 1–2 hours, after which they may be wrapped and placed in a box.

Thawing baked cakes

Undecorated cakes may be thawed in a fairly hot oven for about 10–15 minutes. Leave in the freezer wrappings if these are suitable for the oven, or rewrap in foil. Cakes thawed in the oven more closely resemble the freshly baked product than cakes thawed at room temperature, although they may become dry more quickly afterwards.

Retain the freezer wrappings on cakes whilst thawing at room temperature.

Iced cakes should likewise be thawed in the wrappings, otherwise condensation may occur on the cold surface of the icing which will spoil the appearance. The wrappings should be loosened so that they do not rest on the surface of the cake, or alternatively the cake may be removed from the freezer wrappings and placed in an air-tight container to thaw out.

BATCH BAKING

This would seem to be the best method of making use of the freezer when making cakes. Therefore the procedure followed in this section is to give directions for the preparation of the basic cake mixture in reasonably large quantities, giving in each case the weight of raw mixture to be expected. For each type of cake mixture a table is given setting out the weights of raw mixture required for cake tins of differing sizes. (For tins which have sloping sides the measurement is always taken at the base of the tin and not the top.) Suggestions of suitable additions and variations to the basic cake mixtures are also included in each case.

The second part of this section on cake making is devoted to the preparation of icings and fillings which may be used on and in cakes stored in the freezer, as well as for use with cakes which have been frozen prior to finishing. Some suggestions for the finishing of small cakes and sandwich cakes (or layer cakes) are included.

Plain Cake Mixtures

Here the proportion of fat to flour is not more than half, and so it may easily be rubbed into the mixture.

The weight of raw mixture produced from the basic mixture given below is about 2¾ lb.

The following table gives an indication of the amount of raw mixture required by the different sizes of cake tin.

		Round			Loaf	
Size of tin	5 in.	6 in.	7 in.	8 in.	1 lb.	2 lb.
Weight of raw mixture	12 oz.	1 lb.	1½ lb.	2 lb.	1 lb. plain to 1¼ lb. fruit	2 lb. plain to 2½ lb. fruit

341

Basic recipe for a plain cake mixture

1 lb. flour
¼ teaspoon salt
6 teaspoons baking powder
8 oz. margarine
8 oz. castor sugar

4 eggs
8 tablespoons milk } In small cakes the number of eggs may be reduced from 4 to 2 and an additional ¼ pint milk added to make up the liquid.

A flavouring is required, otherwise the cake is plain and tasteless.

Sift the flour, baking powder, and salt into a bowl.
Place the fat into the flour, cut into small pieces. Rub into the flour with the fingertips until it resembles fine crumbs.
Add the sugar.
Slightly beat the eggs, and mix with the dry ingredients.
Add sufficient milk to give the correct consistency; stiff for the small cakes and slightly slacker for the large cakes (until the mixture will just drop from the spoon).
Place the mixture for the small cakes into paper cases, greased patty tins, or in heaps on a greased baking sheet.
Prepare the large cake tins with a circle of greased greaseproof paper at the base.
Weigh the tin and add the appropriate amount of cake mixture.
Freezing raw plain cake mixtures prior to baking is not recommended.

	Oven temperature	*Time*
Small cakes	400°F (205°C), Regulo 6	10–15 minutes
Large cakes	350°F (175°C), Regulo 4	1–1½ hours

SUITABLE FLAVOURINGS OR ADDITIONS TO PLAIN CAKE MIXTURES

These are sifted in with the flour or added to the mixture together with the sugar:

FRUIT

4 oz. currants
4 oz. sultanas } add with the sugar

LUNCHEON CAKE

½ teaspoon cinnamon ⎫
½ teaspoon nutmeg   ⎬   sift together with the flour
½ teaspoon mixed spice ⎭
grated rind of lemon ⎫
3 oz. currants   ⎪
3 oz. sultanas   ⎬   add with the sugar
2 oz. chopped mixed peel ⎭

COCONUT

4 oz. desiccated coconut
or 6 oz. freshly grated coconut plus coconut milk in place of milk in recipe.

To freeze: Pack large cakes separately in sheet wrapping or polythene bags.
Pack small cakes into bags.

To serve: Thaw in wrappings.
Small cakes—at least 30 minutes at room temperature, or 10–15 minutes in a fairly hot oven covered with foil.
Large cakes—at least 2 hours at room temperature, or 20–30 minutes in a fairly hot oven covered with foil.

RICH CAKE MIXTURES

In rich cakes the amount of fat and sugar is the same as that of flour and the method of mixing used is to cream together the fat and sugar. Large food mixers are able to cream satisfactorily 1 lb. fat plus 1 lb. sugar. Smaller hand models cannot efficiently cream more than 8–12 oz. of fat plus the same amount of sugar. The amount which may be comfortably creamed by hand is 8 oz. each of fat and sugar.

The weight of raw mixture produced from the basic recipe given on the following page is a little less than 4 lb.

The following table gives an indication of the amount of raw mixture required for tins of different sizes, and small cakes, so that it is possible to estimate the yield from a certain weight of mixture in terms of the finished product.

	Sandwich tins	Yorkshire pudding tin	Ring mould	Small cakes	Large cake tin
Size of tin:	6 in. 8 in.	5½ × 9 in.	7 in.		8 in.
Weight of raw mixture:	8 oz. 11 oz.	11 oz.	9 oz.	1 oz. each	2 lb.

Basic recipe for a rich cake mixture

1 lb. margarine or butter
1 lb. castor sugar
8 eggs
½ teaspoon salt
2–3 teaspoons baking powder (the larger amount is used when the mixture is hand creamed).
1 lb. flour (in some cases when a slightly less rich cake is required the amount of flour may be increased).

Stand the margarine in a warm place overnight or for some hours, until well softened, but on no account allow to oil.
Use eggs at room temperature. This makes the cake easier to cream and less likely to curdle.
Beat the margarine and sugar together in a mixing bowl until the mixture is light and fluffy.
Beat the eggs slightly, and gradually add them to the creamed fat and sugar, beating well between additions.
Sift the flour, salt and baking powder and fold lightly into the creamed mixture. Divide the mixture between the tins and paper cases as required; sandwich tins should have a circle of greaseproof paper in the base, and large tins be fully lined.
Freeze the raw cake mixture in tins or cartons, or bake as follows:

	Baking temperature	*Time*
Small cakes	400°F (205°C), Regulo 6	15–20 minutes
Sandwich cake	375°F (190°C), Regulo 5	20–30 minutes
Large cakes	350°F (175°C), Regulo 4	45 minutes to 1½ hours

VARIATIONS OF RICH CAKE MIXTURE

The basic rich cake mixture may be varied by making the following additions. The amounts given are for half the basic recipe, i.e. 8 oz. each fat and sugar.

Victoria Sandwich
The basic mixture is used with no additions or modifications.

Slab or Base Cake—from which small fancy cakes may be prepared, decorated and iced

2 oz. ground almonds	sieve in with the flour

Madeira Cake

4 oz. flour	sieve and fold in the increased quantity of flour as usual
4 tablespoons milk	add the milk when folding in the dry ingredients
sliced candied peel	place the candied peel on the surface of the cake for decoration before baking

Queen Cakes

grated rind of lemon	add the lemon rind to the creamed fat and sugar
3 oz. washed dried currants	add the currants when the flour has been partially folded into the creamed mixture

Orange Sandwich Cake
substitute 1 oz. cornflour for 1 oz. flour

grated rind of 2 oranges	add the grated orange rind to the creamed fat and sugar
1 tablespoon orange juice	add the orange juice when folding in the flour to the creamed mixture

Chocolate Sandwich Cake
reduce the flour by 1 oz.

2 oz. cocoa 2 oz. ground almonds }	sieve the cocoa and ground almonds together with the flour
2 tablespoons brandy, brandy and water, or water	Stir in the brandy when folding in the dry ingredients

Cherry Cakes

8 oz. cherries cut in half 4 oz. angelica chopped }	or 12 oz. cherries	toss cherries in ground almonds and add together with the dissolved tartaric acid when folding in the flour to the creamed mixture

2 oz. ground almonds
½ teaspoon tartaric acid dissolved in 2 teaspoons water

CHOCOLATE CUP CAKES

4 oz. drinking chocolate 4 oz. chopped walnuts 2 dessertspoons coffee essence 8 oz. plain chocolate cut into pea-sized pieces	sieve the drinking chocolate together with the flour; fold into the creamed mixture together with walnuts, essence, and chocolate pieces

To freeze: Freeze *raw cake mixture* in tins or cartons as is convenient. Cool *baked cakes*, separate sandwich cake halves with double thickness of cellophane, and wrap in suitable freezer wrappings. Pack in useful quantities.

To serve: Place *raw cake mixture* for small cakes into a preheated oven directly from the freezer, but thaw cake mixtures for large cakes for about 1 hour at room temperature before baking. Baking will take a little longer than for an unfrozen mixture.

Thaw *baked cakes* in freezer wrappings.

Small cakes—at least 30 minutes at room temperature, or 10–15 minutes in a fairly hot oven.

Large cakes, sandwich cakes—at least 2 hours at room temperature, or 20–30 minutes in a fairly hot oven.

Use or decorate as required.

QUICK-MIX CAKES

These are made using a soft-blended margarine specifically formulated for this type of mixture. After making, weigh into tins as for the basic rich cake mixture.

8 oz. soft-blended margarine
8 oz. castor sugar
4 eggs
8 oz. flour
4 teaspoons baking powder

Sieve the flour and baking powder into a bowl. Add the other ingredients and beat together until well mixed and smooth.

Bake at 335°F (170°C), Regulo 3-4, for 30–40 minutes.
Freeze and use as for the rich cake mixture.

Sponge Cake Mixtures

The amount of raw sponge mixture required for the different sizes of cake tin is as follows:

	Swiss roll	Yorkshire pudding	Sandwich tin		Cake tin	Flan tin
Size of tin:	7 × 11 in.	5½ × 9 in.	2 tins of 6 in. diameter	2 tins of 8 in. diameter	8 in.	8 in.
Weight of raw mixture:	9 oz.	9 oz.	5 oz. per tin	7 oz. per tin	13 oz.	5 oz.

Basic recipe for a sponge cake mixture

6 eggs
6 oz. castor sugar
5 oz. flour

Whisk the eggs and sugar over hot water until the mixture thickens enough to leave a trail over the surface when allowed to run off the spoon.

Remove from the heat and whisk until cold (or whisk the eggs and sugar for about 6 minutes in a food mixer until the foam is thick and there are no large bubbles left visible).

Sift in the flour one-third at a time, and fold in with a metal tablespoon. Pour into the prepared tins and bake as follows.

Freezing sponge mixtures raw is not recommended.

Type of Cake tin	*Baking temperature*	*Time*
Sandwich tins: sides greased and base lined with greased greaseproof paper	375°F (190°C), Regulo 5	20 minutes
Flan tin: well greased and floured	375°F (190°C), Regulo 5	20 minutes
Swiss roll tin: fully lined with greased greaseproof paper	425°F (220°C), Regulo 7	7–10 minutes

CHOCOLATE VARIATION

6 eggs

7¾ oz. castor sugar

¼ oz. vanilla sugar

3 tablespoons water

3 oz. flour

1½ oz. cocoa

mix the water in with the eggs and sugar before beginning to whisk; sift the cocoa in with the flour

To freeze: Cool sandwich cakes, and separate layers with a double thickness of cellophane.

Pack cooled flan cases on a circle of cardboard or in a box for added protection.

Swiss rolls may be frozen before or after filling—for notes on filling with jam and cream see page 356. Turn out Swiss roll to be frozen unfilled on to sugared paper, trim the sides and roll up; when cool, unroll and reroll around cellophane and pack in suitable wrapping.

To serve: Thaw for 1½–2 hours at room temperature. Decorate and use.

GENOESE MIXTURE

This is a sponge incorporating fat.

Basic recipe for a genoese mixture

6 eggs

8 oz. castor sugar

6 oz. flour

4 oz. butter

Melt the butter, clarify if preferred, but it is not essential. Allow to cool slightly.

Sift the flour and put in a warm place.

Whisk the eggs and sugar over hot water until they are thick and leave a trail over the surface of the mixture. Remove from the heat and whisk until cool.

Or whisk in a food mixer for about 6 minutes until a thick foam is formed with no large bubbles visible.

Pour a little of the butter around the edge of the bowl.

Sift one-third of the flour into the foam, and fold in lightly, lifting the butter from the base of the bowl.

Add the remaining butter and flour in thirds.

Mix until the foam is even and the flour and butter have been thoroughly incorporated.

348

This mixture is suitable for cakes, Swiss rolls, sandwich cakes, slab cakes for icing, and flan cases.

Refer to the previous recipe and notes on sponge for details of tin preparation, weight of raw mixture to allow, and baking details. Like the fatless sponge, genoese is not suitable to be frozen as raw mixture.

CAKE ICINGS AND FILLINGS

The following icings and fillings are suitable for use on and in cakes to be frozen.

APRICOT GLAZE

4 oz. apricot jam
⅛ pint water
1 tablespoon lemon juice

Mix together and boil for 5 minutes. Strain. Reheat to use.

BUTTER CREAM OR BUTTER ICING

Any butter icing is suitable. The following recipe gives a creamy icing which is not too sweet.

6 oz. unsalted butter
4 oz. icing sugar
2 egg yolks

Place the butter in a bowl and beat until fluffy. Gradually beat in the icing sugar. Add the yolks and continue beating. Finally mix in the flavouring.

SUITABLE FLAVOURINGS

CHOCOLATE

2 oz. melted plain chocolate
½ teaspoon coffee essence

ORANGE

finely grated rind of 1 orange
1 tablespoon orange juice or orange liqueur

VANILLA

Substitute ½ oz. vanilla sugar for ½ oz. icing sugar

COFFEE
2 tablespoons coffee essence, or strong blended instant coffee

CHOCOLATE BUTTER CREAM II
This icing is excellent with sponge cakes.

3 oz. plain chocolate
10 oz. icing sugar
¼ oz. vanilla sugar
3 tablespoons hot water
3 egg yolks
4 oz. butter

Melt the chocolate. Add the sugars and hot water and beat well.
Add one yolk at a time and continue beating. Add the softened
butter, one ounce at a time. Beat until smooth.

MERINGUE BUTTER CREAM
This is a butter cream icing incorporating beaten egg white.

6 oz. unsalted butter
4 oz. icing sugar
2 egg whites
flavouring

Thoroughly cream the butter. Whisk the egg whites and sugar over
hot water until thick, remove from the heat and continue to whisk as
the mixture cools. Gradually work into the creamed butter. Mix in
the flavouring.

GLACÉ ICING FOR THE FREEZER
4 oz. granulated sugar
⅛ pint water
1 tablespoon liquid glucose or corn syrup
12 oz. icing sugar

Dissolve the granulated sugar in the water and boil for 3 minutes.
Add the glucose, or corn syrup. Cool until the pan is comfortable to
hold. Beat in the sieved icing sugar. Use when warm. Colour and
flavour as required.

350

CHOCOLATE GLACÉ ICING

2 oz. chocolate
2 tablespoons water
4 oz. icing sugar
1 dessertspoon corn syrup or liquid glucose

Break the chocolate into pieces, place into the water over gentle heat until it melts. Bring to the boil. Remove from the heat and add the corn syrup or liquid glucose. Cool the pan until it is comfortable to hold. Add the sieved icing sugar and beat well until smooth. Use when warm, adjusting the consistency with water if necessary.

ORANGE GLACÉ ICING

10 oz. icing sugar
juice 1 orange
1 teaspoon liquid glucose
rind of 1 orange cut into short strips (julienne)—optional

Peel the orange rind very thinly, without taking any of the pith, and cut into very fine strips (julienne). Boil in water for 5 minutes until tender. Drain. Sieve the icing sugar. Warm the orange juice and add sufficient together with the glucose to the sieved icing sugar to give a coating consistency when warm. Mix in the julienne strips and use.

ALMOND PASTE

8 oz. ground almonds
6 oz. castor sugar
6 oz. icing sugar
1 egg, or 2 egg whites, or 3 egg yolks
1 tablespoon sherry—optional
few drops almond essence if necessary

Sieve the ground almonds and sugars. Add the sherry, essence, and sufficient egg to give a crumb-like mixture. The crumbs should hold together when firmly pressed in the hand. If they break away again, a little more egg is needed.

351

CAKE FINISHES—ICING, FILLING AND DECORATING

The cake bases may be Victoria sandwich, quick-mix cake, sponge or genoese. The following suggestions for icing, filling and decorating may be executed before or after freezing.

FINISHING—SMALL CAKES

CUP CAKES

These are small cakes baked and served in paper cases.
Finish the tops with flavoured glacé icing—use lemon, orange, chocolate, according to taste.
3½ dozen cakes require the full quantity of the basic glacé icing mixture (page 350).

BUTTERFLY CAKES

icing sugar
butter cream (page 349, half quantity sufficient for 3½ dozen cakes)
flavouring
Remove a slice from the top of each cake. Cut it into half and dust the top with icing sugar. Pipe a rosette of the butter cream into the centre of the cake and place the cut pieces of cake into this to resemble wings. A thin line of butter cream may be piped between the wings, to give a neater finish.

ICED FANCIES

The preparation of iced fancies is a particularly time-consuming task, since it is necessary first to prepare several different icings. It is easier and more economical to prepare these cakes in large quantities, when it is possible at the same time to achieve a greater assortment. Iced fancies freeze and keep well, but may be thawed quickly at room temperature when required for a tea party.

The slabs of cake for the preparation of iced fancies should be about 1 inch thick. The quantities of icings given are sufficient for two slabs of cake 5½ × 9 inches. The amounts of raw rich cake mixture, quick-mix and sponge cake to produce this amount of cake

13a. Brushing with apricot glaze

PREPARATION OF ICED FANCIES (*page 352*)

13b. Coating with glacé icing

14a. Shaped Loaf, page 384

14b. Croissants, page 390

base are given in the previous section on pages 344, 346, and 347. The slab may be plain or flavoured.

Glacé icing	(page 350, full quantity)	*Yield: about 48 cakes*
Chocolate glacé icing	(page 351, half quantity)	
Butter cream	(page 349, full quantity)	
Almond paste	(page 351, half quantity)	
Apricot glaze	(page 349, full quantity)	

Decorations: glacé cherries, angelica, mimosa balls, frosted petals, dragees, etc.

Cut the cakes into shapes such as circles, crescents, triangles, squares, oblongs, no more than 1½ inches across.

Roll the almond paste to ⅛ inch thickness in icing sugar and cut into shapes corresponding to the cakes.

Any remaining paste can be shaped into balls and placed on top of the cakes under the icing, or coloured and used to decorate the cakes.

Coat the top and sides of the cakes with hot apricot glaze and place the almond paste on top.

Warm the glacé icing, place the cakes one at a time at the end of a palette knife held over the icing and spoon icing across each cake to coat it completely.

Allow to drain and set on a cooling tray.

Decorate with piped butter cream, glacé icing, or cake decorations.

Alternatively, after coating with apricot glaze, spread the sides with a thin layer of butter cream and roll the cakes in chopped nuts, flaked chocolate, or vermicelli.

Coat the top surface with glacé icing and finish as for the chocolate gâteau on page 354.

Or as a further alternative, both the sides and tops of the cakes may be spread with butter cream and coated with nuts or chocolate.

To freeze: Freeze before packing for 2–3 hours. Wrap in suitable freezer wrappings and place inside a box for added protection.

To serve: Thaw in the wrappings, loosening them where necessary to prevent their sticking to the surfaces of the cakes. Or remove cakes to an air-tight container and thaw.

SANDWICH CAKES OR LAYER CAKES

Use bases of rich cake mixture, quick-mix, sponge, or genoese, pages 344–349.

ICED ORANGE CAKE
(coated all over with glacé icing)

For two 6–inch or two 8–inch sandwich cakes the following quantities of icing are required:

Orange-flavoured butter cream   (page 349, half quantity)
Orange icing   (page 351, full quantity)
Apricot glaze   (page 349, half quantity)

Either join the sandwich cakes together with the butter cream or slice each sandwich cake into two and join the four layers of cake with three layers of butter cream.
Brush away any loose crumbs; fill the crevices with more butter cream.
Brush all over with warmed apricot glaze and dry for several minutes. Place the cake on to a cooling tray over a crumb-free surface.
Warm the icing and pour liberally across the cake so that the top is completely covered. Tap the cooling tray lightly on to the table to help the flow of the icing. Scoop icing from beneath the cake to cover any bare patches on the sides.
Decorate with piped butter cream, glacé icing, or cake decorations.

CHOCOLATE GÂTEAU
(finished in butter cream and glacé icing)

For two 6-inch or two 8-inch chocolate sandwich cakes the following quantities of butter cream and glacé icing are required:

Chocolate butter cream   (page 349, full quantity)
Chocolate glacé icing   (page 351, half quantity)
2 oz. chopped walnuts, flaked almonds, or chocolate vermicelli

Reserve 2 tablespoons of butter cream in a piping bag fitted with a small star pipe, for finishing the cake.
Join the sandwich cakes together with butter cream, or slice each sandwich cake in two and join the four layers of cake with three layers of butter cream. Spread a thin layer of butter cream around the sides of the cake and smooth with a palette knife to give an even finish.
Turn the sandwich cake on to its side and roll in the chopped nuts or vermicelli; when evenly coated place on to a cooling tray.

Warm the chocolate icing and pour on to the top of the cake. Ease the icing towards the edges of the cake. If it overflows, allow the icing to harden and lift off.

Finish the top edge of the cake with piped stars, shells or other edging.

These two methods of icing and decorating sandwich cakes which have been described may be used for sandwich cakes of all flavours, providing butter cream and glacé icing of corresponding flavours are used.

To freeze: Place the gâteau uncovered in the freezer until hard. Pack and place inside a box for added protection.

To serve: Loosen packaging from the surface of the cake. Thaw in the packaging or in an air-tight container for 2 hours.

BATTENBURG CAKE

Unless a divided cake tin is available it is more convenient to make two Battenburgs at one time.

The recipe given makes two finished cakes.

1 slab basic cake mixture 5½ × 9 inches
1 slab basic cake mixture coloured pink with food colouring, 5½ × 9 inches

} use rich cake mixture, page 344, or quick mix, page 346

apricot glaze (page 349, double quantity)
almond paste (page 351, as recipe)

Trim the edges of each cake, cut each into quarters lengthwise, giving four strips 1–1½ × 8 inches.

Use two pink strips and two plain strips for each cake.

Place together in two layers, alternating the colours to give the usual chequer-board appearance of a Battenburg cake.

Generously brush the inside edges of each strip with apricot glaze and rejoin them together.

Cut a rectangle of greaseproof paper to fit around the outside of the cake.

Take half the almond paste; sprinkle the paper with castor sugar and roll out the almond paste on top and to fit the paper.

Brush the almond paste with apricot glaze.

Trim the long sides of the paste, place the cake with one edge against the trimmed side. Tightly wrap the almond paste and paper around the cake.

Seal the cut edges of paste together when they meet.

Remove the paper.

Shape the cake by rolling with a rolling pin until square in outline.

Finish the cake by crimping the top long edges, then trim each end.

To freeze: Wrap the Battenburg in suitable wrapping film and freeze.

To serve: Thaw for 1½–2 hours inside loosened packaging material.

SWISS ROLLS

The filling and finishing of Swiss rolls may be completed before or after freezing. If using a Swiss roll which has been frozen, thaw at room temperature for about 2 hours or until it may be unrolled, and proceed as for Swiss rolls freshly baked.

Quantities given are sufficient for 1 Swiss roll base 7 × 11 inches prepared from the sponge cake (page 347) or genoese mixture (page 348).

JAM

4 oz. warm jam	when cooked turn out the Swiss roll on to sugared paper and spread with warm jam (if using a previously frozen Swiss roll base, cold jam may be used); roll up, trim the edges, and sprinkle lightly with castor sugar when serving

BUTTER CREAM FILLING

butter cream (page 349, use half quantity)	turn out cooked Swiss roll on to sugared paper, trim the long sides and promptly roll up; allow to cool; gently unroll, spread evenly with butter cream, re-roll, trim ends, and serve

CHOCOLATE LOG

chocolate butter cream (page 350, use full quantity)	for a richer finish or for chocolate logs, use butter cream to coat the outside of the Swiss roll—roughen the surface with a fork to give the impression of the bark of a tree.

To freeze: Freeze filled Swiss rolls prior to packaging. Then seal in suitable freezer wrappings—support the roll first on a rectangular piece of cardboard.

To serve: Thaw in loosened wrappings or in an air-tight container for about 2 hours at room temperature.

PASTRIES

CREAM HORNS

12 cream-horn cases of puff pastry (page 373)
$\frac{1}{4}$ pint double cream
$\frac{1}{2}$ oz. vanilla sugar
2 oz. raspberry jam

Bake and cool the cases.
Place a little jam at the base of each case.
Pipe the whipped cream flavoured with the vanilla sugar through a $\frac{1}{2}$-inch star piping nozzle into each case, finishing with a rosette on the top.

ECLAIRS AND PROFITEROLES

6–8 éclairs or 18–24 profiterole cases—depending on size (page 379)
$\frac{1}{4}$ pint double cream
$\frac{1}{2}$ oz. vanilla sugar
chocolate glacé icing (page 351, full quantity)
    or coffee glacé icing may be used for profiteroles (page 350, full quantity) or commercial chocolate couverture is also satisfactory for use on éclairs to be frozen.

Whip the cream with the vanilla sugar and place in a piping bag with a small nozzle. Insert the nozzle into the split pastry cases and fill with cream.
Prepare the chocolate or coffee icing and keep warm until used.
Dip the top of the éclairs or profiteroles in the icing.
Put to set on a wire tray.
Serve, or freeze the filled cases until required.

To freeze: Freeze without wrapping until hard, seal in suitable freezer wrappings and place in a box for added protection.

To serve: Loosen wrappings so that they do not bear down on the surface of the icing. Thaw in a refrigerator 4–6 hours or at room temperature for 1–2 hours.

These pastries should be eaten as soon as they are thawed or the pastry will become soggy.

Scones, Tea Breads, Biscuits

SCONES

ALTHOUGH scones may be made quickly and easily, they stale rapidly and should be made and eaten on the same day. However, two or three times the usual mixture may be made at one time with little extra work involved, and frozen in convenient numbers for serving. Thaw in an oven for a fresh-baked smell and taste.

A freshly prepared raising agent should be used to ensure light, well-risen scones. Thoroughly mix together cream of tartar and bicarbonate of soda in the proportions of 2:1—mix together only as much as is required at a time.

OVEN SCONES

8 oz. flour
pinch salt
2 teaspoons freshly prepared raising agent
2 oz. margarine
1½ oz. sugar
4 oz. milk (¼ pint minus 2 tablespoons)

Yield: 18–24 scones
Oven: 475°F (245°C),
Regulo 9.
Time: 10 minutes

Sift the flour, raising agent, and salt, several times.
Cut in the margarine, then rub into the mixture until it is like fine breadcrumbs.
Add the sugar.
Mix to a stiff paste with the milk.
Immediately roll out to ½-inch thickness, cut into shapes with a 2-inch plain cutter; or cut dough in half and shape each half into a round ½ inch thick, and mark into eight or twelve segments.
Place on to a heated baking sheet. Bake in a hot oven for 10 minutes.

VARIATIONS

FRUIT SCONES

2 oz. dried fruit · · · · · · · · · · · mix in with the sugar

CHEESE SCONES

Omit the sugar
1 teaspoon salt
pinch cayenne pepper ⎫
3 oz. grated cheese ⎬ sift in with the flour

mix in before adding milk

Bake at 450° F (230° C), Regulo 8, for about 10–15 minutes

GIRDLE SCONES

Any scones may be cooked on a moderately hot girdle for 4 minutes each side instead of baking in the oven.

To freeze: Cool and pack in useful quantities.

To serve: Thaw at room temperature in wrappings for about 1 hour, or in a fairly hot oven covered in aluminium foil for 10 minutes.
Another alternative method is to split the scones while still frozen and place them under a hot grill.

DROP SCONES

8 oz. flour · · · · · · · · · · · · · · · · · · *Yield: 24–30 scones*
pinch salt
1½ teaspoons cream of tartar
¾ teaspoon bicarbonate of soda
2 oz. margarine
1 oz. sugar
1 standard egg
¼ pint milk

Mix the cream of tartar and bicarbonate of soda, add to the flour and salt, and sift into a bowl.

Cut in the fat, and rub in until the mixture resembles breadcrumbs. Add the sugar. Pour in the beaten egg and some milk, beat well together, add sufficient milk to give a batter which just drops off the spoon.

Pour in tablespoons on to a moderately hot girdle. Cook on one side until bubbles are seen on the surface—about 3 minutes—then turn over and continue cooking.

Place to cool covered with a tea towel so that they remain moist.
Serve whilst warm, or cool to freeze.

To freeze: Pack and seal in suitable quantities.

To serve: Thaw at room temperature in the freezer wrappings for 30
minutes, or spread out frozen scones on to a baking sheet and cover
with foil and place into a cold oven set at 400°F (205°C), Regulo 6.
Warm through for 10–15 minutes.

TEA BREADS, WITHOUT YEAST

DATE AND WALNUT LOAF

This semi-rich cake is excellent thinly sliced and buttered as a tea
bread.

1 lb. flour

$\frac{1}{4}$ teaspoon mixed spice

$\frac{1}{4}$ teaspoon salt

2 oz. sugar

$\frac{1}{2}$ oz. vanilla sugar

3 oz. walnuts

10 oz. whole dates

1$\frac{1}{2}$ oz. butter

1$\frac{1}{2}$ teaspoons bicarbonate of soda

8 tablespoons boiling water

1 large egg

4 tablespoons malt extract

4 tablespoons golden syrup

*Yield: 1 × 2 lb. loaf tin
and 1 × 1 lb. loaf tin*

Oven: 400°F (205°C), Regulo 6

Time: 1 hour

Place the dates, butter, and bicarbonate of soda into a basin and pour
on the boiling water. Mix well together and allow to stand.
Beat the egg with the syrup and the malt extract.
Sieve the dry ingredients together. Pour in the date mixture and stir
well.
Add the egg mixture, and chopped walnuts. Stir until well mixed.
Pour into tins lined with greased greaseproof paper.
Bake for 1 hour, test with a skewer to see that it is fully cooked.

To freeze: Cool and pack in suitable freezer wrappings.

To serve: Thaw at room temperature for 2–3 hours.

Spiced Apple Loaf

4 oz. margarine
8 oz. sugar
2 eggs
1 lb. flour
2 teaspoons cinnamon
good pinch nutmeg
good pinch ginger
½ teaspoon ground cloves
½ teaspoon salt
2 teaspoons bicarbonate of soda
4 oz. sultanas
1 lb. apple purée

Yield: 2 loaves
Size of tin 7½ × 3½ inches
Oven: 350°F (175°C),
Regulo 4 Time: ¾ to 1 hour

Cream together the margarine and sugar.

Whisk the eggs, and beat into the creamed margarine and sugar.

Stir in the apple purée.

Sieve together the dry ingredients and fold into the creamed mixture, together with the sultanas.

Divide between the prepared loaf tins.

Bake in a moderate oven for ¾–1 hour.

To freeze: Cool, pack in suitable freezer wrappings. Seal and freeze.

To serve: Thaw in wrappings at room temperature for 2–3 hours. Ice the top if desired, or spread with butter.

BISCUITS

Although baked and unbaked biscuits freeze well, they need to be very carefully packed if they are to remain intact. If possible choose a tubular carton which permits the biscuits no lateral movement. A well-washed empty plastic detergent bottle may be improvised to make an excellent container. Cut off the tapered top section, and cut the remaining tube in half lengthways—it may be rejoined along one side with tape forming a hinge. The biscuits may be stacked inside one half of the container and the other half be brought on top to form a lid. It may be secured with tape, and placed in a plastic bag.

Another satisfactory solution is to freeze biscuit or cookie mixture in rolls of the same diameter as the required biscuit. The roll may then be partially thawed, and a number of biscuits sliced off directly on to a baking tray and cooked as usual. Several rolls may be prepared from one large batch of mixture and each one flavoured and frozen separately. Any biscuit mixture containing over ¼ lb. fat to 1 lb. flour will freeze satisfactorily.

Soft biscuit mixtures normally piped or spooned out on to a baking tray may be frozen on the tray, then lifted off with a palette knife and packed in polythene bags or containers.

A basic recipe for refrigerator biscuits is given below, which may be varied according to taste and finish. At the same time several recipes are given for biscuits for which the basic method is not suitable.

REFRIGERATOR BISCUITS

4 oz. butter
5 oz. soft brown sugar
1¾ oz. castor sugar
¼ oz. vanilla sugar
1 standard egg
6 oz. flour
¼ teaspoon cream of tartar
¼ teaspoon salt

Yield: 60 biscuits
Oven: 400°F (205°C),
Regulo 6
Time: 8 minutes

Cream the butter well. Add the sugars and beat until light and fluffy.
Beat in the egg gradually.
Sift the flour, cream of tartar and salt.
Add to the creamed mixture and fold well in.
Shape into rolls 3 inches in diameter.

To freeze: Roll up in suitable freezer wrappings. Pack, seal and freeze.

To serve: Thaw sufficiently to enable the roll to be cut into slices ⅛ inch thick. Place on a greased baking tray. Bake at 400°F (205°C), Regulo 6, for 8 minutes.

SUITABLE FLAVOURINGS

SPICE

¼ teaspoon cinnamon or nutmeg   sift in with the flour

CHOCOLATE

2 oz. melted chocolate   add after the egg

NUTS

2 oz. chopped nuts, walnuts, almonds,   add with the flour
or pecans

363

PINWHEEL VARIATION

Divide mixture into two.

Melt 1 oz. chocolate and add to half the mixture.

Chill the doughs, then roll to ⅛ inch thick.

Place one layer on to the top of the other and roll up like a Swiss roll.
Use as for the basic roll.

SUITABLE FINISHES

ICING SUGAR

Dust the surface lightly with icing sugar.

ICING

Coat the surface with white or coloured icing, and decorate with glacé cherry, nuts, coconut, if desired.

CHOCOLATE

Coat the surface with chocolate couverture.

JAM OR BUTTER CREAM

Sandwich two biscuits together with redcurrant jelly or butter cream.

CATS' TONGUES

4 oz. margarine
3½ oz. castor sugar
½ oz. vanilla sugar
2 egg whites
4½ oz. flour

Yield: 3–4 dozen biscuits
Oven: 350°F (175°C), Regulo 4
Time: about 10 minutes

Cream the margarine and sugars together until light and fluffy.

Whisk the egg whites until stiff, and fold into the creamed mixture.

Sift and fold in the flour.

Put the mixture into a piping bag with a ¼-inch pipe.

Pipe in 3-inch lengths spaced at least 2 inches apart on greased baking sheets.

Bake until lightly browned at the edges.

Cool very slightly before lifting off on to wire tray.

To freeze: When cool pack in small cartons or boxes—seal and freeze.

To serve: These biscuits are so thin that they may be served immediately they are removed from the freezer.

SHORTBREAD

Shortbread may be kept some weeks in an air-tight tin, but as a result of the high butter content it has a tendency to develop 'off flavours'. Freezing greatly increases the storage life.

9 oz. flour

3 oz. castor sugar

¼ teaspoon salt

6 oz. butter

Yield: 4 rounds of 6 ins diam.

Oven: 350°F (175°C), Regulo 4

Time: about 15 minutes

Sift together the flour, sugar, and salt directly on to a table or pastry board.

Add the butter in one piece and gradually work it into the dry ingredients.

Knead until smooth.

Form into rounds 6–8 inches in diameter and ¼ inch thick and transfer to a baking sheet. Mark each round into six or eight segments, but do not cut through. Prick with a fork, then crimp the edges.

Alternatively roll to ¼ inch thick. Cut strips 3 inches wide and lift on to baking trays. Prick with a fork and crimp edges as before, then separate into biscuits 1½ inches wide.

Bake until very lightly coloured.

Allow to cool slightly on the trays before lifting on to wire racks.

To freeze: Wrap rounds individually and support with a circle of cardboard.

Pack rectangular biscuits in suitable quantities. Seal and freeze.

To serve: Thaw about 30 minutes in the wrappings at room temperature.

THIN GINGER BISCUITS

8 oz. flour

2 teaspoons ground ginger

1 teaspoon bicarbonate of soda

6 oz. margarine

4 oz. castor sugar

3 tablespoons golden syrup

Yield: about 2½ doz. biscuits

Oven: 325°F (165°C) Regulo 3

Time: 10–15 minutes

Sift together the flour, ginger and bicarbonate of soda.

Rub in the margarine, mix in the sugar.

Mix to a fairly stiff dough with the syrup. Chill in the refrigerator for a few minutes if the dough is a little difficult to handle.

Roll out thinly and cut into rounds with a 2-inch cutter.
Bake until very lightly browned. Cool on a wire tray.
Alternatively this mixture may be formed into a roll, 2 inches in diameter and frozen raw.

To freeze: Pack cooled *baked* biscuits in a tubular carton.
Pack *raw* biscuit mixture in suitable freezer wrappings. Seal and freeze.

To serve: As the *baked* biscuits are so thin, they may be served immediately after removal from the freezer.
Partially thaw the roll of *raw* biscuit mixture for about 1 hour at room temperature before slicing on to a baking tray. Bake at 325°F (165°C), Regulo 3, for about 15 minutes.

Pastry

ALL pastries freeze well whether raw or cooked, and involve no special treatment or problems. Pastries may be frozen raw in bulk, but about 3 hours at room temperature must be allowed for the pastry to thaw sufficiently to be rolled and shaped; it is of greater advantage to roll and shape the pastry before freezing, when it can be placed directly from the freezer into the oven to bake.

SHORTCRUST PASTRY

This pastry may be used widely in dishes frozen raw or after baking. It has a tendency to become shorter in texture after freezing. It may be frozen raw in bulk—this provides little advantage—or after shaping. Shaped raw pastry may be frozen blind in the form of flan cases or pie lids, or in combination with other ingredients such as meat or fruit. Pie lids may be prepared in quantity and several packed together, separated by a double thickness of cellophane—a piece of cardboard cut to shape will give added support. Careful packaging is necessary for all shaped raw shortcrust pastry, particularly when frozen without a filling.

1 lb. flour   *Baking temperature: 425°F (220°C), Regulo 7*
1 teaspoon salt
4 oz. margarine
4 oz. lard
4 tablespoons cold water

Sift the flour and the salt.
Cut the fat into the flour and then rub in with the fingertips until the mixture forms fine crumbs.

Sprinkle the water over the dry mixture, and lightly mix until evenly distributed. Gather together to form a paste.

Freeze in bulk, or after shaping, or after baking.

Use for Flan cases, see below.

> Plate pies, page 301.
>
> Fruit pies, page 299.

FLAN PASTRY

A rich, sweet pastry ideal for flan cases, plate pies, small tartlets, and pastries. It may be frozen equally well in bulk (not considered particularly worth while), after shaping, and after baking.

12 oz. flour *Baking temperature: 400°F (205°C), Regulo 6*
½ teaspoon salt
2 oz. castor sugar
8 oz. butter or margarine
1 standard egg

Sift the flour, salt and sugar on to a board or table.

Make a hollow in the centre of the flour and into this place the butter and egg.

Work the butter and egg into the flour and sugar until a smooth, even paste results. If the pastry is too soft to handle easily, chill slightly in a refrigerator.

Freeze the pastry in bulk, or after shaping, or after baking.

Use for Plate pie, page 301.

> Flan cases, see below.
>
> Mince pie, page 302.

PREPARATION AND SHAPING OF FLANS AND OPEN TARTS

Flan cases may be frozen before or after baking.

Size of flan rings:	6 inch	7 inch	8 inch
Weight of flan pastry:	3–4 oz.	4 oz.	5 oz.
Weight of shortcrust pastry:	4 oz.	5 oz.	6 oz.

Roll out the pastry until it is about 2 inches larger than the flan ring.

Place the flan ring on to a baking tray.

Lift the pastry over a rolling pin and arrange evenly over the flan ring.

Ease the edges of the pastry into the ring, and start moulding into

15a. Pizza, page 387

15b. Currant Loaf, page 390

16. A Selection of Savouries, pages 406–435

(*Left to right*) Shrimp Horns; Sardine Pyramids; Mushroom Bouchées;

shape from the centre outwards. Care should be taken to give a square corner between the base and the sides of the ring.

Trim off the top edges of the flan ring, and lightly prick the base. The pastry can also be shaped in a foil case and this is recommended when freezing raw.

Freeze raw at this stage

or line with a sheet of greaseproof paper weighted down with baking beans and bake until the pastry is set. Remove the beans and continue to cook until the pastry is light golden-brown. Baking pastry without a filling in this way is called 'baking blind'.

To freeze: Freeze *raw flan cases* until hard, then pack in a suitable freezer wrapping. Place in a box for added protection.

Cool *baked flan cases*, pack carefully in freezer wrappings, and place in a box.

To serve: Line *raw flan cases* with greaseproof paper and baking beans. Bake at 400°F (205°C), Regulo 6, for 20–25 minutes. Prick the base of the case after it has just thawed.

Thaw *baked flan cases* at room temperature for about 1 hour. If serving as a chilled sweet, place the filling in the still frozen flan case and keep in a refrigerator until required.

Serve flan cases filled with:　fruit

fruit and cream or custard

chiffon filling

Biscuit Crust

This is an easily made crust for flans. It can be kept in shape after making, either by chilling or baking. The baked crust gives a firmer casing which may be frozen satisfactorily without a filling (i.e. blind). Provided a suitable filling is used, flans made with baked or chilled biscuit crust freeze well.

10 oz. digestive biscuits	*Yield: 1 × 9-inch tin, or*
3 oz. unsalted butter	*2 × 6-inch tins*
3 oz. castor sugar	*Oven: 375°F (190°C), Regulo 5*

Crush the biscuits until they resemble fine crumbs.

Mix the biscuits and the sugar.

Melt the butter and pour over the crumbs.

Mix until the crumbs are evenly coated with fat.

Press into shape around the edges of a Victoria sandwich tin or flan case.
Chill to hold shape before filling, or
Bake for 8 minutes.
Cool and fill.
Use for Chiffon pie, page 305.
Rhubarb cream flan, page 308.

CHEESE PASTRY

This pastry can be used for savoury flans, cheese biscuits, and savouries.

Although the baked pastry will keep quite well for 1–2 weeks in an airtight tin, a much fresher result is obtained when the baked pastry is frozen, and its storage life is greatly increased.

12 oz. flour   *Baking temperature: 350°F (175°C) Regulo 4*
1 teaspoon salt
pepper
pinch cayenne pepper
8 oz. margarine or butter
8 oz. grated Parmesan
3 egg yolks
3 tablespoons milk

Slightly chill the fat to prevent it from oiling during mixing.
Sift the flour and seasonings.
Cut the fat finely into the flour, and rub in with the fingertips to a crumbly consistency.
Mix in the finely grated cheese.
Mix the egg yolks and milk.
Sprinkle the liquid into the dry ingredients and mix lightly. Gather the crumbs together and press into a paste.
Freeze in bulk, or after shaping, or after baking.
Use for Cheese balls, page 419.
        Cheese butterflies, page 417.
        Cheese and vegetable flan, page 402.

SUET PASTRY

As this pastry is so quick and easy to make, there is no advantage gained from freezing suet pastry in bulk. Puddings, sweet or savoury, prepared from suet pastry may be frozen before or after cooking; in either case they may then be placed straight from the freezer into the steamer.

1 lb. flour
1 teaspoon salt
6 teaspoons baking powder
8 oz. suet
½ pint water

Yield: about 2½ lb. finished pastry

Sift the flour, salt, and baking powder.
Use packet suet, or finely chopped or minced block suet.
Mix the suet into the flour, then add the water and mix to a soft dough.
Use for Suet crust puddings, sweet or savoury.

THE FLAKY PASTRIES

All the flaky types of pastry freeze well before and after baking and of all the pastries they are the most worthy of freezer space.

The preparation of the flaky types of pastry is lengthy, as time must be allowed for chilling and relaxing between rollings. Also it is possible to prepare these pastries in larger amounts with little extra work, in fact this is to be recommended, as a more even pastry is produced, which gives a better result on baking.

Flaky pastries may be frozen in bulk, when at least 3 hours must be allowed for them to thaw out before they can be rolled (if possible remove the pastry from the freezer to a refrigerator overnight). As with shortcrust pastry, flaky pastries may be frozen after shaping into pie lids, vol-au-vents, sausage rolls, etc. Shaped raw pastry, whether combined with other ingredients or on its own, may be placed straight from the freezer into the oven.

Baking hint

When baking use a baking sheet with a raised edge to prevent the fat that tends to run from the pastry dripping on to the base of the oven.

FLAKY PASTRY

This is the simplest of these pastries, taking rather less time to prepare than the richer pastries.

1 lb. bread flour (see note on flour, page 383)

Baking temperature: 475°F (245°C)
Regulo 9

1 teaspoon salt
12 oz. margarine
½ pint cold water
2 teaspoons lemon juice

Sift the flour and salt.
Divide the margarine into four equal portions.
Rub one portion of the margarine into the flour until finely mixed in.
Add the lemon juice to the water.
Reserving about 1 tablespoon water, sprinkle the water over the flour and mix to give a soft pliable dough. Add the reserved water if the dough seems a little stiff.
Knead lightly until smooth. Place on a floured plate or board, cover with polythene or damp kitchen paper to prevent the surface from drying out, and leave to relax for 15–30 minutes.
Roll out into a rectangular shape about ¼ inch thick.
Cover two-thirds of the pastry with the second portion of margarine. This is most easily done by cutting small pieces from the block and placing them on to the pastry with a slight spreading action.
Fold the pastry into three, by first folding the uncovered pastry to half cover the fat, then folding the remaining third on top; see Fig. 20 diagram (c).

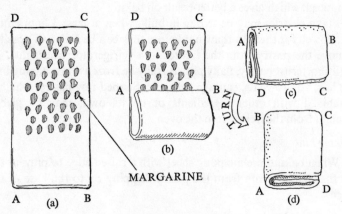

FIGURE 20
PREPARATION OF FLAKY PASTRY

Turn the folded pastry through a right-angle, diagram (d), and roll out into a rectangle along the line of the fold.
Add the third portion of fat by the same process.

Wrap the pastry in polythene or damp kitchen paper and place in a cool place to relax for 15–30 minutes, together with the remaining portion of fat.

Repeat the rolling, add the fourth quarter of fat, then fold as before. Roll and fold once more. The fat should now be evenly distributed, but if any streaks are visible allow the pastry to relax, then roll and fold into three once again.

Allow the pastry to relax for at least 15 minutes before use.

Freeze in bulk at this stage, or after shaping, or after baking.

Use for Steak and kidney pie, page 283.

Mince pies, page 302.

Sausage rolls, page 420.

PUFF PASTRY

This is the richest, flakiest and most even of the pastries, and takes the longest time to make. It keeps in excellent condition when frozen raw, either in bulk or after shaping.

It is a waste of freezer space to freeze puff pastry after baking, as it takes up so much more room and is more fragile.

Since the combined baking and cooling time for the pastry is short, shaped frozen raw pastry may be available for use quite quickly. Puff pastry may be prepared from hard margarine or unsalted butter. Butter gives a very good flavour, but is more difficult to use than a hard or non-cake-making margarine. In any case, very satisfactory results, especially for savoury dishes, are obtained from using margarine.

1 lb. bread flour (note on types of flour, page 383)

Baking temperature: 450°F (230°C) Regulo 8

1 teaspoon salt

1 lb. hard margarine or unsalted butter

½ pint cold water

2 teaspoons lemon juice

Sift the flour and the salt.

Rub 2 oz. margarine into the flour.

Add the lemon juice to the water.

Mix the liquid into the flour until a soft, pliable dough is formed.

Knead very lightly. Place on to a floured board or plate. Cover with polythene or damp kitchen paper to prevent the surface drying out. Leave in a cool place for 30 minutes to relax.

Place the margarine between two layers of greaseproof paper and roll out into rectangle about ¼ inch thick.

Roll the pastry until 1 inch wider and one-third longer than the fat. Place the margarine over two-thirds of the dough, fold the free or uncovered dough half-way across the fat, and finish folding in three. Press the edges lightly together. See Fig. 21.

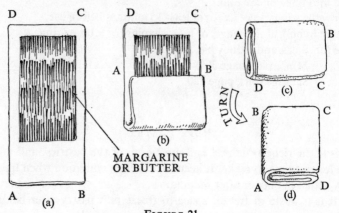

FIGURE 21
PREPARATION OF PUFF PASTRY

Turn the dough around through a right-angle on the board, and roll the pastry to ¼ inch thickness.

Fold into three as before, cover with polythene or damp kitchen paper and allow to relax in a cool place for 30 minutes.

Repeat this rolling, folding, and relaxing, 5 times more. (This method of rolling and folding is the same as that used in the preparation of flaky pastry.)

Freeze in bulk after a total of six foldings and rollings, or proceed to shape the pastry prior to freezing.

Use for Vol-au-vents, page 421.

Gâteau mille feuilles, page 304.

Mince pies, page 302.

SHAPING OF VOL-AU-VENT AND BOUCHÉES FROM PUFF PASTRY

These can be cut in sizes varying from large round or oval cases serving three to four portions, to small cases 2 inches in diameter called bouchées, served filled for cocktail savouries.

1 lb. puff pastry gives   5 round cases of 6 inches diameter,
or 7 oval cases 7 inches by 5 inches,
or 2½ dozen cases of 3 inches diameter,
or 4–5 dozen cases of 1½–2 inch diameter.

Method of making cases

Make the pastry and allow to relax for at least 15 minutes, or thaw
out pastry frozen in bulk.

LARGE CASES, 4 inches and over in diameter

Roll the pastry to ¼–½ inch thick.
Allow to relax for 15 minutes, then cut out.
Place on to baking sheets. Using a knife (or a cutter 2 inches smaller
in diameter), mark all around 1 inch from the edge of the case.
Then cut half-way through the pastry along the mark.
The inner section becomes separated as a lid on baking.
Mark the inner pastry with criss-cross lines to decorate.

Freeze raw at this stage

or brush with egg wash and allow to stand 15 minutes before baking.
Place into the oven at 450°F (230°C), Regulo 8, and cook until the
pastry is risen and set.
The oven temperature can be lowered to 350°F (175°C), Regulo 4, to
finish cooking if the case is browning too rapidly.
The pastry is cooked when the sides of the case are firm.
Remove the lid, carefully take out the uncooked pastry from the
inside of the case.
If the case is not for immediate use, return to a warm oven and dry
out the pastry.

SMALL CASES, 3–4 inches in diameter

These are shaped in the same way as the large cases, but cut out from
pastry between ⅛–¼ inch thick. A lid is marked and cut out in the
same way as for the larger cases, but the cutter selected is 1 inch
smaller in diameter than the case and thus the rim of the case is reduced to ½ inch.

Freeze the cases raw at this stage

or finish and bake as the large cases.

Alternative method of cutting small cases, say of 2-inch diameter

There is often difficulty in removing the uncooked pastry from the inside of the small cases after cooking, so an alternative method avoiding this difficulty is given:

Roll the pastry to $\frac{1}{8}$ inch thick, and allow to relax for 15 minutes.

Cut into rounds with a 2-inch cutter.

Cut out the centre of each with a 1-inch cutter, leaving a ring which

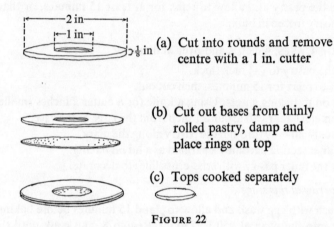

(a) Cut into rounds and remove centre with a 1 in. cutter

(b) Cut out bases from thinly rolled pastry, damp and place rings on top

(c) Tops cooked separately

FIGURE 22

CUTTING OUT BOUCHÉES—ALTERNATIVE METHOD

is to form the top half of the bouchée. It is not wise to retain these small circles cut from the centres for use as lids, as when baked they become too thick, and frequently topple over.

Gather the trimmings of the pastry, including the small circles cut from inside the cases and roll very thinly.

Allow to relax for 15 minutes.

Then with a 2-inch cutter cut out a number of circles corresponding to the number of rings cut from the first rolling to act as bases to the bouchées, and an equal number of 1-inch circles to act as lids.

Prick to prevent rising.

Brush a baking sheet with water, lay on the thinner rounds. Damp each one and fix on the rings cut from the first rolling. See Fig. 22 diagram (b).

Freeze cases and lids raw at this stage

or brush the edges of the rings and the lids with egg wash.

Bake the lids on a separate tray, as they take less time to cook.

Bake in a hot oven until cooked through.

376

Shaping of Cream Horn Cases from Puff Pastry

Roll relaxed pastry $\frac{1}{8}$ inch thick and at least 15 inches long.
Allow to stand for 15 minutes.
Cut into strips $\frac{1}{2}$ inch wide.
Damp one edge of each pastry strip and wrap around a cream horn tin, starting with the pointed end and overlapping the pastry very slightly until the strip is used up. Place with the end of the strip against the baking sheet.

Freeze horn cases raw at this stage

or brush with water and sprinkle with castor sugar, then bake in a hot oven until the pastry is set.
Remove the horn tins, lower the temperature to dry out the pastry cases.

Miniature Horn Cases

It will be necessary to prepare small horn tins from heavy-gauge tinfoil to correspond to the size required.
Roll out the pastry $\frac{1}{8}$ inch thick as when preparing large cream horn cases.
The pastry need be only 8 inches long.
Cut strips $\frac{1}{4}$ inch wide, damp one edge and wrap around tinfoil horn cases.
Place on baking trays as before, but do not brush with sugar and water.

To freeze: Freeze *raw* vol-au-vent cases, bouchées and horn cases prior to packing. Seal in polythene bags. Place horn cases in a box for further protection.
Package *baked* cases carefully in a box, seal and freeze.

To serve: Place the still frozen *raw* cases into an oven at 450°F (230°C) Regulo 8, to bake.
Thaw *baked* cases in a hot oven for about 10 minutes, or at room temperature for 1 hour, before filling.

Choux Pastry

This is a very adaptable pastry with many uses in the preparation of sweet and savoury dishes. It freezes well raw either in bulk or after shaping. The latter is preferable, as shaped frozen choux pastry

377

may be placed directly from the freezer to the oven, whereas raw choux pastry frozen in bulk must be thawed some hours before it may be piped.

Choux pastry also freezes well after baking.

3 oz. butter or margarine   *Baking temperature: 400°F*

8 oz. water ($\frac{1}{2}$ pint minus 4 tablespoons)   *(205°C), Regulo 6*

$\frac{1}{2}$ teaspoon salt

4 oz. flour

4 large eggs (7–8 fluid oz. beaten egg)

Cut the butter into small pieces and place into the water.

Bring slowly to boiling-point, so that the fat melts before the water boils.

Remove from the heat and immediately pour in the sifted flour and salt.

Beat well with a wooden spoon until a smooth paste is formed, leaving the sides of the pan clean.

If the mixture is soft, the beating can be continued over heat until the correct consistency is obtained.

Allow the pan to cool slightly.

Beat the eggs, and add gradually to the mixture in the pan, beating well, between each addition until a smooth paste is formed. Continue until all the egg has been added, giving a soft paste of piping consistency.

Freeze choux pastry in bulk, after shaping, or after baking.

If freezing choux pastry in bulk, pack in a rigid container from which it may easily and cleanly be transferred to a piping bag.

Baking hint

The main faults with this pastry occur in cooking, as an uncooked case immediately on its removal from the oven can still look brown and feel crisp, only to collapse and become soggy on cooling. It is essential that the pastry is cooked undisturbed for about 25–30 minutes, according to size. When risen, brown and crisp, remove from the oven. Puncture the cases, return to an oven at a lower heat to completely dry out the inside. Cool before freezing.

Use for   Chocolate éclairs, page 357.

         Cream profiteroles, page 357.

         Choux puffs, page 220.

         Savoury profiteroles, page 426.

         Choux canapés, page 415.

         Cheese aigrettes, page 427.

PREPARATION OF ÉCLAIR AND PROFITEROLE CASES FROM CHOUX PASTRY

Choux pastry cases may be frozen raw after piping, or after baking.

SWEET ÉCLAIRS

Place the choux pastry into a piping bag and use a plain ½-inch nozzle. Pipe into 5-inch lengths leaving about 2 inches between each one to allow for rising.

Freeze éclair cases raw at this stage
or bake until crisp and golden brown, about 30 minutes.
Split the sides and allow to dry out.

SAVOURY ÉCLAIRS—CAROLINES

These are usually smaller than sweet éclair cases.
Use a plain ¼-inch piping nozzle and pipe into 2½-inch lengths, allowing space between them for rising.

Freeze éclair cases raw at this stage
or bake as for sweet éclairs.

PROFITEROLES

Use a plain ½-inch nozzle and pipe into small rounds about 1 inch in diameter, allowing space between them to allow for rising.

Freeze profiterole cases raw at this stage
or bake as for éclairs.

To freeze: Freeze *piped raw pastry* for 2–3 hours on a greased tin, then lift off with a palette knife and pack in suitable containers.
Freeze *baked cases* before or after filling. Use a suitable freezer wrapping and place in a box for added protection.

To serve: Return the *frozen piped raw pastry* to a greased tin and bake at 400°F (205°C), Regulo 6. Cooking direct from the frozen state will take about 5 minutes longer than for freshly made pastry.
Thaw *baked cases* in wrappings at room temperature for about 1 hour or in a fairly hot oven for 10 minutes, or fill immediately the cases are removed from the freezer—the addition of the filling will hasten thawing.
Thaw *filled cases* in a refrigerator at least 6 hours, then use immediately.

RAISED PASTRY (HOT WATER CRUST)

Pies made from raised pastry may be frozen before or after baking. As they are more frequently served cold, there is greater advantage in baking prior to freezing. This pastry cannot be frozen in bulk (raw) because it may only be satisfactorily moulded whilst warm.

1 lb. flour
2 teaspoons salt
6 oz. lard
¼ pint milk
2 egg yolks

Baking temperature: Begin at 450°F (230°C) Regulo 8, lowering to 350°F (175°C), Regulo 4

Cut the lard into pieces and place with the milk in a saucepan.
Heat until the milk boils and the fat melts.
Sift the flour and salt.
Pour in the boiling liquid and mix in slightly.
Add the beaten yolks and continue mixing.
Knead the pastry lightly until smooth.
Use whilst still warm, and before it sets and becomes brittle.
Use for Veal and ham pie, page 284.
   Game pie.

Yeast Mixtures

MOST yeast mixtures may be frozen before or after baking.

BREAD DOUGHS

AFTER BAKING

For the housewife who likes to bake her own bread, freezing the baked loaves may seem the best way to utilize the services of the freezer. A batch of loaves, sufficient for the family's requirements, may be baked together and then frozen. A loaf may then be removed from the freezer when required. If it has been packed in foil, it may be placed directly in the oven to thaw, and the resulting bread will resemble that which has been freshly baked.

BEFORE BAKING

Uncooked yeast dough may be frozen at various stages during its preparation with varying degrees of success:

After rising

This is usually referred to as *risen white dough* and describes the yeast dough ready for shaping; i.e. after it has been mixed and kneaded, and risen (or proved) once.

The amount of yeast used in a dough to be frozen raw should be doubled, otherwise the second rising (after frozen storage) is liable to be very slow. The dough should be kept as cool as possible during the mixing prior to freezing. For this reason it is recommended to put the dough to rise in a domestic refrigerator. The subsequent kneading should be performed as quickly as possible, so that the dough is not given a chance to warm up before freezing. A warm dough placed in

381

a freezer will continue to rise until its temperature is substantially reduced, and may even burst through the freezer wrappings.

The use of low temperatures for yeast cookery may be new to some readers, but in fact slow rather than rapid rising has been found to result in a better volume and bread with better keeping qualities. For further information on the subject of yeast cookery the reader is referred to *Let's Cook With Yeast*, by Patty Fisher.

After shaping; and after shaping and rising

Successful results from the freezing of shaped, and shaped and risen rolls and loaves cannot be guaranteed. Satisfactory results have been achieved in both cases when the period of frozen storage is restricted to about a week. However, time must be allowed for thawing and rising in the first case, and for partial thawing in the second. Freezing the bread dough in these forms may be found a useful form of advance preparation on some occasions, but if the requirement is for freshly baked rolls at breakfast, 'parbaked' rolls have been found to give the best results.

Parbaked rolls

As their name suggests, these are rolls which have been part-baked, after the final rising. The rolls are put into the oven set at 300°F (150°C), Regulo 1½, until the dough is set, but the rolls still pale in colour; this takes about 20 minutes. After this the rolls are cooled, packed in usable quantities, and frozen. When required, the parbaked rolls are placed uncovered and without thawing in an oven set at 450°F (230°C), Regulo 8, for a further 20 minutes.

OTHER YEAST MIXTURES

AFTER BAKING

Other yeast products which may be frozen after baking include the richer mixtures such as savarin and tea breads which have added eggs and sugar. These products keep well for about three months. Baked yeast pastries (which have fat incorporated in them in the same way as for flaky pastry) should not, however, be kept longer than two to three weeks.

BEFORE BAKING

On the whole, the richer yeast mixtures do not give such good results in terms of volume and texture when frozen raw, and at the same time they are not available as quickly as those mixtures frozen after baking.

Yeast pastry, however, may be successfully frozen raw, and in view of the length of time it takes to prepare this will prove a considerable advantage. The best results are obtained from freezing this pastry in bulk, although for short periods of frozen storage it may be pre-shaped into croissants and Danish pastries.

SPECIAL NOTE

Freezing Yeast

It is very useful to keep a supply of yeast in the freezer. Contrary to popular belief, yeast cells are not killed by low-temperature storage. Buy yeast in ½ lb. or 1 lb. packages, and cut into 1 oz. cubes.
Wrap the cubes individually in moisture-vapour-proof film, and pack into a preserving jar.
When required the individual yeast cube is removed from the preserving jar and allowed to thaw (about ½ hour at room temperature). For immediate use the yeast may be grated.

Type of Flour Used

All the recipes for yeast mixtures call for bread flour. This flour is milled from hard wheat. The soft flour suitable for most other kinds of baking is milled from soft wheat, and cannot be used successfully in bread making. If bread flour is not readily available at the grocer's, it may usually be obtained on request from the bread roundsman, or direct from those shops where bread is baked on the premises. Bread flour should also be used for the preparation of flaky pastries in order to obtain the best results.

Rising or Proving Yeast Mixtures

The time taken for a yeast dough to rise is as follows:

about 30 minutes in a warm place (for example an airing cupboard),
1 hour at room temperature (kitchen),
4 hours in a cool place (larder),
overnight in a domestic refrigerator.

Where time permits, better results are obtained with yeast mixtures risen slowly.

WHITE BREAD

½ pint warm water
1 teaspoon honey or sugar
1 oz. fresh yeast or ½ oz. dried yeast
3 lb. bread flour
1 oz. lard or 2 tablespoons oil
1 tablespoon salt
1 pint warm water or milk and water

Yield: 5 lb. dough, or
4 × 1 lb. loaves
Oven: 450°F (230°C),
Regulo 8
Time: 40–45 minutes

Place the warm water, sugar or honey, and 1 teaspoon of flour into a small basin. Crumble in the fresh yeast, or if using dried yeast sprinkle this over the surface of the water. Leave until the surface appears frothy, about 30 minutes.

Sieve together the flour and salt and either rub in the lard or stir in the oil. Combine together in a large bowl the frothy yeast mixture, the flour mixture, and the pint of water (or milk and water) and work to a soft firm dough which leaves the bowl clean. More water may be added at this stage if the mixture feels too dry.

Knead the dough on an unfloured board, or in a mixer set at a low speed and fitted with a dough hook attachment, until the dough is firm and elastic.

Place the dough inside an oiled polythene bag, secure the mouth of the bag, and put to rise (see note on page 383) until it has doubled in size and springs back when pressed lightly with the fingers.

Knead or 'knock back' the dough for a further 5 minutes by hand or with a mixer.

At this stage it may be called *risen white dough*. It is now ready for shaping into loaves.

Divide the mixture into four pieces, each weighing about 1¼ lb. Shape each piece and place in a greased and floured tin. Brush the top with oil and put to rise in a polythene bag until it has doubled in size.

Brush the surface with salty water for a crusty loaf.

Bake in the centre of a hot oven for 40–45 minutes.

To get some steam in the oven and improve the rise, pour ½ pint boiling water into a meat tray and place in the oven.

To test when the loaf is done, turn out on to an oven cloth and tap the base; it will sound hollow when the loaf is cooked.

SHAPING LOAVES AND ROLLS

If desired, the dough may be formed into loaves of various shapes or made into rolls.

Use the dough after it has been kneaded for the second time.

PLAIT

1–1½ lb. dough

Form three rolls about 14 inches long. Pinch the ends together and plait loosely until half the roll has been used.

Turn over and continue plaiting to the end and pinch ends together to finish.

Place on a greased and floured tray, brush with oil, and prove inside a polythene bag.

Sprinkle with poppy seed if liked.

Bake as usual.

COTTAGE LOAF

1–1½ lb. dough.

Divide the dough into two portions, one twice the size of the other. Knead each and shape into a ball.

Place each ball of dough on a floured board, brush with oil, and prove inside a polythene bag until risen to twice the original size.

Place the larger ball on a greased and floured baking tray and thoroughly dampen with salt water; place the smaller one on top. Make a hole through the centre of both rounds to the bottom and so fuse them together.

Notch the sides of both rounds with scissors.

Bake as usual.

ROLLS

1 lb. dough makes about eight to ten rolls.

DINNER. Divide the dough into ten portions, knead each one, and form into a round ball against the hollowed palm of the hand.

KNOTS. Divide into eight portions, and form each into a roll about 10 inches long. Tie a knot.

WINKLES. Form into 10-inch long rolls as for knots. Wind each into a coil.

Brush the rolls with oil, rise to double the size on a greased and floured baking tray inside a polythene bag.

Bake at 450°F (230°C), Regulo 8, for about 20–30 minutes.

To freeze: Thoroughly cool the rolls and loaves, and pack in aluminium foil or other freezer wrappings. Seal and freeze.

To serve: Either thaw in the freezer wrappings at room temperature for about 30 minutes for rolls, and up to 2 hours for loaves, or in a fairly hot oven, wrapped in foil, for 10–30 minutes.

Soft Bread Rolls—Parbaked for Breakfast

½ pint water—slightly warm
2 teaspoons sugar
2 oz. fresh yeast,
    or 1 oz. dried yeast
2 lb. bread flour
4 teaspoons salt
½ pint milk
4 tablespoons cooking oil

Yield: 24 breakfast rolls
To parbake: 300°F (150°C), Regulo 2,
    for 20 minutes
To finish baking: 450°F (230°C),
    Regulo 8, for 20 minutes

Dissolve the sugar in the warm water and crumble in the fresh yeast or sprinkle on the dried yeast; leave about 10 minutes until frothy.

Sieve the salt and flour together in a large bowl and mix in the frothy yeast mixture, the milk, and the oil.

Work to a firm dough which leaves the bowl clean.

Knead for a further 5–10 minutes on an unfloured board.

Rise in an oiled polythene bag until the volume is doubled.

Knead again and divide into twenty-four pieces.

Roll each piece of dough into a round ball against the hollowed palm of the hand.

Place the rolls on to a greased and floured baking tray and place the tray inside a polythene bag. Put to rise for about 30 minutes in a warm place.

For parbaked rolls: Bake rolls in a preheated oven at 300°F (150°C), Regulo 2, until set but not coloured, about 20 minutes.

To freeze: Cool thoroughly and pack in required quantities.

To serve: Return unthawed rolls to a floured baking tray. Bake at 450°F (230°C), Regulo 8, for 20 minutes or until cooked.

For fully baked rolls: If desired the rolls may be completely baked at 450°F (230°C), Regulo 8, for 30–40 minutes before freezing. They can then be cooled, packed, and frozen in the usual way.

To serve: These rolls may be thawed and reheated in foil wrapping in a fairly hot oven for about 10 minutes.

Freezing Risen Dough

This dough may be prepared in quantity and frozen. It may then be removed from the freezer, thawed, and used as required.

Use the recipe given on page 384 for white bread, but increase the quantity of the yeast to 2 oz.

Combine the ingredients together as directed, and after the first kneading place the dough in a polythene bag and put to rise in a domestic refrigerator.

(With the increased quantity of yeast in the dough, rising will take 2–3 hours.)

When the dough has risen, knead it thoroughly but quickly, so that it is not given a chance to warm up.

Divide the dough into four or five portions and pack individually in suitable containers.

Seal and freeze.

To use: When required, thaw out the dough in a domestic refrigerator overnight or at room temperature for 2–3 hours. Knead and shape as required.

Prove in a warm place.

Bake in the usual way.

Frozen risen dough may be used in the following recipes for pizza, lardy cake, tea breads, and yeast pastries.

Pizzas

These flat savoury tarts are traditionally made from a yeast dough as in the following recipe, although puff pastry, shortcrust pastry, or a plain scone mixture may also be used with good results. All these mixtures may be satisfactorily frozen after baking. Thawing and re-heating a baked pizza in the oven takes about 30 minutes, so that it may be quickly available for lunch or supper.

Pizza

1 lb. bread dough, freshly made (page 384), or previously frozen (page 386)

7 tablespoons olive oil

2 tablespoons chopped onion

2 tins tomatoes (net weight 14 oz.)
    or 2 lb. fresh tomatoes

6–8 oz. cheese (either sliced Bel Paese or grated Cheddar)

20 anchovy fillets, or 10 rashers streaky bacon

20 black olives

1 teaspoon mixed dried herbs—basil, oregano, marjoram

Yield: 4 × 8-inch pizzas each serving 6 portions

Oven: 450°F (230°C), Regulo 8, for 20 minutes

Use the dough after it has been proved once (risen white dough).
Knead well and divide into four portions.
Knead each portion and press or roll into a round about 8 inches in diameter.
Grease 8-inch sandwich tins, or baking sheets.
Place the dough in the tins and press well to give a good shape.
Brush over the surfaces with 2 tablespoons olive oil.
Gently fry the onions in 1 tablespoon oil without colouring, until they are tender.
Slice the tomatoes.
Spread the onions over the bases. Cover with the sliced tomatoes, to within ½ inch of the edges.
Season well with salt and pepper.
Slice the Bel Paese very thinly. Arrange slices over the tomatoes, or sprinkle the grated Cheddar over the surfaces.
Decorate with the anchovy fillets or streaky bacon to give the effect of a cartwheel. Remove the bacon rinds and cut each rasher in half lengthways; usually five fillets or five strips of bacon are arranged in this way on each pizza.
Place the stoned black olives between each fillet or rasher.
Sprinkle the dried herbs over the surface of the pizzas.
Finally sprinkle 1 tablespoon oil over each of the pizzas.

The following filling, although not classical, gives an appetizing result.

ALTERNATIVE FILLING FOR PIZZAS

2 lb. sausage-meat	combine
8 tablespoons tomato sauce	together
or condensed soup	in a
2 teaspoons basil	basin

8 oz. processed cheese or Cheddar—cut into thin strips ½ inch wide
16 rashers of streaky bacon—each one stretched and made into 2 bacon rolls.

Brush the pizzas with oil as before, and spread the sausage-meat over each one.
Use the cheese and bacon rolls to make the cartwheel pattern.

To bake pizzas

Put to prove in a warm place for 20–25 minutes.
Bake in an oven at 450°F (230°C), Regulo 8, for 20 minutes (for the

sausage-meat alternative reduce the temperature to 375°F (190°C), Regulo 5, and continue to cook for a further 20 minutes).

To freeze: Cool rapidly. Pack, seal, and freeze.

To serve: Place the pizza straight from the freezer into a fairly hot oven for 20 minutes, or thaw for 2 hours at room temperature, before placing in a fairly hot oven for 10–15 minutes to heat through. The second method gives a more moist finish.

LARDY CAKE

1½ lb. risen white dough, freshly made (page 384), or previously frozen (page 386).

4 oz. lard

4 oz. brown or castor sugar

½ teaspoon mixed spice ⎫
6 oz. sultanas   ⎬ if liked

Tin: 10 × 8 inches
Oven: 400°F
(205°C), Regulo 6
Time: 30–40 minutes

Roll out the yeast dough into a strip ¼ inch thick, the length to be three times the width.

Distribute one-third of the lard in small pats over the strip, and then sprinkle with one-third of the sugar, some sultanas and spice.

Fold in three as for flaky pastry, and seal edges.

Turn the dough around on the board through a right-angle.

Roll out into a strip as before.

Repeat additions of lard, sugar, sultanas, and spice, twice more.

Roll out into a strip again, and fold loosely into three.

Place in a well-oiled cake tin, and push down until about 1½ inches thick.

Brush the top with oil and sprinkle with castor sugar.

Score the top lightly in a criss-cross pattern.

Put in an oiled polythene bag to rise until the size has doubled—it should come to the top of the tin.

Bake for about 40 minutes.

Cool in the tin, and finally serve, cut in squares, on a hot dish.

To freeze: Cool, shape a piece of aluminium foil into a tray to fit the lardy cake, cover completely in foil. Seal and freeze.

To serve: Thaw and reheat in foil container and wrappings in a fairly hot oven for about 20–30 minutes.

TEA BREADS

A variety of tea breads may be prepared from risen white dough, freshly made (page 384), or previously frozen (page 386). The additional ingredients, fruit, sugar, and fat, are incorporated into the yeast mixture by the same method of rolling and folding as used for the preparation of lardy cake in the previous recipe.

Two recipes are given below, although the additions of fruit and other flavourings may be varied to taste.

The quantities given are sufficient for two 1 lb. loaves.

RAISIN AND HONEY LOAF

1½ lb. risen white dough
6 oz. stoned raisins
2 oz. honey
3 oz. margarine

CURRANT LOAF

1½ lb. risen white dough
4 oz. currants
grated rind and juice of a lemon
4 oz. sugar
3 oz. margarine

Bake at 400°F (205°C), Regulo 6, for 30 minutes.

To freeze: Cool, wrap in foil, seal, and freeze.

To serve: Place wrapped loaf in a moderate oven for 15–20 minutes, or thaw at room temperature in freezer wrappings for 2 hours.

The loaves may be iced with lemon-flavoured icing; and spread with butter.

YEAST FLAKY PASTRY

As well as for the preparation of croissants for which directions are included below, yeast pastry may also be used for making mince pies, eccles cakes, sausage rolls.

1 lb. risen white dough, freshly made (page 384),   *Yield: 12 croissants*
    or previously frozen (page 386)   *Oven: 425°F (220°C),*
4–6 oz. hard margarine   *Regulo 7. Time: 20 minutes*

If using previously frozen dough, thaw but keep chilled; freshly prepared dough should also be chilled in the refrigerator before adding the fat. Roll out the chilled dough into a rectangle 18 inches by 6 inches.

Divide the margarine in half, and cover two-thirds of the yeast dough with small pats of margarine as when making flaky pastry. Fold, in three, seal all the edges by pressing with a rolling pin.

Turn the pastry at right-angles to the original position and roll out into a rectangle again, and add the remainder of the margarine as before. Fold and seal.

Roll out the pastry and fold as before twice more.

Wrap in polythene and chill in the refrigerator before shaping.

If desired the yeast pastry may be frozen in bulk at this stage.

SHAPING CROISSANTS

See Fig. 23.

Roll out the pastry into a rectangular strip about ⅛ inch thick and 5 inches wide.

From this strip cut triangles which are 5 inches high and about 9 inches along the base. (It is possible to cut them alternately from the strip of pastry.)

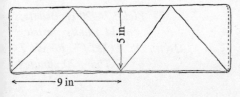

(a) Roll out the pastry into a rectangular strip and cut into triangles

(b) Damp the top of each triangle and roll up from the base

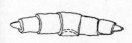

(c) Roll to give number of edges as shown; arrange with the tip underneath

(d) Shape into a crescent and twist the ends to prevent them unrolling

FIGURE 23

SHAPING CROISSANTS

Roll up each triangle beginning at the base, damp the top of the triangle, and seal.

Form the roll into the traditional crescent shape, give a twist to the points, and place on a baking tray. The sealed tip of the roll should lie next to the tray to prevent the crescent unrolling.

Relax the shaped croissant at room temperature for about 30 minutes.

Brush with salted beaten egg white.

Bake for about 20 minutes.

To freeze: Cool, wrap the croissants individually, and pack in a box for added protection.

To serve: Place on a baking tray in a fairly hot oven for 10 minutes.

DANISH PASTRIES

Danish pastry is made from an enriched yeast dough (i.e. with added eggs and sugar). Although slightly sweet, this pastry may be used for the preparation of croissants (details of shaping given in the previous recipe) as well as the more elaborate filled pastries.

1 oz. yeast	*Yield: 14–20 croissants;*
$\frac{1}{4}$ pint milk	*or 14 Danish pastries*
1$\frac{1}{4}$ oz. sugar	*Oven: 425°F (220°C), Regulo 7*
2 eggs	*Time: 20 minutes*
1 lb. bread flour	
1 teaspoon salt	
1$\frac{1}{4}$ oz. margarine	
9 oz. hard margarine	

Scald and partly cool the milk; sweeten with 1 teaspoon of the sugar. Crumble in the fresh yeast, and leave for about 10 minutes until the mixture is slightly frothy.

Add the eggs.

Sift the flour into a bowl together with the salt and the remainder of the sugar.

Rub in the margarine (1$\frac{1}{4}$ oz.).

Pour the yeast and egg mixture into the flour and mix together to form a light dough.

Knead on a lightly floured board for 5 minutes.

Chill in a refrigerator for up to 30 minutes.

Roll out the dough into a rectangle about $\frac{1}{4}$ inch thick.

Roll out the 9 oz. of margarine between greaseproof paper until half the size of the rolled-out dough.

Place the margarine on to one half of the dough and fold over the other half so that the margarine is sandwiched between them.

Relax for 10 minutes in the refrigerator.

Turn the dough through an angle of ninety degrees so that rolling commences along the direction of the fold made in the pastry. Roll into an oblong. Fold into three as for flaky pastries.

Repeat twice, turning the pastry through a right-angle between each rolling.

Chill for 30 minutes.

SHAPING DANISH PASTRIES   *14–16 pastries*

Various fillings can be used and many shapes are formed.

Roll the pastry as thinly as possible.

Cut into strips 5 inches wide, then at intervals of 5 inches to produce squares and diamonds. See Fig. 23.

FILLINGS—each sufficient for fourteen pastries

ALMOND

1 egg	mix the fat into the sugar; beat
3 oz. castor sugar	in the egg;
3 oz. ground almonds	fold in the ground almonds
almond essence to improve flavour	
½ oz. margarine	

CUSTARD

1 egg	beat the egg and sugars until
1 oz. castor sugar	thick; fold in the flour; scald
¼ oz. vanilla sugar	the milk, beat into the egg
½ oz. flour	mixture gradually; return to a
½ pint milk	low heat and stir until boiling-point; cook for 2–3 minutes; cool

CURRANT MIXTURE

3 oz. currants
1½ oz. sugar

CHERRY MIXTURE

4 oz. chopped cherries
2 oz. sugar

393

From a 5 in. square or diamond prepare the following shapes:

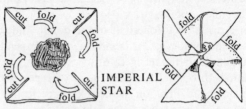

IMPERIAL STAR

Make a diagonal cut from each corner half-way to the centre. Place the selected filling in the centre. Bring alternate points over the filling in rotation. Points just overlap in the centre

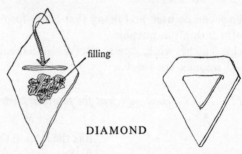

DIAMOND

Make a cut of about 2 in. as shown. Place the filling in the centre. Fold the top point underneath and bring it up through the slit to cover the filling

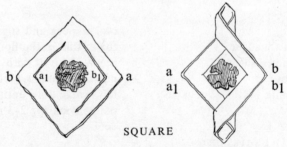

SQUARE

Cut slits as indicated 1 in. from the edge of the square with a 1 in. gap between the ends. Place the selected filling in the centre. Fold the outside corners as shown: a to a_1, b to b_1

FIGURE 24

SHAPING DANISH PASTRIES

APPLE PURÉE

Thick sweetened apple purée can also be used.

To finish pastries: After baking, coat with thin glacé icing and browned nuts.

To freeze: Freeze the completed pastry *in bulk* after the final rolling. Wrap in foil or other suitable freezer wrapping.

Freeze *shaped raw pastries* for 2–3 hours before packaging. (Freezing pastries in this form is not particularly recommended except for very short periods of frozen storage.)

Cool *baked pastries* and freeze before or after icing, according to preference. Freezing without icing is easier, as there is less risk of spoiling the surface of the icing, and the pastries can be refreshed in the oven when thawing if desired.

To serve: Thaw pastry frozen *in bulk* overnight in a refrigerator, or for about 3 hours at room temperature before rolling and shaping according to the directions on the previous page.

Thaw *shaped raw pastries* about an hour at room temperature, 65°F (18°C), or longer if the ambient temperature is below this. Bake on a tray with sides, as there is a tendency for the fat to leak out during baking. Bake at 425°F (220°C), Regulo 7, for 20 minutes. Cool, ice, and finish with nuts.

Thaw *baked pastries, frozen without icing,* in wrappings at room temperature for about 30 minutes, or in a fairly hot oven for 5 minutes. Loosen the packaging and thaw *iced Danish pastries* at room temperature for 30–45 minutes.

Croissants prepared from Danish pastry should be thawed in the oven, as they are normally served warm.

YEAST BATTER

This batter is very easy to use and coats food thinly. When fried it gives a crisp and puffy finish. It keeps well deep frozen and regains its crispness when placed in a hot oven directly from the freezer.

Left-over batter can be cooked in tablespoonfuls on a girdle, and used with a sweet or savoury sauce, or in place of potatoes on top of casseroles, or fried with bacon as a breakfast dish.

8 oz. bread flour
1 teaspoon sugar
½ oz. yeast

½ pint milk
1 egg
1 teaspoon salt

Warm the milk.
Sift the flour into a bowl.
Cream the yeast with the sugar, gradually add the lukewarm milk.
Pour into a well in the flour, together with the yolk of the egg.
Gradually beat together until smooth.
Allow to stand in a warm place until frothy—about 30 minutes.
Whisk the egg white until stiff.
Sprinkle the salt over the batter, then fold in with the egg white.
Use as required.
Use for Kromeskies, page 429.
        Anchovy puffs, page 430.
        Fruit fritters, page 311.

CHEESE BREAD

8 oz. milk (½ pint minus 4 tablespoons)   *Yield: 2 × 2 lb. loaves,*
1 oz. butter   *or 4 × 1 lb. loaves*
2 teaspoons salt   *Oven: 400°F (205°C), Regulo 6*
1 oz. yeast   *Time: 40 minutes*
2 teaspoons sugar
½ pint water
½ lb. grated Cheddar cheese
1½ lb. bread flour
olive oil

Heat the milk to scalding-point, add the butter and the salt. Cool until lukewarm.
Pour the water, which should also be lukewarm, into a basin. Add the sugar and 1 teaspoonful of the flour. Crumble the yeast on to the surface and leave until it appears frothy—about 30 minutes.
Add the milk to half the flour and mix well in.
Add the cheese, then the yeast mixture.
Pour in the remaining flour and mix thoroughly.
Knead the dough until it is smooth.
Place in a bowl, cover with polythene sheeting, and prove in a warm place for about 1 hour, or until the volume is doubled.
Knock back and knead for 5 minutes.
Cut into portions for shaping and cover. Allow to relax for 10 minutes.

Shape the loaves. Place into greased and floured tins.

Prove for 1 hour. Brush with olive oil.

Bake for between 40 minutes and 1 hour.

To freeze: Cool, pack in foil or other suitable wrapping, seal and freeze.

To serve: Thaw in wrappings for 2 hours at room temperature, or if wrapped in foil transfer to a fairly hot oven for 20 minutes.

CONTINENTAL FAMILY CAKE

This all-purpose mixture is very useful. It can be baked and served as a cake, baked in a loaf tin and served spread with butter, or steamed and served as a pudding.

8 oz. bread flour	*Yield: 2 lb. loaf tin or 8-*
2 oz. semolina	*inch square cake tin*
½ teaspoon salt	*or greased 2-pint basin*
3 oz. margarine	*Oven: Cakes and loaves*
1 oz. yeast	*at 400°F (205°C), Regulo 6*
2 oz. castor sugar	*for 35–40 minutes*
6 oz. currants	*Steam pudding: 30–40 minutes*
minced rind and flesh of ½ orange	
2 eggs	
⅛ pint milk	

TOPPINGS FOR BAKED MIXTURES

JAM AND CRUMBLE

2 oz. jam —spread on to the surface before adding the crumble

1 oz. sugar
1½ oz. flour
1 teaspoon cinnamon
1 oz. butter or margarine

sift dry ingredients together; rub in the butter.

STREUSAL

2 oz. flour
2 oz. sugar
½ teaspoon cinnamon
½ oz. chopped nuts
1 dessertspoon melted butter

sift dry ingredients together; add nuts and melted butter.

397

Sift the flour, semolina, and salt.

Rub in the margarine until the mixture resembles breadcrumbs.

Crumble the yeast over the mixture and rub in evenly.

Add the currants and sugar.

Scald the milk and cool. Add to the beaten eggs.

Pour the egg mixture on to the dry ingredients, add the orange pulp, and mix in thoroughly to a soft dough.

Turn the mixture into the greased tin or pudding basin, as required.

Finish the baked mixture with the topping selected.

Cover the tin or basin with polythene sheet and put in a warm place to rise until double the size, 30–60 minutes.

To bake: Bake cake and loaves for 35–40 minutes.

To steam: Place puddings in steamer over actively boiling water for 30–40 minutes; or if desired, the unrisen pudding may be proved whilst cooking at the same time. Place the unrisen pudding in the top of the steamer over cold water and bring up to boiling-point slowly, taking about 30 minutes, during which time the pudding is sufficiently proved. Continue cooking as for proven pudding.

To freeze: Divide a *baked loaf* into smaller portions or slices before freezing if desired. Wrap in suitable freezer wrappings.

Cover the top of a *steamed pudding* with foil and place in polythene bags.

To serve: Thaw cakes and loaves at room temperature for 2–3 hours according to the size of the portion, or covered by foil in a fairly hot oven for 20 minutes. Separated, covered slices may be thawed 30 minutes at room temperature.

Steam yeast pudding rapidly for about 20 minutes before serving.

LEMON SAVARIN

5 tablespoons scalded milk
1 teaspoon honey
1 oz. fresh yeast
2 large or 3 small eggs
8 oz. bread flour
4 oz. butter
1 oz. sugar
1 teaspoon salt
grated rind and juice of 1 lemon

Yield: 2 × 1 pint ring moulds, each one 4–6 portions
Oven: 400°F (205°C), Regulo 6 for 30 minutes, or,
Steam: for 1 hour

Place the warmed milk and honey in a basin and crumble in the yeast. Leave until frothy.

Add the unbeaten eggs.

Sift the flour, sugar, and salt into a large bowl, and rub in the butter.

Add the yeast liquid and eggs to the flour together with lemon rind and juice to make a soft mixture.

Beat for 3 minutes with a wooden spoon.

Pour into greased moulds, half filling them. Cover with oiled paper. Stand to rise until barely at the top of the mould.

Bake or steam.

To freeze: Remove the cooked savarin from the moulds, cool, pack in suitable freezer wrappings, seal and freeze.

To serve: Thaw in wrappings for 2 hours at room temperature. Complete the savarin to serve as follows:

1 pint water
8 oz. sugar
juice of 1 lemon
2 tablespoons honey
macedoine of mixed fruits
   (allow ½ lb. for each mould)
4 tablespoons rum
apricot glaze (page 349.)
blanched almonds

dissolve the sugar in the water, bring to the boil and allow to boil for about 5 minutes; add the lemon juice and honey; place the fruit in the hot syrup; cool a little; add 4 tablespoons rum; strain the syrup into a jug; arrange the savarins on a cake rack over a dish and pour the syrup over them several times to saturate them; place the savarins on to a serving dish; fill the centres with the fruit; brush the sides with hot apricot glaze, and spike the sides with split almonds; serve with whipped cream

RUM BABA

Rum babas may be prepared from the same recipe as for lemon savarin, on the previous page.

Omit the juice and rind of one lemon from the basic mixture

Add 2 oz. cleaned currants

Yield: 16 dariole moulds, or
8 individual rum baba moulds
Oven: 425°F (220°C), Regulo 7,
for 15 minutes, or
Steam: for ¾ to 1 hour

Grease the moulds and place some of the cleaned currants in each mould.

Prepare the yeast mixture as for savarins.

Half fill the small dariole moulds or the larger ring moulds.

Stand in a warm place until the mixture has doubled in size.

Bake or steam.

To freeze: Remove from the moulds. Stand on an aluminium foil tray and seal over the top with aluminium foil.

To serve: Thaw the foil-covered babas in a fairly hot oven for 10–15 minutes, or 45 minutes at room temperature. Saturate the babas in rum syrup as follows, and serve hot or cold.

8 oz. sugar	dissolve the sugar in the water; boil
1 pint water	a few minutes and add the lemon
juice of 1 lemon	juice; the rum may be added to the
rum	syrup or sprinkled directly over the babas after saturating them in the hot syrup.

Miscellaneous Dishes

PANCAKES

Some pancakes become tough and leathery on freezing and subsequent thawing. However, the enriched pancake batter given below has been found to give satisfactory results.

4 oz. flour
¼ teaspoon salt
1 egg
1 egg yolk
½ pint milk
1 tablespoon oil or melted butter

Yield: 12 × 8 inch pancakes
18 × 4 to 5 inch pancakes

Sift the flour and salt. Make a well in the centre of the flour.
Place in the egg and egg yolk and some milk. Gradually work the flour into the well until a smooth batter is formed. If the mixture is lumpy, beat until smooth while still thick.
Beat in the remaining milk.
Finally add the melted butter or oil.
Make the pancakes very thin, in a hot greased pan; cool in a tea towel.

To freeze: Separate with a double layer of cellophane, although they normally separate quite easily after thawing slightly. Pack, seal, and freeze.

To serve: Separate the pancakes. Place on a baking sheet and cover with foil. Place in a hot oven for 10–15 minutes.
Serve filled with sweet or savoury fillings.

FILLINGS

For suitable sweet fillings see page 310 in Hot and Cold Sweet section.

401

Use savoury fillings given for vol-au-vents (page 421) and carolines (page 426).
Larger savoury pancakes may be served as a supper dish.
Smaller pancakes may be served as an hors-d'oeuvre or savoury.

CHEESE AND VEGETABLE FLAN

This may be frozen completed, or assembled from previously frozen ingredients.

baked 6–inch flan case of cheese
  pastry (page 370).
about 8 oz. cooked diced vegetables
  (carrots, turnips, potatoes,
  corn, peas, green beans, etc.)
tomatoes can be added for colour
⅓ pint cheese sauce
1 oz. grated cheese—add on serving

Yield: 4 portions as a supper dish

garnish: sliced tomatoes and peas

Mix sufficient cheese sauce with the vegetables to bind them well, normally about ¼ pint is sufficient.
If tomatoes are included in the mixture, skin them first and use the flesh only, cut into dice.
Place the vegetable mixture in the flan case.
Coat with the remaining cheese sauce,

Freeze at this stage
or sprinkle the remaining cheese on top.
Heat through in a hot oven; finally place under the grill to brown the top.
Garnish and serve.

To freeze: Pack, seal, and freeze the completed flan. Retain the grated cheese for the top until serving.

To serve: Place in a fairly hot oven straight from the freezer. Heat through for 45 minutes (sprinkle on the cheese after the first 30 minutes).
Place under the grill before serving to brown the top. Garnish and serve.

QUICHE LORRAINE

Pastry case

Use 4 oz. shortcrust pastry (page 367)

Filling

½ oz. butter	*Yield: 6-inch flan ring serving 4 portions*
2 oz. onion	*Oven: 425°F (220°C), Regulo 7,*
2 oz. streaky bacon	*for 5 minutes then 350°F (175°C),*
1 egg	*Regulo 4, for 30–40 minutes*
1 egg yolk	
2 oz. grated cheese	
½ teaspoon salt	
pinch cayenne pepper	
mustard	
¼ pint single cream	

Melt the butter in a pan. Finely slice the onion and cook in the butter without colouring until tender.

Remove the rinds from the bacon, then blanch it and chop finely.

Add to the onion and continue cooking for a few minutes.

Beat the eggs and add the grated cheese. Add the seasoning and the cream.

Add the onion and bacon to the egg mixture and stir thoroughly.

Pour the mixture into the prepared flan case and bake towards the bottom of a hot oven for about 5 minutes, then lower the temperature for a further 30–40 minutes.

To freeze: Cool rapidly. Place on a foil dish and pack in a rigid container for added protection. Seal and freeze.

To serve: Thaw in a refrigerator for at least 6–8 hours if to be served cold, or place in a moderate oven for about 20 minutes if to be served hot.

HAM SOUFFLÉ

8 oz. lean ham	*Yield: 5-inch soufflé dish*
1 tablespoon gelatine	*8 portions*
3 tablespoons water	
¼ pint béchamel sauce	
¼ pint double cream	
1 teaspoon tomato paste	

seasoning
2 egg whites

garnish: aspic jelly and gherkin

Finely chop or mince the ham, or blend in a liquidizer with a little of the béchamel sauce if a very fine texture is required.

Mix with the tomato paste.

Place the gelatine in a small basin with the water and stand this in a pan of hot water until the gelatine dissolves.

Prepare a 5-inch soufflé dish by securing a band of foil around the outside of the dish to stand 3–4 inches higher than the rim.

Mix the ham with the béchamel sauce and season well.

Mix in the gelatine.

Cool until it begins to thicken.

Fold in the partly whipped double cream.

Finally whisk the egg whites and fold them into the mixture.

Pour the completed mixture into the prepared soufflé dish.

Allow the mixture to come 2 inches above the rim.

Place the soufflé into the freezer and freeze until hard.

Fold down the foil over the top of the soufflé, pack and seal.

To serve: Stand the foil upright again, place the soufflé in a large polythene bag or other suitable container, and thaw in a refrigerator overnight. Garnish with aspic jelly and pieces of gherkin.

STUFFINGS

Stuffings should be prepared and packed separately in suitable containers; they should not be left in the cavity of the bird when frozen.

FORCEMEAT STUFFING

Suitable for chicken and fish.

3 oz. fresh breadcrumbs
2 teaspoons chopped parsley
¼ teaspoon mixed herbs
seasoning
1 oz. melted margarine
little lemon juice or grated lemon rind
egg or milk to bind

Mix together all the dry ingredients, add the melted margarine, lemon juice and rind, bind with the egg, or egg and milk.

If preferred, the forcemeat stuffing may be formed into small balls and cooked in the tin beside the bird.

CHESTNUT STUFFING

Suitable for turkey.

1 lb. peeled chestnuts
2 oz. breadcrumbs
1 oz. melted butter
2 teaspoons fresh herbs or 1 teaspoon dried mixed herbs
2 eggs
salt and pepper
French mustard

Simmer the chestnuts in a little milk, drain and push them through a sieve. Mix in the breadcrumbs.
Add the melted butter and herbs, beaten eggs, seasoning, and mustard. Pack in a suitable carton and freeze.

To use: Thaw out in a domestic refrigerator for about 12 hours.

SAUSAGE FORCEMEAT

Suitable for turkey or goose.

1 lb. sausage meat
2 oz. finely chopped streaky bacon
liver from the bird
3 oz. lamb's liver
1 onion
1 egg
stock if necessary
2 oz. fresh white breadcrumbs
salt and pepper
2 teaspoons fresh herbs or 1 teaspoon dried mixed herbs

Mince the two kinds of liver and the onion.
Mix the liver, onion, bacon, sausage-meat, breadcrumbs, seasoning and herbs, and bind together with the beaten egg and a little stock if necessary.
Pack in suitable cartons and freeze.

To use: Thaw out in a domestic refrigerator for about 12 hours.

Savouries

The preparation of savouries, whether to be served at a special tea party, a formal cocktail party, or with some pre-luncheon drinks, takes a considerable time. In fact, many people find them rather tedious, especially when required in any quantity. Fortunately most varieties of savouries may be prepared in advance, when there is time to spare, and frozen until required. Small savouries can be removed from the freezer and are ready to serve within the hour, although more time must be allowed for the stacks of party sandwiches.

The varieties described in the following section are suitable to be handed on a dish. They can be broadly divided into: those prepared from fresh bread or toast; those prepared on a pastry base; and those coated in batter.

PARTY SANDWICHES

These are suitable for cocktail parties and special tea parties.

To prepare party sandwiches in quantity it is easier to use slices of bread cut along the length of the loaf, rather than across it. Thus the following recipes give the number of lengthwise slices needed in each case. To prevent wastage and to increase the variety, it is recommended that two or three varieties of party sandwiches be prepared at the same time. For example, from one large white and one large brown loaf it is possible to prepare both a stack of Neapolitan and a stack of mosaic sandwiches.

NEAPOLITAN OR RIBBON SANDWICHES

Slices of white or wholemeal bread are spread with butter and delicately coloured fillings, and piled one on top of the other to produce

406

a stack containing seven to nine slices of bread. Weights are placed on top of the stack to compress the slices. Sandwiches cut from the stack are attractive to serve and delicious to eat.

5 slices white bread ⎫ cut lengthwise from a one- *Yield: about 36–*
4 slices brown bread ⎰ day-old sandwich loaf *40 small sandwiches*
4–6 oz. softened butter or butter spread (page 410)
2–3 oz. each of four fillings which complement each other in flavour and colour (page 411)

Slice the bread thinly, as near to ⅛ inch as possible, along the length of the loaves.
A stack of nine slices requires four fillings (a stack of seven slices requires three fillings).
Arrange the slices in the order of two white, two brown, one white, two brown, two white.
Brush or spread with softened butter, on one side only, the slices which are to form the top and bottom (or outside) slices of each pile. Brush the other slices (middle slices) with butter on both sides, taking the butter to the edge of the slice in each case.
Place an outside piece of bread on a board with the buttered surface uppermost and spread with the first filling.
Cover with a middle slice (buttered on both sides).
Spread on the second filling, and cover with another middle slice.
Repeat all the fillings twice and use the slices of bread in the order specified.
Trim off the crusts and neaten the edges of the stack with a sharp knife.
Tightly enclose in foil and seal.
Weight between two boards for 30 minutes.
Seal in a polythene bag, and freeze.

To serve: Thaw in wrappings in the refrigerator for at least 12 hours, or at room temperature for 4–6 hours. (Thawing may be hastened by replacing the foil wrappings with others of waxed paper.) Cut the stack in half lengthwise, then cut into slices ¼–½ inch thick to serve.

MOSAIC SANDWICHES

3 slices white bread ⎫ cut lengthwise from *Yield: about 60 small*
3 slices brown bread ⎬ a one-day-old sandwich *sandwiches*
              ⎭ loaf
4 oz. butter spread (page 410)
See Fig. 25 for the preparation of mosaic sandwiches.

MOSAIC SANDWICHES

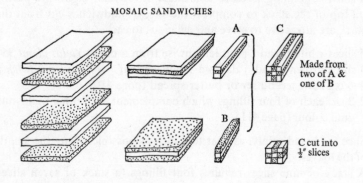

PINWHEEL SANDWICHES

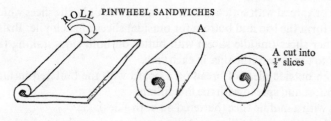

FIGURE 25

PREPARATION OF PARTY SANDWICHES

Cut three slices of white bread and three slices of wholemeal bread ½ inch thick along the length of a sandwich loaf.

Spread a slice of white bread with a butter spread.

Cover with a slice of wholemeal, spread this also with a butter spread, and finally cover with a slice of white bread.

Using two wholemeal slices and one white slice, repeat the process.

Press each stack lightly for 30 minutes.

Trim away the crusts and make the two stacks even.

Cut each into slices ½ inch thick through the layers along the length of the loaf (each stack will probably give about nine layered slices).

Spread each slice with a layer of butter spread and put together two slices from one stack with one slice from the other, which gives a mosaic effect when cut into ¼–½–inch slices to serve.

Wrap the sandwich blocks firmly in foil or waxed paper and weight for 30 minutes.

Seal and freeze in a polythene bag.

To serve: Thaw as for Neapolitan sandwiches, and cut blocks into $\frac{1}{4}$–$\frac{1}{2}$ inch slices to serve.

PINWHEEL SANDWICHES

Slices of bread are cut lengthwise from the loaf, spread with butter and a suitable filling, and rolled up. The filling chosen should if possible contrast with the colour of the bread, i.e. cream cheese in brown bread and salmon in white.

Allow 8 oz. of filling to a 2 lb. loaf.
Any filling which freezes well may be chosen, cheese and anchovy and smoked salmon are two examples given in the following recipe.
1 × 2 lb. loaf (new)
4 oz. creamed butter
(i) 4 oz. cheese and anchovy filling, page 412, }
    stuffed green olives   or any two
(ii) 4 oz. thinly sliced smoked salmon   of the fillings
    or finely mashed tinned salmon   listed on
    1 lemon   page 411
    paprika

Cut the loaf lengthwise into $\frac{1}{8}$–$\frac{1}{4}$ inch slices, remove the crusts.
For bite-size pinwheel sandwiches cut the slices 6 inches long:
for larger ones use the whole length of the loaf.
Spread or brush each slice with melted butter.
See Fig. 25 for the preparation of pinwheel sandwiches.

For cheese filling: Spread with the cheese and anchovy filling. Arrange a row of stuffed green olives across the width of the slice at one end and begin rolling from this point.

For salmon filling: Arrange the thinly sliced salmon along the length of the buttered slices of bread, stopping $\frac{1}{2}$ inch short of one end. Sprinkle with lemon juice and paprika, make a small fold along the end of the slice not covered with the salmon and use this fold to begin rolling the sandwich.
Wrap the filled rolls of sandwiches in foil, seal, and freeze.

To serve: Thaw in wrappings in a refrigerator for about 8 hours, or at room temperature for 4–5 hours. Cut into $\frac{1}{4}$–$\frac{1}{2}$ inch slices and serve.

ASPARAGUS ROLLS

24 fresh asparagus tips, or tin of asparagus tips (net weight 10½ oz.)
1 small Hovis or wholemeal loaf (new)
4 oz. butter

Soften or cream the butter without melting it, until it will brush on to the bread.

Cut the loaf in paper-thin slices, slice across the loaf for small to medium spears, lengthwise for thicker ones, and brush with butter.

Trim off any woody stalk from the asparagus spears and cut into 3-inch lengths.

Place an asparagus spear diagonally across the slice and roll up.

If the bread is very new the crusts may be left on.

To freeze: Place the rolls close together on an aluminium foil tray to prevent them unrolling. Cover closely with aluminium foil. Seal and freeze.

To serve: Thaw in wrappings in the refrigerator for at least 8 hours, or at room temperature for 3–4 hours. (Thawing may be hastened by replacing the aluminium foil covering to the tray containing the rolls with waxed paper.)

BUTTER SPREADS AND SANDWICH FILLINGS

BUTTER SPREADS

If desired a flavoured butter spread may be used in place of the butter in the preparation of any of the party sandwiches, giving added flavour.

Cream the softened butter with sufficient of one of the following flavourings:

Lemon juice.
Anchovy essence.
Grated horseradish.
Roquefort cheese.
Chopped parsley.

CHEESE BUTTER

Blend together equal quantities of butter and grated Cheddar or cottage cheese. Add seasoning, mustard, and moisten with a little béchamel sauce if too dry. This cheese butter may be coloured by the addition of chopped parsley, anchovy essence, or tomato purée.

PEANUT BUTTER

This may be used where the flavour is appropriate.

SANDWICH FILLINGS

The following fillings are suitable for any of the party sandwiches as well as for use in a packed meal. They may be prepared in advance and frozen in small cartons. This is a very useful way of using up small quantities of chicken or ham.

Mayonnaise may be used to moisten the filling, but it should not make up more than one-third of the total volume.

MEAT- AND POULTRY-BASED FILLINGS

Mince or finely chop the meat or poultry used, and combine with the other ingredients as follows:

Chopped ham, highly seasoned, moistened with cream.

Chopped chicken or turkey, crumbs of crisp bacon, moistened with mayonnaise.

Minced ham, with sour cream and horseradish.

Minced chicken with crumbled crisp bacon, finely chopped mushrooms, pepper and cream.

DEVILLED HAM

4 oz. finely minced ham
1 oz. butter.
sprinkling of cayenne pepper
sprinkling of black pepper
⅛ teaspoon curry powder (or less to taste)
pinch ground ginger
1 teaspoon Worcestershire sauce

cream the butter and combine with the other ingredients

CHICKEN LIVER PÂTÉ (page 200)

FISH-BASED FILLINGS

Finely flake or mash the fish with added ingredients:

Salmon, mashed with butter, seasoning, and lemon juice.

Sardine, finely mashed, mixed with a little soft butter, lemon juice, and seasoning.

Smoked haddock, finely mashed, mixed with a little cream cheese.

CHEESE-BASED FILLINGS

Finely grated Cheddar or soft cottage cheese moistened with cream or mayonnaise may be used in combination with other ingredients:

Finely chopped chicken or ham.

Finely chopped chives.

Chopped preserved ginger.

Crushed pineapple.

Chopped shrimps.

Finely chopped green and red peppers.

Finely chopped parsley.

The following cheese spreads are a little more elaborate:

CHEESE AND ANCHOVY

2 tablespoons butter	cream the butter and cheese and beat
4 oz. grated Cheddar	in the other ingredients
1 teaspoon vinegar	
salt, paprika, mustard	
anchovy essence	

FRUIT AND NUT CHEESE

2 oz. demisel or other soft cheese	soak the apricots, and finely
1 oz. dried apricots	chop or mince; combine with the
1 oz. chopped blanched almonds	cheese and almonds

LIPTAUER CHEESE

4 oz. cream cheese	mix the cheese and butter to-
2 oz. butter	gether and beat in the other
1 teaspoon chopped capers	ingredients
1 teaspoon paprika	
$\frac{1}{2}$ teaspoon caraway seeds	
$\frac{1}{2}$ teaspoon salt	
1 teaspoon chopped chives	

AVOCADO CHEESE

4 oz. avocado purée
1 oz. crumbled Roquefort
1 teaspoon lemon juice
salt and pepper

TOAST CANAPÉS

Canapés using toast as a foundation may be prepared in advance for a party and frozen. The butter used to spread the croûtes must be brought to the edge to provide a waterproof layer and prevent the toast becoming soggy. The completed canapés may be coated with aspic prior to freezing, when, although the glaze may not be as clear as usual after thawing, it prevents the ingredients used on the canapé from becoming dry.

Satisfactory results cannot be guaranteed with canapés prepared from fried croûtes. This is because the croûtes, which are not insulated from the topping by a layer of butter (as with toasted croûtes), have a tendency to become soggy.

TOAST CANAPÉS

slices of bread ¼–½ inch thick   *Yield: 1 slice yields 3–4 canapés*
butter spread (page 410)
Suitable toppings:
Thinly sliced ham, chicken, tongue, other cooked meats.
Liver pâté, salmon pâté.
Thinly sliced smoked salmon.
Herring fillet, anchovy fillet.
Cheese-based fillings (page 412).

Garnish of:
walnuts, stuffed olives, gherkins, picked shrimps, pimento pith
aspic jelly

Cut the bread into ¼–½-inch slices, and toast on each side.
Cut into circles, triangles, or fancy shapes no bigger than 2 inches in diameter.
Allow to dry before spreading with one of the butter spreads.
Arrange the toppings on the prepared shapes—the thinly sliced meats and fish may also be cut into shapes, or thin strips, and arranged decoratively.
Finish with a suitable garnish.
Place on a wire tray, and brush a little aspic over the topping.
Allow to set.
Place the completed canapés on a baking tray and freeze before packaging.
Canapés prepared in this way keep one to two weeks satisfactorily.

To serve: Thaw in a refrigerator in loosened wrappings for about 3 hours, or ½ hour at room temperature. Serve immediately after thawing.

SARDINE PYRAMIDS

2 slices of bread   *Yield: 16 small croûtes*
1 tin sardines
2 oz. butter
little lemon juice
Worcestershire sauce
salt and pepper
aspic to glaze

Cut bread ¼ inch thick, toast on both sides, and allow to dry.
Spread a little of the butter in a thin layer over the surface, bringing it to the very edge.
Cut each slice into eight triangles.
Pound the sardines until finely divided, beat in the rest of the butter, a few drops of lemon juice, Worcestershire sauce to taste, and seasoning.
Pile the sardine mixture on to the croûtes to resemble a pyramid.
Use a palette knife, dipped in hot water, to get a good shape and sharp edges.
Brush a thin layer of aspic over the sardine mixture.

To freeze: Place on a baking tray in the freezer until hard. Pack in layers, seal, and freeze.

To serve: Loosen the wrappings and thaw in a refrigerator for about 3 hours, or at room temperature for about ½ hour.

WELSH RAREBIT

6 slices of bread   *Yield: 24 small canapés*
2 oz. butter
6 oz. grated cheese
2 teaspoons cornflour
salt, pepper, mustard
4 tablespoons milk

Cut the bread into ¼-inch slices and toast on each side.
Remove the crusts and spread with butter to the edges.

Mix the grated cheese, cornflour, and seasoning, to a stiff paste with
the milk.
Spread on the buttered toast.
Cut into fingers or squares.
Brown lightly under the grill.

To freeze: Separate the layers between a double thickness of cellophane.
Seal and freeze.

To serve: Place on a baking tray and put into a fairly hot oven for
about 5–10 minutes.

CHOUX CANAPÉS

These are toast canapés the topping of which is made of uncooked
choux pastry blended with a cheese or ham mixture. When baked the
topping puffs up and is very light and tasty. They should be frozen
before baking.

CHEESE TOPPING

2 oz. choux pastry (page 377)
2 tablespoons single cream
1 oz. Parmesan
½ teaspoon salt
cayenne pepper

*Yield: 8 slices of bread gives
32 hors-d'oeuvre pieces
or 64 bite-size pieces
Oven: 425°F (220°C), Regulo 7
Time: 15 minutes*

HAM TOPPING

2 oz. choux pastry
1 oz. finely chopped ham
¼ teaspoon French mustard

Cut the bread about ¼-inch thick and toast on one side.
Spread the untoasted side with a generous layer of flavoured choux
pastry, and cut each slice into suitable shapes.

Freeze the canapés raw and package when hard
or bake in a hot oven to serve immediately.

To serve: Place *raw* canapés in the oven at 450°F (230°C), Regulo 8,
and cook for 15 minutes (a slightly higher temperature is used when
cooking the canapés straight from the freezer). Serve hot.

PASTRY SAVOURIES

For a note on the 'weight of pastry' given in recipes, see page 299.

USING SHORTCRUST PASTRY

MINIATURE QUICHE LORRAINE

The recipe on page 403 makes twelve to fifteen miniature Quiche Lorraine, 2 inches in diameter.

Bake at 425°F (220°C), Regulo 7, for 5 minutes.

Reduce to 350°F (175°C), Regulo 4, for a further 10–15 minutes.

Roll the pastry thinly and line the patty tins.
Finely chop the onions instead of slicing them when preparing the filling.
Begin baking in a hot oven, lowering the temperature after the first 5 minutes. Cook 15–20 minutes in all.
Cool, pack, and freeze.

To serve: Thaw and reheat in a fairly hot oven for 5–10 minutes.

FILLED TARTLETS AND BARQUETTES (PASTRY BOATS)

8 oz. shortcrust pastry (page 367) yields: 36 round cases 2 inches diameter
or 30 round cases 3 inches diameter
or 36 pastry boats.

Oven 425°F (220°C), Regulo 7, for about 10 minutes.

FILLINGS

SPINACH TARTLETS
(sufficient for 24 cases)
4 oz. creamed spinach (page 248)
2 oz. very thinly sliced smoked salmon

fill the tartlets with the spinach and decorate with a curl of smoked salmon on top

CHICKEN CREAM
cooked chicken
double cream

make a purée of the chicken with a little stock, blend with an equal quantity of double cream

416

Chicken Liver Pâté

chicken liver pâté (page 200)	pipe the liver pâté in boat-shaped moulds; decorate with chopped truffles
chopped truffles	

Some of the fillings suggested for sandwiches on page 411 may also be used to fill these pastry tartlets and boats.

Roll the prepared pastry as thinly as possible, and line the patty tins and small boat-shaped moulds.

Cut the round tartlet cases with a cutter ½ inch larger than the patty tins. Line the tins and prick the bases with a fork.

When lining the boats, place all the moulds close together, lift the pastry over the rolling pin and lay it over the boats. Run the rolling pin across the tops of the moulds to trim off the surplus pastry.

Ease the pastry into each case, prick lightly with a fork to prevent the base rising.

Although tedious it is worth placing greaseproof paper and beans or rice in each tartlet and pastry-boat case to prevent the pastry rising during baking.

This should be removed after the pastry has set, and the base allowed a few minutes longer to dry out.

Cool and fill with one of the fillings.

Decorate with gherkins, capers, walnuts.

To freeze: Place on a baking tray and freeze. Pack in waxed boxes, separate layers with cellophane. Seal.

To serve: Remove from the box and spread out on trays. Cover lightly with greaseproof paper for 30 minutes at room temperature. Serve immediately, or keep in the refrigerator.

USING CHEESE PASTRY

Cheese pastry may be used in place of shortcrust pastry to produce the pastry boats and small tartlets in the previous recipe. It may also be used to make a variety of savouries as in the following recipes.

Cheese Butterflies

6 oz. cheese pastry	*Yield: 2 dozen cheese butterflies*
6 oz. cream cheese	*Oven: 350°F (175°C), Regulo 4*
2-3 tablespoons single cream	*Time: about 10 minutes*

finely chopped parsley
paprika $\Big\}$ add on serving

Prepare the pastry and roll out ⅛ inch thick.
Cut into four dozen rounds, a little less than 2 inches in diameter.
Place on a baking tray and cut two dozen of the small rounds into two across the centre.
Bake the whole and half biscuits until very lightly browned.
Cool on a wire tray.
Mix the cream cheese with a little cream until sufficiently smooth for piping.
Using a star-shaped pipe, pipe a little cheese on to each complete biscuit and place two half-biscuits into the cheese to form butterfly wings.
Place on a baking tray to freeze, pack carefully in a waxed box.

To serve: Loosen the packaging material, thaw 2–3 hours in a refrigerator or 30 minutes at room temperature. Decorate with chopped parsley and paprika.

TALMOUSES OF SMOKED FISH

These savouries freeze equally satisfactorily before or after baking.

6 oz. cheese pastry	*Yield: 30*
1 oz. butter	*Oven: 375°F (190°C), Regulo 5*
3 oz. cooked smoked haddock (weight after removing bones and skin)	*Time: 10–15 minutes*
1½ oz. grated cheese	
salt and pepper	
3 tablespoons white sauce	
egg wash—before baking	

Prepare the pastry and roll out very thinly. Cut into 2-inch rounds.
Finely chop the haddock and mix with the butter, cheese, seasonings, and white sauce.
The consistency of the mixture should be such that it can be rolled into small balls.
Divide the mixture into the same number of balls (the size of marbles) as rounds of pastry, and place one on each.
Dampen the edges of the pastry rounds, and fold and seal above the ball of fish in the form of a three-cornered hat.

418

Freeze at this stage

or brush with egg wash and bake for about 10–15 minutes.

To freeze: Freeze the *raw talmouses* before packaging.
Cool the *baked talmouses* and wrap in foil or other suitable packaging.
Seal and freeze.

To serve: Brush the *raw talmouses* with egg wash and place in an oven at 375°F (190°C), Regulo 5, for 20–30 minutes.
Thaw and reheat the *baked talmouses* in a fairly hot oven for 10 minutes.

CHEESE BALLS

These may conveniently be made from scraps of pastry.

They may be frozen before or after baking. On the whole it is better to freeze the mixture raw, as in this form it takes up less space in the freezer and there is no risk of the biscuits becoming broken during storage.

Bake at 350°F (175°C), Regulo 4, 10–15 minutes.

Roll out into a rectangle about 7 inches long.
Place pimento-stuffed olives across the width of the pastry (this can be of any measurement) and roll up as for pinwheel sandwiches.

Freeze in the roll at this stage

or slice into biscuits ⅛ inch thick, place on a baking tray and bake for 10 minutes.

To freeze: Pack and freeze *raw mixtures* in the roll before slicing.
Cool *baked biscuits*, pack carefully, seal and freeze.

To serve: Thaw *unbaked roll* 1 hour at room temperature, slice and bake at 350°F (175°C), Regulo 4, for 10–15 minutes.
Thaw *baked cheese balls* in wrappings for ½–1 hour at room temperature.

CHEESE SHORTBREADS

These thick, short biscuits can be served hot or cold, and are ideal just by themselves. They may be frozen before or after baking.

5 oz. butter	*Yield: 3½ dozen biscuits 2 inch diameter*
6 oz. flour	*Oven: 375°F (190°C), Regulo 5*
6 oz. grated Cheddar	*Time: about 15 minutes*

2 hardboiled egg yolks
seasoning, $\frac{1}{2}$ teaspoon salt; pepper, cayenne, mustard

Place the flour, butter, and grated cheese into a bowl.
Sieve the egg yolks, add to the other ingredients together with the
seasonings and mix into an even dough.
Roll out to $\frac{1}{2}$ inch thickness, cut into various shapes (rounds, squares,
triangles, etc.) about 2 inches in diameter.

Freeze raw at this stage

or brush over with egg wash and decorate with caraway seeds or
chopped nuts, and bake in a fairly hot oven for about 15 minutes.
To freeze: Freeze *raw shortbreads*, pack, and seal.
Pack *baked shortbreads* in suitable quantities, seal, and freeze.

To serve: Brush *raw shortbreads* with egg wash and decorate with
chopped nuts or caraway seeds, bake at 375°F (190°C), Regulo 5, for
15–20 minutes.
Thaw *baked shortbreads* in wrappings at room temperature for 30
minutes.

USING FLAKY PASTRY

SAUSAGE ROLLS

Sausage rolls may be frozen before or after baking. The keeping
time of unbaked rolls is determined by the keeping time of the sausage-
meat, and should not be longer than 1 month.
Sausage rolls may also be prepared from puff pastry.

8 oz. pastry   *Yield: 16 large or 36 cocktail size rolls*
12 oz. sausage-meat   *Oven: 475°F (245°C), Regulo 9, lowering*
                        to 375°F (190°C), Regulo 5, after 15 minutes

Roll the pastry to $\frac{1}{8}$ inch thickness and into a rectagular shape wide
enough for two rolls to be made side by side—about 8 inches wide.
Divide the sausage-meat in half and roll each half to the length of the
pastry.
Place the sausage-meat rolls on to the pastry about $1\frac{1}{2}$ inches in from
the long edge of the rectangle.
Damp the outside edges of the pastry and fold them in towards the
centre of the rectangle over the sausage meat to enclose it.
Press down to seal the pastry. Cut into two long rolls. Trim and flake
the cut edges of the rolls.

Freeze sausage rolls raw at this stage

or brush with egg wash and cut into even-sized pieces.
Decorate by cutting the top surface with a knife or scissors, and bake in a very hot oven, lowering the temperature after the first 15 minutes.

To freeze: Divide *raw sausage rolls* into even pieces on the trays, freeze until hard, pack and seal.
Freeze *baked sausage rolls* in suitable quantities.

To serve: Place *raw sausage rolls* on to a baking tray with raised sides, brush with egg wash, and bake at 475°F (245°C), Regulo 9, for 20 minutes, lower to 375°F (190°C), Regulo 5, to complete cooking.
Thaw *baked sausage rolls* in wrappings in the freezer for 6–8 hours, or if to be served hot thaw and reheat in a fairly hot oven.

USING PUFF PASTRY

VOL-AU-VENTS, BOUCHÉES, AND MINIATURE HORN CASES

The best results are obtained with pastry freshly baked. The simplest method of achieving this is to prepare the puff pastry and cut out the shapes required some days or weeks ahead, or whenever convenient. These shapes may be frozen raw, they can be placed straight into the oven from the freezer when needed, and they will be ready for use in less time than it takes to prepare the fillings.

For preparation and shaping of cases, see page 374.

For fork-luncheons and supper dishes, serve vol-au-vents cases 3–4½ inches in diameter.
For cocktail party savouries, serve bouchées 1½–2½ inches in diameter, and miniature horn cases 2 inches in length.
They may be served hot or cold depending upon the filling selected.

FILLINGS FOR VOL-AU-VENTS, BOUCHÉES AND HORN CASES

A slightly more moist filling is required for the vol-au-vent and bouchée cases than for the horn cases.
The quantities given are for 18 cases served as an entrée,
or 3 dozen cases served at a party.

Bouchées à la Reine

½ pint velouté or béchamel sauce
6 oz. chopped chicken
3 oz. chopped mushrooms
1 oz. butter
juice ½ lemon

cook the mushrooms in butter and lemon juice over gentle heat for 5–7 minutes covered with a lid; add the chicken and sufficient velouté sauce to bind and soften; heat the mixture and fill into the cases just before serving

or substitute:

1 tin condensed chicken soup
    (net weight 10½ oz.)
2 oz. chopped cooked chicken
3 oz. chopped mushrooms cooked
    in ½ oz. butter and juice of
    1 lemon as before

} combine and heat together

Bouchées aux Champignons

8 oz. chopped mushrooms
2 oz. butter
juice 1 lemon
seasoning
½ pint velouté or béchamel sauce

} make as à la Reine, omitting the chicken

or substitute:

1 tin condensed mushroom soup
(net weight 10½ oz.)
2 oz. mushrooms cooked in ½ oz. butter
juice of 1 lemon

Bouchées aux Crevettes

1 pint prawns or 8 oz. frozen
    prawns
½ pint béchamel sauce
2 oz. chopped mushrooms
1 oz. butter
juice ½ lemon

prepare and chop the fresh prawns; cook the chopped mushrooms in butter and lemon juice as for à la Reine; add sufficient béchamel sauce to the mixture of prawns and mushrooms to bind it; if fresh prawns are used, reserve the heads and use to garnish each case

COOKING FOR THE FREEZER

Bouchées de Homard

lobster meat from 1 small cooked
  lobster (see page 175 for
  cleaning)
¾ pint velouté sauce made from
  fish stock
3 oz. lobster butter
  (page 234)
cayenne pepper

cut up the lobster meat—
it needs to be quite finely
chopped if to be used in
small cases, but larger
pieces are used for vol-au-
vents; add the lobster to
the sauce and heat
thoroughly without boil-
ing; remove from the heat
and beat in the lobster
butter; season with cay-
enne; fill into cases just
before serving

The following ingredients can also be diced and combined in a
sauce:

Aubergine in sauce suprême (page 225).
Celeriac blanched, then cooked in butter until soft, in sauce suprême
(page 225).
Chicken, mushrooms, cooked as for à la Reine, in sauce Italienne
(page 227).
Crayfish, in béchamel sauce (page 222).
Sweetbreads, truffles, and mushrooms in Madeira sauce (page
228).
Foie gras or minced sautéed chicken livers in Madeira, port or sherry
sauce (page 228).
Game in thick rich brown sauce (page 225).
Ham in thick rich brown sauce (page 225).
Truffles or mushrooms pounded with chicken purée.

Allumettes

Although allumettes freeze equally well raw or cooked, they take
up less space when frozen before baking.

Puff pastry (page 373).
(Trimmings from vol-au-vents and
  bouchées may be used.)

Oven: 450°F (230°C), Regulo 8
Time: 10–15 minutes

FILLINGS

SPINACH

3 oz. creamed spinach	mix together and spread on to
1 oz. grated Parmesan	puff pastry; sprinkle a little
	Parmesan over the surface.

HADDOCK

cooked smoked haddock	finely flake or purée the had-
little béchamel sauce	dock, add seasoning; mix with
salt and pepper	sauce to give spreading con-
	sistency

HAM AND ONION

2 oz. choux pastry	mix together as a spread
1 oz. minced lean ham	
½ teaspoon paprika	
1 oz. finely chopped onion, cooked	

Roll out the puff pastry into a square 12 × 12 inches, and ⅛ inch thick.
Cut four strips 3 inches wide and 12 inches long.
Spread the strips with one or other of the mixtures.
Cut into ribbons 3 × 1 inches.
Place on to baking sheets.

Freeze allumettes raw at this stage

or bake in a very hot oven for 10–15 minutes.

To freeze: Freeze *raw allumettes* before packing.
Pack *baked allumettes* in a waxed box, separate layers with double thickness of cellophane, seal, and freeze.

To serve: Bake *raw allumettes* at 450°F (230°C), Regulo 8, for 15 minutes.
Thaw and reheat *baked allumettes* in a fairly hot oven for 5–10 minutes.

D'ARTOIS

These are similar to allumettes, but the filling is covered by a second layer of pastry. Freeze before or after baking.

puff pastry (page 373)   *Oven: 450°F (230°C), Regulo 8*
fillings as for allumettes, mixed a little   *Time: 20–25 minutes*
    softer, also those suitable for puff
    patties (page 425) may be used
egg wash

Cut eight strips of puff pastry 3 × 12 inches, four strips to form the bases and four strips to form the lids of the pastry sandwiches.

Spread four strips with the prepared mixtures, taking it to within ⅓ inch of the edge.

Dampen the edge, cover with the pastry lids, and seal.

Mark with a knife at 2-inch intervals along each strip, but do not cut. Decorate the tops with a criss-cross pattern.

Freeze raw at this stage

or brush the tops with egg wash and bake as for allumettes. Separate at 2-inch intervals after baking.

Freeze and serve: As for allumettes.

PATTIES

Although patties may be frozen after baking, they take up much less space when frozen raw.

puff pastry (page 373) *Oven: 450°F (230°C), Regulo 8*
Time: 20 minutes

FILLINGS

Cooked white fish, chopped mushrooms and truffle.

Fish moistened with velouté sauce, well seasoned.

Chopped or minced game in rich brown sauce.

Anchovy fillets.

Sausage-meat mixed with chopped green and red pepper, and finely chopped cooked onion.

Sausage-meat and chopped parsley.

Equal quantities of sausage-meat and duxelles (page 245).

Liver pâté, or foie gras and chopped truffle (or tinned equivalent).

Roll the pastry to ⅛–¼ inch thick. Cut into rounds 2 inches in diameter.

Arrange half the circles 2 inches apart on a baking tray.

Dampen the edges.

Fill with a teaspoonful of one of the suggested fillings.

Place the second rounds on top and seal the edges.

Mark the top lightly with a criss-cross design.

Freeze raw at this stage

or bake about 20 minutes until well risen and fully cooked. The risen sides should be quite firm.

425

To freeze: Freeze *raw patties* before wrapping.
Baked patties are cooled, packed, and then frozen.

To serve: Bake *raw patties* at 450°F (230°C), Regulo 8, for 20 minutes; reduce to 400°F (205°C), Regulo 6, for a further 10 minutes.
Reheat *baked patties* in a hot oven for 10 minutes.

USING CHOUX PASTRY

SMALL SAVOURY ÉCLAIRS (OR CAROLINES) AND PROFITEROLES

The best results are obtained with choux pastry cases filled just prior to serving. Although it is possible to freeze the cases complete with a savoury filling, unless they can be served promptly after thawing, the tendency is for the pastry to become soggy.

For preparation and shaping of choux pastry cases, page 379.

FILLINGS

Allow 8 oz. prepared filling for 15–18 carolines or profiteroles.

HAM OR CHICKEN

8 oz. diced chicken or ham
4 oz. chopped celery
4 oz. grated Cheddar
2 oz. chopped stuffed olives
salt and pepper
mayonnaise to moisten

CREAM CHEESE

8 oz. cream cheese
2 tablespoons chopped chives or parsley
   or 2 oz. finely chopped prawns

PURÉED VEGETABLES

green peas
beans
asparagus tips
celeriac
whipped cream

cook the selected vegetables in butter; blend in a liquidizer and mix in the cream

CHOUX PASTRY PIECES—DEEP FRIED

CHEESE AIGRETTES

9 oz. choux pastry (page 378)   *Yield: 50 aigrettes*
3 oz. grated cheese—Cheddar or Parmesan
½ teaspoon salt
cayenne pepper

Mix the cheese and seasonings into the choux pastry.
Drop in teaspoonfuls into oil at 360°F (180°C)—an easier method is
to place the prepared mixture into a piping bag with a ½-inch plain
pipe and cut off ½–1inch pieces of choux pastry directly into the fat.
Cook for 5–10 minutes, until golden brown and cooked in the centre.
Drain well on absorbent paper.
Cool rapidly.

To freeze: Pack and seal in suitable quantities.

To serve: Place on a tray in a fairly hot oven for 10–15 minutes until
heated through and crisp. Drain on absorbent paper. Toss in finely
grated Parmesan cheese. Serve hot.

The following beignets soufflés may be prepared by making certain
other additions to the choux pastry, in place of the cheese as in cheese
aigrettes. The method of frying and freezing is the same as for cheese
aigrettes.

PIGNATELI

9 oz. choux pastry
2 oz. chopped ham or tongue
2 oz. finely diced Gruyère or grated Cheddar
1 oz. shredded almonds
seasoning

ANCHOVY

6 oz. choux pastry
1 tablespoon finely chopped anchovies
seasoning
little anchovy essence if required

SAVOURY

6 oz. choux pastry
2 tablespoons chopped cooked onion
½ teaspoon paprika

427

HAM

9 oz. choux pastry
2 oz. Parmesan
2 oz. chopped lean ham
2 oz. chopped mushrooms—cooked in butter
   or 1 oz. chopped truffles
seasoning

BATTER-DIPPED SAVOURIES

Savouries dipped in batter and served piping hot are delicious, but are not normally easy things to serve at a party if the hostess is not to spend much of her time in the kitchen.

YEAST BATTER (page 395) is used throughout this section.

All the fillings used have been pre-cooked prior to being coated with the batter.

Prepare the savouries and dip in batter.
Fry in fat at 360°F (180°C) until the batter is crisp and golden brown.
Drain well on absorbent paper, and cool rapidly.
Pack in suitable quantities, seal, and freeze.

To serve: Reheat in a fairly hot oven for 15–20 minutes. Drain on absorbent paper. Serve hot.

SPICED MEAT BALLS

Highly spiced meat balls bound with egg are cooked, then dipped in batter.
The following recipe is suitable or any favourite recipe can be used:

1 lb. finely minced lean meat   *Yield: 3 dozen small balls*
1 egg
⅛ pint tomato juice
¼ teaspoon pepper
1 teaspoon prepared mustard
few grains nutmeg

Mix all the ingredients well together and shape into bite-size balls.
Fry gently tossing frequently; do not overcrowd the pan so much that they are not able to roll around during frying, and so keep their round shape.

Cook until well browned outside and fully cooked in the centre, about 20 minutes.

Cool.

Coat in batter.

Fry in the fat bath, drain, cool, and freeze.

Thaw and serve as above.

KROMESKIES

These all consist of chopped meats and vegetables in a sauce, rolled into balls and dipped into batter. They may first be rolled in a small pancake, or in streaky bacon, before being dipped in the batter. They are excellent hot savouries, hot hors-d'ouevres, or party snacks. The sizes can be varied.

Yield: 5 dozen bite-size pieces; 2½ dozen hors-d'oeuvre portions

CHICKEN

4 oz. chopped cooked chicken
4 oz. chopped mushrooms
½-1 oz. butter
2 oz. chopped ham
4 tablespoons béchamel sauce
1 egg yolk
optional—16 rashers streaky
   bacon spread out thinly
   with a knife

chop the mushrooms and cook in the butter until tender; mix in the chicken and ham, and add the hot sauce; beat in the egg yolk; allow to cool; divide into equal portions; roll into cork shapes or balls; if desired wrap around with half a rasher of bacon; dip in batter and fry

FLORENTINE

4 oz. creamed spinach, page 248
2 oz. grated Parmesan
thin unsweetened pancakes, page
   401

cut pancakes into strips; mix ingredients together; if the mixture is rather soft, wrap up portions in a strip of pancake; dip in batter and fry

GAME

4 oz. cooked game meat
rich brown sauce

finely chop meat and add sufficient sauce to bind portions together; dip in batter and fry

KIDNEY   *Yield: 28 bite-size savouries; 14 hors-d'oeuvre portions*

4 kidneys, 3–4 oz. prepared weight
1 oz. chopped mushrooms
¼ pint thick brown sauce
salt and pepper
14 rashers of bacon spread out thinly with a knife—optional

melt the butter and cook the chopped mushrooms; lift out of the pan; add prepared and roughly chopped kidney; cook over gentle heat without toughening; finely chop kidney, mix with the mushrooms and just sufficient sauce to bind them together; season well; dip in the batter and fry. (wrap in streaky bacon before coating in batter if desired)

VEGETABLE PUFFS

Suitable vegetables:

artichoke hearts quartered
white asparagus tips
baby Brussels sprouts
cauliflower sprigs

olive oil
lemon juice
seasoning
chopped parsley   } for marinade

Cook selected vegetable until tender in salted water.
Drain and marinade for 30 minutes.
Drain, dip in yeast batter, and fry.

ANCHOVY PUFFS

thin slices of bread
anchovy fillets

Cut out circles from the bread 1½ inches in diameter.
Place a rolled anchovy fillet between two circles of bread and sandwich them firmly together.
Dip into the yeast batter and fry.

430

GRUYÈRE CHEESE SLICES

thin slices of Gruyère cheese
mixture of béchamel sauce and minced ham

Cut the cheese into circles 1¼ inches in diameter.
Sandwich the slices together with the mixture of minced ham and béchamel sauce.
Dip in the batter and fry.

FISH PUFFS

white fish—cooked
olive oil
lemon juice   } for marinade
salt and pepper
chopped parsley

Cut fish into bite-size pieces and marinade for 30 minutes.
Drain, dip in batter, and fry.

CREAM CHEESE PANCAKES, IN BATTER

4 oz. cream cheese
4 oz. softened butter
salt and pepper
grated nutmeg
1 egg
small, thin, unsweetened pancakes, 4 inches in diameter (page 401)

Work the cheese and butter together.
Beat in the egg and season well.
Divide into portions of 1 tablespoon.
Place on the centre of each pancake.
Fold the edges to seal the filling.
Dip into batter.
After frying and draining, sprinkle with salt. Garnish with chopped or fried parsley.

COCKTAIL KEBABS

SAVOURY HAM KEBABS

8 oz. cooked ham
2 tablespoons tomato ketchup

3 teaspoons lemon juice
½ teaspoon pepper
2 oz. nuts finely chopped

Mince the ham finely.
Mix with the ketchup, lemon juice, and pepper.
Form into small cork shapes.
Roll in nuts.
Chill.

To freeze: Pack in small cartons, separate layers with a double thickness of cellophane. Seal and freeze.

To serve: Thaw in a refrigerator for 6 hours. Serve spiked on skewers, or cocktail sticks, with sliced cucumber and cocktail onions.

CHICKEN LIVER KEBABS

chicken livers
streaky bacon

Wrap pieces of chicken liver in streaky bacon.
Bake at 400°F (205°C), Regulo 6, until cooked.

Freeze and serve as for savoury ham kebabs.

HAM CORNETS

thin slices of cooked ham
3 oz. cream cheese
6 chopped olives   } or chicken liver pâté, page 200
1 teaspoon horseradish
2 tablespoons cream

Combine cheese with chopped olives, horseradish and cream.
Trim slices of cooked ham, and shape into horns or rolls.
Fill with cheese mixture or the liver pâté; or spread the ham with these fillings before shaping.
Secure with cocktail stick if necessary.
Pack in layers, seal and freeze.
To serve: Thaw in a refrigerator for 6 hours before serving.

CREAM CHEESE BALLS

4 oz. cream cheese
1 oz. chopped walnuts

Mix the cheese with chopped walnuts.
Form into small balls, roll in finely chopped nuts.
Freeze and serve as for ham cornets.

DIPS

Many savoury dips may be prepared in advance and frozen in cartons. Transfer from the freezer to the refrigerator to thaw the day before required. Serve cold dips well chilled. Reheat hot dips in a water bath.

It may be decided to serve three or four dips of contrasting flavours, when the quantities given below would be sufficient for a party of fifteen to twenty. If only one dip was to be served, then the quantities would need to be doubled or trebled to cater for the same number.

MUSHROOM DIP

$\frac{1}{2}$ lb. mushrooms   *Serves 15–20*
2 oz. butter
1 teaspoon flour
$\frac{1}{4}$ pint double cream
salt and pepper
pinch nutmeg

Chop the mushrooms finely and cook in the butter for about 5 minutes.
Add the flour, stir in the cream.
Cook until thick.
Add the nutmeg, and season well.
Serve: Hot with toast or crackers.

CHEESE DIP I

3 egg yolks   *Serves 10–15*
$\frac{1}{4}$ pint single cream
$2\frac{1}{2}$ oz. butter

433

2½ oz. grated Cheddar
¼ teaspoon paprika
pinch salt

Beat the egg yolks.
Scald the cream and beat into the yolks.
Return to the pan and cook until it is thickened.
Cool rapidly with frequent stirring.
When cool beat in the cheese and butter in small pieces. Season well.

Serve: Chilled with cheese biscuits, celery, potato crisps.

CHEESE DIP II

This is a very highly seasoned dip.

1 tablespoon tomato ketchup	*Serves 10–15*
1½ teaspoons lemon juice	
5 drops tabasco	Mix and allow to stand for 2
¼ teaspoon salt	hours
sprinkling of pepper	

4 oz. cream cheese, or other suitable soft cheese to taste
2 tablespoons soured cream
1 teaspoon chopped celery
1 teaspoon chopped green pepper
¼ teaspoon chilli powder
1 teaspoon mayonnaise
sprinkling of cayenne pepper

Mix the cheese with the soured cream until smooth.
Add the ketchup mixture.
Add the spices and seasoning to taste.
Add the mayonnaise, chopped celery, and pepper.

Serve: With any small dunks.

GUACAMOLE

1 avocado	*Yield: 1 medium-sized avocado*
2 teaspoons lemon juice	*gives about 6 oz. of dip*
¼ teaspoon onion salt, or onion juice	*Serves 10*
+ ¼ teaspoon salt	
sprinkling of pepper	

2 teaspoons tomato ketchup
1 teaspoon mayonnaise
¼ teaspoon chilli powder
sprinkling of cayenne pepper

Cut the avocado in half and remove the stone. Scoop out the flesh. Mash the avocado.

Season well with the onion salt, lemon juice and pepper.

Beat in the tomato ketchup and mayonnaise until it resembles thick cream.

Chilli powder can be added if a highly spiced dip is required, and the seasoning can be heightened with a little cayenne pepper.

Serve: Chilled with small shapes of brown toast, potato crisps, or small savoury biscuits.

Entertaining

THE freezer can be of particular use for all types of parties. Many dishes can be prepared in advance, leaving the hostess with little to do on the day of the party except for thawing and arranging the various dishes. Only unfreezable garnishes, such as salad vegetables, need be prepared at the last minute.

Party foods are usually stored for short periods of, say, one to two weeks; therefore it is worth while to be a little more generous with packaging space. Where possible use waxed boxes to give added protection.

THE COCKTAIL PARTY

The cocktail party is open to widely differing interpretations and obviously the selection of savouries to be served will be based on the function the party is intended to serve. If it is the prelude to a dinner party, then the quantities and varieties served will be few. If it is to be a large cocktail party lasting two or more hours, the savouries served should be more varied and substantial.

Several points, however, remain the same, whatever the size and form of the party. The savouries must be well presented; on the whole large trays and dishes display the savouries better than smaller plates. The savouries selected should be varied in texture, colour, and flavour. Avoid overworking one base, whether it be a type of pastry, or bread, or toast; however varied the fillings may be, the effect is still inclined to be monotonous. Batter-coated savouries, kebabs, a well-flavoured dip, all serve to make good contrasts.

A selection of freezable savouries is found on pages 406–435.

By serving the majority of foods straight from the freezer a little time can be set aside for the preparation of some of the more choice salads and fruits, and those savouries which cannot be frozen, for example:

STUFFED CELERY PIECES

Cut inner white celery stalks into 2-inch pieces and fill with Stilton cheese.

STUFFED CUCUMBER RINGS

Peel cucumbers and cut into rings ⅜ inch thick; slightly hollow out the centres to hold a small ball of finely mashed salmon or tuna fish (blended with mayonnaise).

STUFFED GRAPES

Remove the pips and fill the centres with well-seasoned cream cheese.

THE BUFFET PARTY

To many people a buffet party is more popular than a cocktail party, both to give and to attend. A variety of dishes both hot and cold, sweet and savoury, may be prepared and arranged on an attractively decorated table, from which the guests may help themselves.

The dishes should be so prepared and served that it is possible for the guests to be able to eat the food on their plates using a fork only—without embarrassment, unless, accommodation can be provided in the form of small tables and chairs, when dishes requiring a knife and fork may be served.

The variety of dishes can be chosen to suit the occasion, the season of the year and, not least, the guests.

The following list includes dishes suitable to serve at buffet meals, whether for an evening party or a fork-luncheon.

SAVOURY DISHES

Served cold:

Ham Soufflé, page 403.
Salmon Mousse, page 256.
Savoury Meat Loaf, page 286.
Terrine, page 201.
Quiche Lorraine, page 403.
Veal and Ham Pie, page 284.

Serve with: lettuce, tomatoes, celery, cucumber, cole slaw, beetroot, sweetcorn in mayonnaise.

Served hot:

Chicken à la King, page 293.
Crêpes à la Marinière, page 265.
Curried Veal, page 275.
Large or Individual Vol-au-Vents, page 421.
Meat Balls in Savoury Sauce, page 278.
Paella, page 264.
Pizza, page 387.

Suitable hot vegetables include: Spiced red cabbage, page 240.
Potato croquettes, page 241.

SWEETS

Almond Cherry Flan, page 305.
Baked Cheese Cake with Fruit—pears or strawberries, page 307.
Cassata, page 335.
Charlotte Russe, page 318.
Chiffon Pie, page 305.
Chilled Cheese Cake, page 321.
Cold Soufflé, page 319.
Frozen Apricot Mousse, page 336.
Fruit Gâteau, page 314.
Fruit in Wine Syrup, page 322.
Gâteau Mille Feuilles, page 304.
Lemon Savarin, page 398.

TEENAGE PARTY

If catering for a teenage party, the choice of menu might be made from the following dishes.

SAVOURY DISHES

Baked Stuffed Potatoes, page 243.
Fried Chicken Joints, page 290.
Individual Steak and Kidney Pies, page 283.
Meat Balls in Sauce, page 278.
Sausage Meat Pizza, page 388.

Served with: salad vegetables, chips (reheat frozen chips on baking trays, page 244).

SWEETS

Blackcurrant Flan, page 308.
Biscuit Tortoni, page 337.

Fruit Pie and Cream, page 299.
Neapolitan Ice Cream, page 334.
Rum Babas, page 399.

ROLLS AND CROISSANTS

Serve freshly baked rolls (parbakedrolls, page 386) and hot croissants, page 390.

ICE

The owner of a freezer has no excuse for running short of ice. It can be made in quantity days before the party.

Cocktail cherries, mint leaves, pieces of lemon and fruits, may be set in ice cubes and served in wine and fruit cups.

THE DINNER PARTY

By utilizing the services of the freezer, the preparation of a dinner party need no longer be an exhausting occasion; the various courses may be prepared when time and inclination exist, and they need not all be prepared on the same day. At the same time, it is possible to include those dishes made from foods currently out of season, and some of those dishes which would otherwise need to be cooked at the last minute such as potato croquettes. It is often worth while to prepare and freeze a dish for just a few days if by this means the problems of last-minute preparations are overcome.

These points are illustrated in the following dinner party menu:

Consommé Madrilène, page 208 (note a).

Coquilles St Jacques à la Parisienne, page 257 (note b).

Veal Escalopes, page 269 (note c).
Duchesse Potatoes, page 240 (note d).
Creamed Spinach, page 280.
Tomatoes Stuffed with Duxelles, page 245.

Strawberry Sorbet, page 330 (note e).

Baked Cheese Cake (orange flavoured), page 307.

Notes

(*a*) A consommé is time consuming to prepare; after thawing serve en gelée.

(*b*) Scallops are not always in season and thus make a welcome change in the fish course.

(*c*) Frozen escalopes can be reheated in the oven; no need for last-minute frying.

(*d*) Duchesse potatoes reheated from the frozen state are as good, if not better, than those freshly made.

(*e*) The low temperature of a freezer is necessary for the preparation of this excellent sweet.

The need to work out a timetable when preparing and serving a dinner composed entirely or largely of pre-frozen dishes is just as important as when working with fresh raw ingredients. Otherwise dishes may be served apparently hot and steaming on the outside, but still containing ice crystals in the centre.

A timetable for the sample dinner menu to be served at eight o'clock is given below:
Transfer the *Consommé Madrilène* to the domestic refrigerator twenty-four hours before required.

2.0 p.m. Transfer the *Strawberry Sorbet* to the ice compartment of the domestic refrigerator.
Transfer the *Baked Cheese Cake* to the domestic refrigerator (these tasks may be done in the morning if away from home during the day).

6.0 p.m. Prepare the garnish for the *Escalopes* and *Consommé Madrilène*. Look at the *Cheese Cake*, and bring out to room temperature if necessary to accelerate thawing.

7.30 p.m. Place the *Coquilles St. Jacques* into a cold oven set at 400°F (205°C), Regulo 6.

7.45 p.m. Place the *Escalopes* and the *Duchesse Potatoes* into the oven at the same setting as for the *Coquilles St Jacques*.

7.50 p.m. Place the *Spinach* in the top section of a double boiler over gently simmering water.

7.55 p.m. Remove the *Coquilles St. Jacques* from the oven, sprinkle with browned breadcrumbs, add a few knobs of butter, and place under a grill until lightly browned. Keep hot.

Lower the oven temperature so that the *Escalopes* and *Duchesse Potatoes* are kept hot, but not allowed to dry.

8.0 p.m. Serve the *Consommé Madrilène*.

GUESTS FOR THE WEEK-END

This must surely be the most taxing of all forms of entertaining and it will require careful planning if the hostess is to be able to enjoy the company of her guests. It is on this occasion perhaps more than any other that the freezer comes into its own.

The following menu suggestions are made for a week-end visit extending from the Friday evening to the Sunday tea-time. (Obviously the arrangement of the meals will be varied to suit the times of arrival and departure of the guests, as well as the week-end's activities.)

In the event of the guests taking their host and hostess out to a meal, then at least the food stored in the freezer will not be wasted, but will remain to be served on another occasion.

Friday Supper
French Onion Soup, page 210.
Pizza, page 387.
Green Salad.
Fresh Fruit.

Saturday Breakfast
Grapefruit, page 199.
Kedgeree, page 255.
Hot Rolls, page 386.
Coffee.

Saturday Lunch
Veal Sauté Marengo, page 274.
French Beans.
Potato Croquettes, page 241.
Almond Cherry Flan, page 305.

Saturday Dinner
Spinach Soup, page 212.
Chicken Kiev, page 291.
Corn.
Cauliflower with Sauce Duxelle, page 224.
Frozen Apricot Mousse, page 336.

441

Sunday Breakfast
Fruit Juice.
Savoury Pancakes, page 401.
Croissants, page 390.
Coffee.

Sunday Dinner
Vichyssoise, page 215.
Noisettes of Lamb, page 271.
Tomato Sauce, page 229.
Asparagus.
New Potatoes.
Stuffed Peaches, page 324.

Sunday High Tea
Ham Soufflé, page 403.
Salad.
Cheese Bread, page 396.
Fruit Gâteau, page 314.
Tea.

ENTERTAINING AT CHRISTMAS

Entertaining at Christmas can be a combination of all the previous types of entertaining; guests staying for several days, a cocktail party, a teenage party, and Christmas dinner itself. Formidable as the task appears to be, if a thoroughly comprehensive plan is worked out, and then systematically worked through, the Festive Season can be enjoyed by all.

The following Christmas dinner menu can be almost entirely pre-prepared and pre-frozen.

Consommé Julienne, page 207.

———

Roast Turkey, (note a).
    with stuffings, page 405.
Bread Sauce, (note b).
Cranberry Sauce, page 237.
Bacon Rolls, (note c).

442

Duchesse Potatoes, page 240.
Boiled Sprouts (frozen).

———

Christmas Pudding.
Brandy Butter, page 234.

———

Mince Pies, page 302.

Notes

(*a*) Remove the raw bird (bought at economical prices some months previously) from freezer two to three days before Christmas—depending on size. Prepare and freeze chestnut and sausage-meat stuffings up to two weeks before Christmas—transfer to the refrigerator twenty-four hours before required.

(*b*) Prepare bread sauce from frozen fresh breadcrumbs as follows.

BREAD SAUCE

½ pint milk
1 small onion stuck with 3 cloves
1 small bayleaf.
seasoning
2–3 oz. fresh breadcrumbs (frozen)
knob of butter

infuse for 15 minutes; remove onion and bayleaf
stir in and simmer for 3–5 minutes
add butter and adjust the seasoning

If preferred, this sauce can be completed and frozen in advance.

(*c*) Prepare bacon rolls up to two weeks before Christmas.

Recipe Index

To assist in menu planning, the recipes in this book have been listed in alphabetical order in their culinary categories, as follows:

RECIPE INDEX

447

RECIPE INDEX

General Index

Accelerated freeze drying, 118
Ageing carcass meat, 158
Alarm system for freezers, 74
Allumettes, 423
Almond cherry flan, 305
Almond paste, 351
Aluminium, 91
 in summary, 97
Anchovy butter, 234
Anchovy puffs, 430
Anchovy sauce, 223
Apple loaf, spiced, 362
Apple sauce, 236
Apples, freezing, 137
Apricot glaze, 349
Apricot mousse, frozen, 336
Apricot sorbet, 329
Apricots, freezing, 138
Ascorbic acid, as an antioxidant, 44
Ascorbic acid, quantities added to
 fruit, 134
Asparagus, freezing, 150
Asparagus rolls, 410
Aubergines, freezing, 150
Aubergines à la Provençale, 247
Aurore sauce, 223
Avocados, freezing, 138

Babyfoods, freezing, 191
Bacteria—see micro-organisms, 46
Baked cakes, packing, freezing and
 thawing, 340
Baked cheese cake, 307
Baked stuffed marrow rings, 249
Baked stuffed potatoes, 243
Baking tins, use in the freezer, 99
Bananas, freezing, 138
Batch baking, 341
Battenburg cake, 355

Batter-dipped savouries, 428–31
Bavaroise, 317
 Chocolate, 318
 Coffee, 318
 Fruit, 318
Beans, freezing, 150
 Broad, 150
 French, 151
 Runner, 151
Béarnaise sauce, 233
Béchamel sauce, 222
Beef, cuts of, 159
Beef, Flemish, 277
Beef goulash, 277
Beef râgout, 276
Beetroot, freezing, 151
Beignets soufflés, 427–8
 Anchovy, 427
 Ham, 428
 Pignateli, 427
 Savoury, 427
Beurre manié, 194
Bigarde sauce, 226
Bilberries, freezing, 139
Birdseye, 21
Biscuit crust, 369
Biscuit tortoni, 337
Biscuits, 362–6
Blackberries, freezing, 139
Blackcurrant flan, 308
Blanch, to, 194
Blanching fruits in steam, 135
 in syrup, 136
Blanching, nutritive losses during, 38
Blanching vegetables in water, 145
 in steam, 146
Blast freezing, 111
Block freezing, 129
Boil-in-the-bag products, 89

449

Vitamin A, 37, 39
Vitamin B, 37, 39
Vitamin C, 35
  as a quality indicator, 37
Vitamin D, 39
Vol-au-vent, shaping, 374
  fillings, 421

Waxed containers, 85
  in summary, 96
Weekend guests, suggestions, 441
Welsh rarebit, 414
Whisked fruit ice cream, 334

White bread, 384
White sauces, 221

Yeast batter, 395
Yeast flaky pastry, 390
Yeast freezing, 383
YEAST MIXTURES, 381–400
Yeasts—*see* micro-organisms.

Ziegler process for polythene, 88
Zone of maximum crystal formation,
  26